IEE Control Engineering Series 28
Series Editors: Prof H. Nicholson
Prof B.H. Swanick

Robots and automated manufacture

Robots and automated manufacture

Edited by J. Billingsley

Peter Peregrinus Ltd. on behalf of the Institution of Electrical Engineers

Published by: Peter Peregrinus Ltd., London, UK.
© 1985: Peter Peregrinus Ltd.

ISBN 0 86341 053 7

Printed in England by Short Run Press Ltd., Exeter

Contents

List of contributors

Paper 1

D.J. Todd
Jurrassic Robots
Swindon
Wiltshire SN1 4JZ

Paper 2

R.K. Stobart
Cambridge Consultants Ltd
Cambridge CB4 4DW

Paper 3

M. Edkins & C.R.T. Smith
Department of Mechanical Engineering
Manufacturing and Machine Tools Division
UMIST
Manchester M60 1QD

Paper 4

D. Harrison, J. Billingsley & F. Naghdy
Department of Electrical and
Electronic Engineering
Portsmouth Polytechnic
Portsmouth PO1 3DJ

Paper 5

M.A. Woollett
Cranfield Robotics and Automation Group
Cranfield Institute of Technology
Cranfield
Bedford

Paper 6

B.M.S. Zainol Anvar & P.D. Roberts
Control Engineering Centre
School of Electrical Engineering
and Applied Physics
The City University
London EC1V 0HB

Paper 7

D.J. Rhodes, E.H. Stenson & P. Blanchfield
Department of Electrical and
Electronic Engineering
University of Nottingham
Nottingham NG7 2RD

Paper 8

J.A. Lock, I. McLeod & G.J. Smith
Department of Industrial Engineering
Napier College
Edinburgh EH10 5DT

Paper 9

D.J. Drazan, S.H. Hopkins & C.J. Bland
Department of Mechanical Engineering and
Engineering Production
UWIST
Cardiff CF1 3XE

Paper 10

A.H. Redford
Department of Aeronautical and
Mechanical Engineering
University of Salford
Salford M5 4WT

Paper 11

F. Naghdy, J. Lidbury & J. Billingsley
Department of Electrical and
Electronic Engineering
Portsmouth Polytechnic
Portsmouth PO1 3DJ

Paper 12

E.R. Davies
Department of Physics
Royal Holloway and Bedford Colleges
Egham
Surrey TW20 0EX

Paper 13

R.E. Jones & P.M. Hage
Joyce-Loebl
Gateshead
Tyne and Wear NE11 0QW

Paper 14

J. Weston
Research Laboratory
Abington Hall
Cambridge CB1 6AL

Paper 15

M.A. Browne, E.R.D. Duckworth,
R.A. Wainwright & J. Ashby
Department of Instrumentation and
Analytical Science
UMIST
Manchester M60 1QD

Paper 16

A.M. Dean
Department of Electrical Engineering
University of Bradford
Bradford BD7 1DP

Paper 17

P.D. Chuang and J.P. Cosmas
Department of Mechanical Engineering
Imperial College
London SW7 2BX

List of referees

A.P. Ambler (University of Edinburgh)
M.C. Bonney (University of Technology, Loughborough)
G.C. Bryan (Mars Electronics)
A.K. Forrest (Imperial College, London)
L. Goldstone (Department of Trade and Industry)
J.R. Hewit (University of Newcastle upon Tyne)
J.J. Hunter (National Engineering Laboratory)
C. Lamont (Unimation)
A.C. McMillan (Alan Bradley International)
R.J. Poppleston (University of Edinburgh)
A. Pugh (University of Hull)
I.C. Pyle (University of York)
J. Rees Jones (Liverpool Polytechnic)
T.J. Stoneham (Brunel University)
J.A. Weaver (Philips Research Laboratories)
R. Weston (University of Technology, Loughborough)

Editor's Introduction

Five years ago the SERC Robotics Initiative was launched. The word 'robot' then held the same excitement that 'microprocessor' had possessed a year or two before. Now it has lost its glamour, not because robotics is in any way waning, but because it has been accepted into everyday technology. That is not to say that its adoption by industry is without effort — just the reverse.

The first rush of enthusiasm was dominated by a concern to study the robot itself, its sensors, coordination, gripper, accuracy and agility. Applications in the UK and abroad were often notable for their novelty — the boning of bacon backs, shearing of sheep and even brain surgery. Now research has matured to concentrate on the harder task of integrating the mechanical manipulator into a production environment. Not only must it cooperate with the process and machining equipment, it must be made compatible with the abilities of the industrial programming team.

The papers gathered here cover a variety of topics, but a recurring theme is the achievement of the programming task, especially by the use of off-line methods. 'Solid modelling' is presented as one solution to the need to predict potential collisions between a robot and its workspace. An 'error map' helps to reduce positional inaccuracies without resorting to the measurement of point-by-point corrections. Rule-based methods allow a loose verbal description of a task to be applied to implement on-line control. The addition of a sensory probe is suggested as an aid to location editing. Collision avoidance is considered again, this time with the complication of a number of robots within a single work cell.

Component assembly is an important application area, and a pallet based technique is described to aid assembly in a flexible manufacturing system. Other papers discuss the use of compliance in the assembly task and its influence on programming methods. Another distinct robot application area is to welding, and papers include a general survey and a more particular investigation of the use of vision methods.

Vision and other sensing techniques are well represented. The use of motor drive monitoring to deduce the robot force has its problems; statistical processing methods are put forward as a solution. Vision systems are considered with their abilities to recognise shapes and to determine locations of regular profiles to fine tolerances. Their applications are studied for position control during assembly, and for the recognition of a particular motor car headlamp among a dozen variations.

One paper falls outside all these categories; it is concerned with a novel steering mechanism for a multilegged walking robot. With two- four- and six-legged robots on show in the Japanese Tsukuba Expo 85 and a growing body of research in the UK and in the United States, this could be an area to watch in future.

It is no accident that the words 'Automated Manufacture' have been aded to the title of this year's meeting. The earlier conferences on 'UK Robotics Research' were inspired by the annual SERC gatherings of the recipients of grants under the Robotics Initiative, although the papers were by no means restricted to that source. Now the Robotics Initiative has been absorbed into ACME, a programme for the 'Application of Computers to Manufacturing Engineering,' while even greater efforts have been made to encourage papers from industry.

Although industry is still shy to publish, it will find much useful material within this book. The academic authors have worked closely with industrial partners to formulate and solve their mutual problems. Robotic development is challenging, stimulating and fascinating, but it is only in its exploitation by industry to enhance productivity and profitability that any real success can be measured.

Warm thanks are due to Brian Davies of Imperial College, Peter Pugh of the Institution of Mechanical Engineers, Peter Smith of the Science and Engineering Research Council and Peter Walters of Thorn EMI Central Research Laboratories, members of the hard-working committee who brought the meeting and papers together. We are all grateful to the many referees who assessed the submissions and suggested numerous improvements which have been incorporated by the successful contributors. Especial thanks are due to Miss Beverley Preston of the Institution of Electrical Engineers for so ably coordinating the frantic traffic of papers between contributors, referees, the committee and printers.

John Billingsley,
Chairman, Organising Committee.

A novel steering mechanism for legged robots

D.J. Todd

1   INTRODUCTION

Many walking machines have six legs, since this allows an efficient and stable gait in which the machine is supported by two tripods of legs alternately. It is common to give each leg three powered joints so each foot can be moved up and down, backwards and forwards, and from side to side. This requires a total of 18 actuators, each with a driving device such as an amplifier or servovalve, and the associated cables or hoses. The weight, cost and complexity of all this equipment presents an obstacle to the widespread use of legged vehicles.

Further, some manoeuvres such as turning, or even straight-line walking in some designs, requires the co-ordinated action of several joints, presenting a considerable problem of computing, in real time, the joint angles and velocities needed, and driving the actuators to produce them.

For many purposes there is no alternative to the fully servo-controlled 18 degree of freedom vehicle. However, at this early stage in the development of walking robots some investigations can be done with cruder machines. For example, there has been little research on the interaction of legged vehicles with different surfaces such as sand, mud and loose earth. Studies of the interaction of vehicle loading and foot design with surface conditions do not need a high degree of adaptability. Another example is the development of aids to stable locomotion such as balancing devices.

This paper describes such a simple vehicle. The next sections cover the methods used to reduce complexity: of these the most unusual is the steering geometry and mechanism. There follows a summary of the design of the current version of the machine.

2   METHODS OF SIMPLIFICATION

2.1 Leg Geometry

It is commonly held desirable to decouple the degrees of freedom of a leg so that precisely horizontal motion of the machine needs the use of only the propulsion actuators

(Hirose (1), Waldron (2)). This implies the use of a prismatic joint or an equivalent linkage. The most promising designs at present use pantographs to transfer the motions of orthogonal sliding joints to the foot.

The hexapod described here is intended to use pantograph legs, but their engineering is demanding, so this phase of the project has been deferred until the rest of the system is satisfactory. The legs presently fitted achieve similar results, but with a limitation to fairly smooth and obstacle-free ground. The design is shown in Figs. 1 and 2.

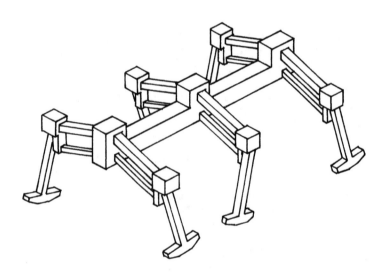

Fig. 1   General arrangement of hexapod

It separates the functions of lift, propulsion and turning. The parallelogram linkage constituting the 'femur' constrains the knee to move in a vertical circular arc of large radius so that the sideways travel of the foot while on the ground is limited to less than 1 cm, an amount easily taken up in the elasticity of the structure and slackness of the joints. In effect, when a leg is raised or lowered the foot travels very nearly in a vertical straight line.

The horizontality of the propulsion stroke is produced by the rolling action of the foot, the profile of which is a circle centred on the knee. This requires long feet, so the machine is unsuitable for stepping over obstacles.

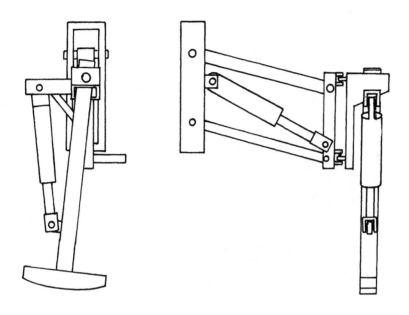

Fig.2   Side and front views of left front leg

2.2   Turning

2.2.1   Previous research. When a legged vehicle turns while
walking, each foot must move sideways relative to the body
as well as backwards, with a velocity depending on the
situation of the particular foot. The difficulty of computing
this velocity and servo-controlling it has led several
investigators to devise vehicle designs which generate the
correct sideways movement automatically. One simple approach,
that of Kessis et al. (3) is to make the legs laterally
flexible and elastic so they deform by the needed amount
during a turn. This is unlikely to work well on slopes as
there is nothing to stop the whole vehicle flopping down the
slope in a rather uncontrolled manner. A cumbersome but
effective way of turning is to divide the body into two
sections which can rotate relative to each other about a
central vertical axis, the steering joint, with one set of
three or four legs attached to each half of the body. This
method is used by the Komatsu underwater octopod (Ishino (4))
and by a small pressure vessel inspection robot (Kemmochi and
Katsuoka (5)). An elegant method is that of Sutherland (6),
in which a hydraulic connection between the sideways actuators of the three supporting legs allows the differential
movements occurring in a turn.

2.2.2   Turning geometry. The steering method described here
is based on that of wheeled vehicles in which each wheel is

pivoted about a vertical axis, through an angle such that all the wheels roll in concentric circles. There are differences, since the feet do not move (apart from a limited rolling from heel to toe) but the knees describe circular arcs. Also, the arrangement is not kinematically perfect, some of the legs being forced to slip or deform slightly, but not noticeably in practice.

Fig.3 shows a plan view of the hexapod at the start and

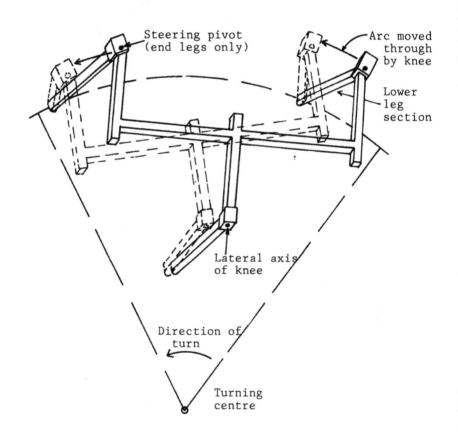

Fig.3 Plan view of hexapod at start (solid lines) and end (broken lines) of a left turn. Only the supporting legs are shown.

finish of a left-turning stride. At all times it walks with the alternating tripod gait. As shown in Figs.1 and 2, during a stride the foot swings in a vertical plane, pivoting about the lateral axis of the knee, rather like a wheel. For a left turn the foot-swinging plane of the right end legs is rotated about a vertical axis so as to face the cetre of the turning circle. The plane of the third supporting leg, the left middle, is not rotated since it already faces the turning centre. Its stroke during the turn is less than that of the right end legs since it is nearer the turning centre.

In the alternating tripod gait, the vehicle is supported alternately by the front left, back left and right middle legs (leg set A) and by the front right, back right and middle left legs (leg set B). Left turning can only take place during the half cycle when leg set B is supporting, and right turns only when leg set A is supporting. Therefore a sustained turn in either direction consists of an alternating series of short arcs and short straight lines. This causes the effective turning radius to be larger (in this case about 4m) than the turning radius during an individual stride, which is 1.6m. It would be possible to turn on both halves of the cycle at the cost of slightly more complexity.

2.2.3 Steering actuation. Since this steering method does not require the leg yaw angles to be controlled individually it is possible to use a single actuator and valve instead of

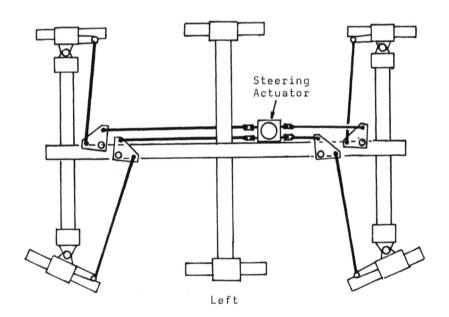

Steering
Actuator

Left

Fig.4   Plan view of hexapod showing steering linkage. A right turn is illustrated: the left end legs have been rotated. The bell cranks are located under the spine. The push rods have ball joints to allow the legs to swing vertically.

six. This actuator is a hydraulic cylinder of 150 mm stroke.

At the centre of the stroke (75 mm) all the legs are in the
longitudinal (sagittal) plane and the vehicle walks in a
straight line. When the piston extends beyond 75 mm a linkage of bell cranks and push rods (Fig.4) swivels the right-
hand legs for a left turn. When it retracts to less than
75 mm the left-hand legs rotate. These two separate actions
are made possible by the shape of the slots in the cam
plates (Fig.5).

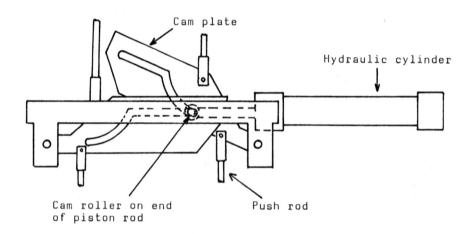

Fig.5   Side view of steering actuator. It is mounted
        with its long axis vertical, on the right-hand
        side of the spine, so the cam plates are in a
        vertical longitudinal plane.

2.2.4 Hydraulic circuit modifications. In straight-line
walking the knees of all the supporting legs are powered to
make the machine move forward, and they all have the same
stroke. During a turn, however, the middle leg (on the inside
of the turn) must make a shorter stroke, in the same time.
To avoid active velocity control the middle knees are fitted
with valves which in their 'off' state do not isolate the
two sides of the piston but connect them together. During a
turn this knee is not powered but allowed to coast; oil
flows freely from one side of the piston to the other.

2.3 Actuator Control Using Directional Valves

    Accurate velocity control of a hydraulic ram needs the
use of either a servovalve or a variable displacement pump,
both being very expensive. A lesser accuracy can be achieved
using a cheaper proportional solenoid valve. Finally, if
the velocity is fixed, or controlled by some auxiliary
device, an actuator can be controlled by a simple directional

solenoid valve, at a cost approaching a tenth that of a servovalve.

One of the purposes of the machine described here is to see how well a hexapod can be controlled using directional valves only, or directional valves together with very few proportional valves. A similar but rather more advanced approach was adopted by Sutherland (6), who used directional valves and a few variable displacement pumps.

The hydraulic circuit is shown in simplified form in Fig.6. At present a total of nine three-position, four-way

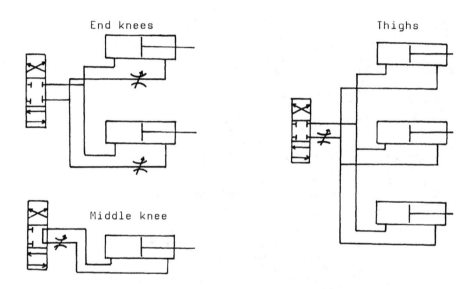

Fig.6   Hydraulic circuit for one leg set. The valve
        used for the middle knee allows the knee to
        coast when not powered.

solenoid valves are fitted. Each set of three thighs shares a valve, as do the end knees. If all thighs and knees had individual valves a total of 15 would be used: six thighs, six knees, one steering and two bypass valves. One bypass valve is manually operated and allows two walking speeds; the other is automatically switched on whenever no cylinder is active, reducing the load on the motor and the dissipation of power in the relief valve.

The cylinder velocity is limited by restrictors but is not precisely controlled. A problem with this approach is that all movements start and stop suddenly. It is proposed that in later tests a combination of directional and proportional valves will be used.

The present arrangement of shared valves makes it

impossible to control individual legs, so the gait cannot be
adapted to deal with slopes or obstacles. It is seen as a
temporary expedient to simplify the machine.

3   DESIGN OF THE HEXAPOD

3.1 Mechanical

The basic form of the machine is as in Fig.1. It is
constructed of aluminium alloy, mostly in the form of square
sectioned tube. The legs are attached to a spine from which
hang two platforms, the front carrying the hydraulic supply
and the rear the valves. The hydraulic supply consists of a
fixed displacement gear pump delivering 13 litres/min at
3000 rpm. It is driven by a 2 kW 240 V single phase mains
induction motor. A reservoir, relief valve, pressure gauge
and filter are also fitted. The robot is 1.8 m long, 1.3 m
wide, 1.1 m high and weighs 140 kg.

3.2 Electronics And Computing

Only a brief summary is given, as this subject is
covered in more detail in an earlier paper (Todd (7)) on a
pneumatic hexapod whose control system was used, with some
modifications, for the vehicle described here.
The walking cycle is generated by a computer running a
program which implements a state machine. A 'state' is a
unit in which some self contained function occurs, such as
raising a set of legs. A state begins with the turning on of
the appropriate solenoid valves. The joint angles are then
monitored until the end condition for that state is detected,
when the valves are turned off. The program then checks for
commands from the user before looking up the next state in a
table. When walking is first commanded the program carries
out a set of leg movements which ensures that the robot is
standing on a valid leg set, with the other set raised, so
it is in a correct condition for the state machine to start.
The computer, which can be on board or external, is
connected to the interface electronics by a serial link. The
computer used at present is based on a Z80 microprocessor.
The interface circuits are Eurocard modules in a rack
on the robot's spine. A controller, communicating with the
computer over the serial link, addresses the interface
modules, loading outputs and interrogating inputs. The main
interfaces are optically isolated solenoid valve drivers, an
8 bit 16 channel analogue to digital converter and a 64
channel binary multiplexer.
The user's interface consists of a hand-held box
carrying switches to select the direction of turning and
to select forwards or backwards walking. The interface rack
contains displays showing the state of the program.

4   RESULTS AND CONCLUSIONS

The steering mechanism has proved capable of a reasonably smooth turn. The plane of the leg pivots through 20°
.

and the vehicle turns through 10° in one stride. A tighter
turn is desirable, and in principle is not difficult to
achieve; in practice the linkages would need strengthening.
    As remarked earlier, straight horizontal motion is
produced by using a wheel-like foot unsuitable for rough
ground. The leg being designed to replace it is shown in
Fig.7. A pantograph transfers the motion of a horizontal

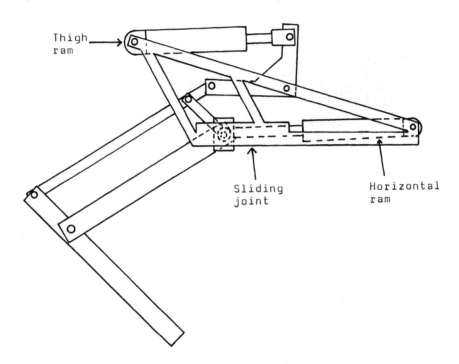

Fig.7   Pantograph leg with one sliding joint, and a
        swinging link to generate the vertical movement

slide to the foot, but there is no need to make the lift
(thigh) motion exactly vertical, so a swinging link is used
instead of a slide. This simplifies the engineering and
allows a reasonably compact design while using standard
cylinders.
    The experiments with directional solenoid valves
indicate that although they are adequate for the thighs, the
knees need some control of velocity, particularly at the end
of the stroke. They will be fitted with proportional valves.
    The overall conclusion is that, while the initial hopes
for a very simple and cheap machine were over-optimistic,
especially in the area of joint velocity control, it is still
possible to design a usable hexapod at only a moderate level
of engineering sophistication.

REFERENCES

1. Hirose, S., 1984,
    'A study of the design and control of a quadruped
    walking vehicle', Int. J. Robotics Res., 3, 2, 113-133.

2. Waldron, K. J., Vohnout, V. J., Pery, A., McGhee, R. B.,
    1984,
    'Configuration design of the adaptive suspension
    vehicle', Int. J. Robotics Res., 3, 2, 37-48.

3. Kessis, J. J., Rambaut, J. P., Penne, J., 1983,
    'Walking robot multi-level architecture and implement-
    ation', 4th CISM-IFToMM Symp. on Theory & Practice of
    Robots & Manipulators, PWN, Warsaw, 297-304.

4. Ishino, Y., Naruse, T., Sawano, T., Honma, N., 1983,
    'Walking robot for underwater construction', ICAR 1983,
    107-114.

5. Kemmochi, S., Kazuoka, S., 1983,
    'Mechanized devices for the inservice inspection of
    nuclear power plants', ICAR 1983, 83-90.

6. Sutherland, I. E., Ullner, M., K., 1984,
    'Footprints in the asphalt', Int. J. Robotics Res.,
    3, 2, 29-36.

7. Todd, D. J., 1984,
    'An experimental study of pneumatic walking robots',
    Digital Systems for Industrial Automation, 2, 4.

The use of simulation in the off-line programming of robots

R.K. Stobart and C. Dailly

The general requirements of a robot simulation system are discussed and the benefits of integrating such a system with a solid modeller are indicated. The status of a robot simulator integrated with the BUILD solid modelling system is examined. The application example cited is the Condition Number analysis for optimum positioning of a workpiece in a robot's workspace. The robot simulator is used to verify that no interference takes place between the robot and workpiece in the optimum position.

1.   INTRODUCTION

Simulation is an essential component of a process which must work in practice without major modifications. This requirement may apply to large projects in manufacturing or a complex robot operation which must be integrated with a larger production process.

Robot simulation has been used in both of these situations and may find wider applicability with falling software costs or the increasing demands placed on robot design and manufacture.

This paper describes work done to implement a robot simulation system within the BUILD research solid modelling program. The robot simulator co-exists with a number of other CAE tools such as NC data generation and process planning facilities. All the BUILD tools draw on a common solid modeller and so the passing of information between them is greatly facilitated.

Many computer based manufacturing systems require the use of computer generated solid models. The integration of techniques demonstrated in BUILD points to the plausibility of a common solid modeller supporting a diverse range of CAE tools and thus indicates progress towards computer integrated manufacturing.

Simulation may be directed at planning a robot installation or at producing a working program for a robot. In the latter case there is a need to produce a manipulation language program to be transferred to the 'target' robot. This program must take into account the inaccuracies in the robot which will produce differences between the achieved and simulated tool positions. The aim of the simulation is to allow the programmer to program the robot model as if it were the real robot. The simulated conditions allow him to check for errors before transferring the manipulation program to the robot for execution of the task.

2.   THE BUILD SOLID MODELLER

A solid modeller may be defined as a computer program which generates and manipulates models of solid objects.

The BUILD solid modeller is such a program and has been developed over a period of about 10 years, primarily at the University of Cambridge Computer Laboratory. BUILD is described by Jared (1).

BUILD version 4.0 now supports NC data generation, Parkinson (2), feature recognition, Jared (3), double quadratic surface handling, Jared & Varady (4) and draughting, as well as the robot simulation described in this paper. The relationship of these facilities to the core solid modeller is illustrated in Figure 1. Each of the CAE tools draws on the solid generation and handling facilities of the solid modeller.

3.   THE FORM OF A ROBOT SIMULATION SYSTEM

A robot simulation system requires a collection of facilities which together provide the capability to model robots in a work place. This section attempts to introduce the facilities and how each contributes to the usefulness of the simulation system. The components of the system are illustrated in Figure 2.

3.1   The Robot Model

The robot model is fundamental. The representation of the robot should permit modification of joint parameters and easy access to the data structures describing the form of the robot. The robot should be easily constructed and once made should be able to be stored.

3.2   Moving the Robot Model

To allow the programming of a tool position or path in cartesian space, the simulator should be able to relate a cartesian co-ordinate set to the robot model's joint angles. This process is known as the inverse kinematic transformation. For generality, no special software should be required to handle a particular robot.

3.3   A Language Interface

To facilitate programming of the simulator, a manipulation language interface should be available. The language may be the same as that of the robot system for which the simulation is being prepared.

3.4   Collision Detection

When programming off line, the programmer may request an illegal operation in which the robot collides with an object in the work place. This collision should be reported as quickly as possible in order to allow interactive programming. To enhance this facility, minimum distance checks should also be performed to compensate for uncertainty about the true robot path and to ensure that adequate clearances are maintained.

If a collision occurs the objects involved should be highlighted in the graphics display of the work place.

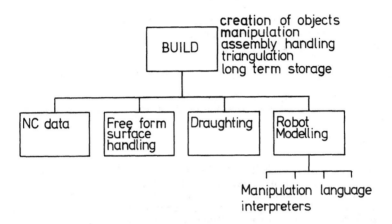

Fig. 1   The Structure of the BUILD Solid Modelling System

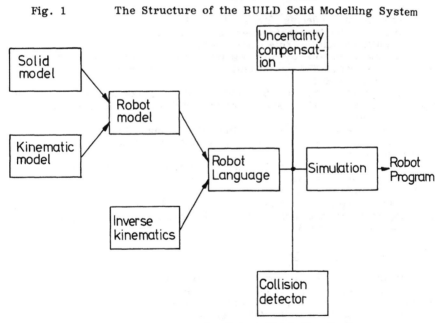

Fig. 2   The Structure of a Robot Simulation System

3.5   Uncertainty in Robot Operation

However well the robot is modelled there will be differences between actual robot performance and the modelled robot performance. The difference will depend on the assumptions made when designing the robot model.

If the true robot performance is to be simulated then the errors must be accommodated by the simulation system. The errors may be measured by a calibration system.

4.   THE COMPONENTS OF A ROBOT SIMULATION SYSTEM

This section expands on the descriptions of the simulation system components and describes the implementation in the BUILD simulation system.

4.1   Robot Model

The robot model is the basic component of the simulation system. In the BUILD simulation system the model is formed from solid models created using the solid modeller and is then assembled in the datum configuration. The assembly process produces a data structure known as a binary tree and this is illustrated in Figure 3. This structure contains all the information about the boundaries of the objects and the way they are disposed relative to each other.

The kinematic description is made up of joint rotation and position information and is formed as a series of linked lists, Figure 4.

The robot model is formed by linking the kinematic and solid descriptions. The joint parameters may then be modified to change the robot configuration. This re-configuration process replaces the transformations in the original binary tree with those describing the new configuration. This new configuration can then be manipulated like any other solid object.

The model formed by combining the kinematic and solid descriptions of the robot is a complete model and can form the basis of further developments.

The models supported by the BUILD simulation system include serial and simple parallel types and some simple examples are shown in Figures 5 and 6.

The detail of the model can be varied to reflect the required accuracy in the robot model. For example, Figure 6 shows the detailed intersections in the Workmaster base.

Using BUILD's viewing facilities the robot model may be viewed from anywhere in the workspace in either axomometric or perspective view. It may be transformed, scaled or stored for later retrieval.

4.2   Inverse Kinematics

In order to make a simulator of any practical use there is a need to be able to relate the tool position and orientation to joint parameters. Furthermore, the technique should be generalised so that a specific routine does not have to be implemented to deal with each robot type. We may then list the basic requirements of an inverse kinematics handler as:

      i)   handling of a multiplicity of robot types without
                   specific routines;

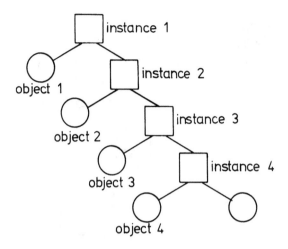

Fig. 3   Data Structure of the Robot Solid Model

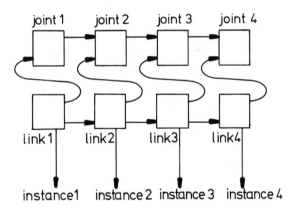

Fig. 4   Data Structure of the Robot Kinematic Model

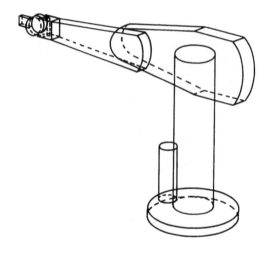

Fig. 5   A Puma Model

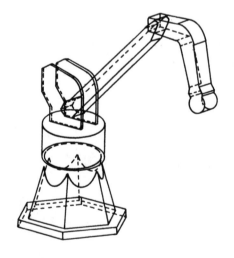

Fig. 6   A Workmaster Model

ii)   handling alternative configurations and eliminating those which are impossible or not allowed in the real robot;

iii)   reporting singularities and reacting in a sensible fashion.

The inverse kinematics processor in the BUILD simulator generates a kinematic class for the robot during the process of linking the kinematic and solid descriptions.

The kinematic classification identifies the type of inverse kinematic process to be applied. When the robot is 'solved' for a given tool position a sequence of procedures is called dependent on the classification. To illustrate the process the methods of solving the Workmaster and a 6 degree of freedom Puma will be considered.

To solve the Puma requires the following steps: (axis numbers refer to Figure 7).

i)   From the tool position and orientation calculate the wrist position;

ii)   calculate the angle of rotation of axis 1;

iii)   calculate angles 2 and 3 from the two link mechanism connecting the shoulder and the wrist;

iv)   calculate the remaining joint angles using spherical trigonometry.

The corresponding steps for the Workmaster are:

i)
ii)   as above

iii)   calculate the position of joint 4 using the wrist geometry; calculate angles 2 and 3 from the two link mechanism connecting the shoulder and link 4.

iv)   as above.

The similarity between the techniques is apparent and indicates the generality that may be applied to the inverse kinematics problem.

If the robot configuration is close to a singularity then this is reported and no further movement is attempted.

4.3   The Manipulation Language Interface

A robot is usually programmed offline using a manipulation language. The programming language is then a natural means of driving a robot simulation.

The language definition for the BUILD simulator is outlined briefly in Appendix A. It is a necessarily simple language and is designed to demonstrate that a simulation system of the type described in this paper can also accommodate a manipulation language interface.

A language interface is provided in the ROBOCAM system described by Craig (5) where programming is done in one language, and when complete the program is translated into the 'target' language ready for downloading to the robot.

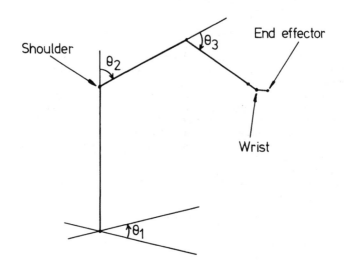

Fig. 7   Axis Descriptions for the Puma Model

An alternative philosophy might be that the simulator should be programmed in the target language so as to avoid translation errors, and allow a programmer to 'debug' a program on the target robot system with which he is already quite familiar.

Manipulation languages generally process sensory inputs. The simulation of such inputs would require the inclusion of the random processes reflecting the variation which may occur in the robot's environment. For this aspect of operation the robot simulator may be able to borrow techniques from FMS simulation.

4.4   Collision Detection

Automatic collision detection is an essential part of a robot simulation system. Without the facility, the programmer must adopt the correct viewpoints for graphic displays of the workplace so as to observe potential collisions. This viewpoint will probably need to constantly change during robot motion and even with care the programmer may miss a collision.

Two principal demands of an automatic collision detection system are that it should be reliable and fast; speed is essential for interactive use in a programming environment.

In offline programming there is a degree of uncertainty about the configuration which the robot will take up in response to a tool position demand. This problem is discussed more fully in section 4.5. For the purposes of collision detection the problem of uncertainty means that a report of a collision is no longer adequate. A check on minimum distance is required throughout critical stages of the proposed robot motion. If this value falls below a specified value based on the uncertainty in the robot's motion then the programmer should be informed. To make an effective correction the programmer should be aware of which parts of the objects involved were about to collide.

4.4.1 The Build Simulator Collision Detection System

A collision detection system has been implemented which uses the solid descriptions contained in the basic data structure (Figure 3). During the creation of the robot model each of the objects making it up is boxed, i.e. a cuboid most closely approximating to the shape of the object is calculated. This boxing process may also be applied to any other objects in the robot workspace. As the robot is subsequently transformed, the boxes are transformed with it and the boxing process never needs repeating. The advantage of the box as a means for observing the spatial relationships in a robot's environment is the computational efficiency with which they may be handled. As most robot members are cuboid or cylindrical the box is generally very close to the object's true boundaries and so can be considered to be a 'good' representation. The dimensions of the boxes may be exaggerated to compensate for uncertainties. This approach, though simple, may be adequate for a first check of a robot motion in the presence of uncertainty.

To check for a collision, the intersection of two boxes is investigated. This is a simple and fast process and would require about 100μs CPU time on a VAX 11/780.

A more sophisticated approach based on Cameron (6) has been designed and can be summarised as follows.

    i)   check for a collision between each of the boxes representing robot members and those representing the components of the machine tool;

    ii)   if there is a collision report it, otherwise using approximate techniques select the pairs of objects most likely to collide;

    iii)   do a detailed minimum distance check on these pairs of objects and so produce an overall minimum distance.

So as to track the minimum distance trends while maintaining computational efficiency a parameter identification technique is used to fit a model to the minimum distance data. This model may be used in a predictive fashion for minimum distance and collision checking.

4.4.2. A Programming Interface for Collision Detection

A manipulation program needs to have access to collision detection facilities. The most appropriate form is that of a monitor, activated at the start of the program. This is a small concurrently executing program which passes a message back to the main manipulation program in the event of minimum distance falling below a specified value or a collision occurring.

In the BUILD robot simulator where there are no facilities for multi-processing the monitor has been designed to simulate this effect.

4.5   Uncertainty Handling

Robot simulators are not generally accurate as they do not allow for dynamic effects, deficiencies in the robot control systems or systematic errors in either trajectory calculations or inverse kinematic processing.

The implementation of a full dynamic model including the control system would be out of the question for a general simulator where each robot would be found to have a unique set of characteristics. Instead methods based on calibration data would be quite appropriate for many applications.

Calibration methods are attracting increasing interest. The work described by Hunter and Kay (7) uses an optical technique for determining the position of the robot end effector.

We might categorise robot errors in two ways,

i)   static errors resulting from control deficiencies and systematic errors in the inverse kinematics,

ii)   dynamic errors in path following resulting from control deficiencies and treatment of 'via' points.

Static errors may be calibrated across the range of the robot using a method which produces the end effector position in the robot's base co-ordinate system. The robot is moved to a series of set-points in cartesian space and the resulting achieved position compared with the demanded position. The error vector may then be deduced over the robot's range of movement and then stored.

For any robot set point programmed off-line the error vector can be found by interpolation and the required robot setpoint deduced.

Static error correction is useful where only the trajectory end points are of interest.

Dynamic error correction is far more difficult to implement because of the large number of variables which influence the path of the tool. One method may use calibration procedures to establish the general form of the path following quality in various parts of the robot's range of movement. This 'path following quality' may then be applied to a desired path to obtain the true path. This method is applied in a simple form in the ROBOCAM simulation system (5).

Mismatches may appear between simulation and the real robot operation. If the simulator is to produce a robot program as output then as many choices as possible should be passed on to the run time control system of the robot. This technique avoids uncertainty by deferring decisions until the appropriate stage. For example, Lozano-Perez (8) describes a path definition method in which the robot run time system is supplied with a nominal cartesian path definition and an allowed deviation from it. The computation of such information is very appropriate for a solid modelling system in which methods for analysing space are well developed.

No attempt has yet been made to implement an error handler in the BUILD simulator. The first step of such a development would be the formulation of an interface to a calibration system. Calibration systems are not yet well developed and so this step has not yet been taken.

5. APPLICATION EXAMPLE: CONDITION NUMBER ANALYSIS AS AN AID TO WORKPIECE POSITIONING

5.1 Introduction

Experience with industrial robots shows that there are some areas in the robot workspace which are unfavourable from the control point of view. In the proximity of these areas, a robot's motion may be fast and unpredictable, regardless of how slowly the robot is programmed to move. This results in actuator saturation and potentially catastrophic consequences for the workpiece or anything else which happens to be in the vicinity at the time. This phenomenon is independent of the complexity of the control system.

Condition number analysis seeks to identify these uncontrollable areas and provide a quantitative measurement of the performance attainable by the robot in operating on a particular workpiece in a given location. It also seeks to maximise performance by optimisation techniques using a figure of merit based on the configuration of the robot over the whole workpiece. In this way, the workpiece is placed at the location in the robot's envelope which allows best performance from the robot.

Collision detection is then used to ensure that the optimised workpiece location is not invalidated due to the robot linkages intersecting with some part of the workpiece in order that the end effector may access some other part. This is a purely geometric process and returns an unambiguous result in that intersection either has, or has not occurred.

This application is seen to require the construction of a robot model; the use of inverse kinematics to direct the robot end effector to a given point; and the use of static collision detection to check on the intersection of the robot and workpiece. The solid modelling facilities of BUILD are used to provide the description of the workpiece. This illustrates a powerful advantage of the co-existence of the solid modeller and robot simulator.

Another use of condition number analysis would be in the aid of collision detection software. During a robot trajectory, it cannot be assumed that the actual and demanded trajectories will coincide. Trajectory following will break down in certain robot configurations. These configurations, called singular points, occur in every robot and so are best avoided if control is to be maintained. The proximity to a singular point which the robot can controllably and predictably go, depends upon the quality of the controller, the available actuator power, and the demanded direction of movement through the volume surrounding the singular point, Daniel and Cook (9).

Therefore, in order to effectively predict a collision, it is necessary to consider not only the geometry of the problem, but also the unpredictability in trajectory following, and to be able to quantify this.

5.2 Relation Between Condition Number and Physical Robot Movements

We may relate robot joint space to cartesian space by the transformation

$$d\theta = J.dX \qquad .....(5.1)$$

where dx is a vector of differential cartesian position errors.
dΘ is the corresponding vector of differential joint angle errors.
J is the matrix of first order partial derivatives.
This equation only holds when J is not singular.

The Condition Number, (Salisbury (10)), may be defined as

$$C(J) = Pmax(J) / Pmin(J) \quad \ldots\ldots(5.2)$$

where Pmax(J) and Pmin(J) are the maximum and minimum principal gains of J respectively.

The Condition Number of the Jacobian matrix of the robot can be thought of as an indicator of the robot's worst dynamic accuracy in a given configuration. It therefore gives an indication of the accuracy with which forces can be exerted on the workpiece, since forces and velocities are closely related. In fact:

$$dF / F \quad <= C(J)* \; dT / T \quad \ldots\ldots(5.3)$$

where dT / T is the relative joint torque error.
dF / F is the corresponding relative force error.

Since J is a configuration dependent matrix, the principal gains of J will be configuration dependent i.e. they will vary as the robot makes a trajectory. For the best control, we desire to have the spread of principal gains as small as possible, and therefore Pmax(J) = Pmin(J). Physically, this means that the robot has adopted some configuration where, regardless of the demanded cartesian direction, the magnitude of the joint movements will be identical. In this case, the Condition Number is 1.

Conversely, at a singular point of the Jacobian, Pmin(J) = 0. Physically, there is at least one cartesian demand direction which will cause some vast joint movements, and this accounts for the loss in control. Now the Condition Number is infinity. In this way the Condition Number, or better, the reciprocal of it, provides the basis for a figure of merit for the ease of controllability of a robot in a given configuration.

Mathematically, it may be shown that the upper bound on relative angular error in terms of the relative cartesian error is

$$d\Theta / \Theta \quad <= C(J)* \; dX / X \quad \ldots\ldots(5.4)$$

where dX / X is the relative cartesian error.
dΘ / Θ is the corresponding relative joint angle error.
The duality between equations 5.3 and 5.4 is evident. Equation 5.4 states that regardless of the size of relative cartesian error, at a singular point, when C(J) is infinite, the relative angular error may be very large.

The process of optimising workpiece location so that at no point will the robot attempt to attain a singular configuration is as follows:

The workpiece shape is expressed as a set of spatial co-ordinates. At each co-ordinate point the Jacobian matrix for the manipulator is calculated and the condition number derived from it. A cost function is evaluated using the cost number derived for all the points on the workpiece. This cost function indicates how well the robot configuration is suited to the current workpiece position. Using optimisation techniques described by Bazarra & Shetty (11) the workpiece is translated and rotated until the best location is found.

Essentially, the object of the exercise is to locate the position in the robot envelope at which the robot may contact the workpiece in the most controlled and balanced way. This is done by keeping the Condition Number of all discretised points as close to unity as possible. Large deviations from unity are penalised and so too are large changes in Condition Number. The result is a location allowing the full performance of the robot to be realised. The performance will also be relatively constant over the whole workpiece area.

5.3 Experimental Results

Figure 8 shows the reciprocal Condition Number plot of an aircraft body panel, approximately 1.5m * 1m in size, before and after optimisation. Similarly, figure 9 shows a 1m square plane. The cost function, which may be assumed to be proportional to the dynamic accuracy of the robot, shows nearly a twofold improvement. Up to a tenfold improvement was observed, depending upon the quality of the initial position chosen for the workpiece. This is usually a random choice.

It was observed that different starting positions could return different optimised locations. This is because the 6 dimensional space of the robot in question had many singular positions and also a number of local maxima of the cost function. There is no easy way to guarantee finding the global maximum on exit from the optimisation procedure. The number of local maxima will depend on the size of the workpiece and the complexity of the robot. For small workpieces there may be very many maxima, and this may cause problems. However, in the majority of cases, small workpieces do not demand large and fast movements from the robot and a sensible location can be fairly intuitively found. The optimisation becomes valuable for larger workpieces, which by virtue of their size, relative to the robot, preclude a great number of local maxima.

6. CONCLUSION

A robot simulation system has been implemented in the BUILD solid modeller alongside other CAE tools. It demonstrates a variety of capabilities including collision detection to validate workpiece positions produced by condition number analysis. A means of accommodating robot calibration data has not yet been designed and awaits a generally accepted calibration method.

ACKNOWLEDGEMENTS

We would like to thank Graham Jared of for his help with the BUILD solid modeller.

24

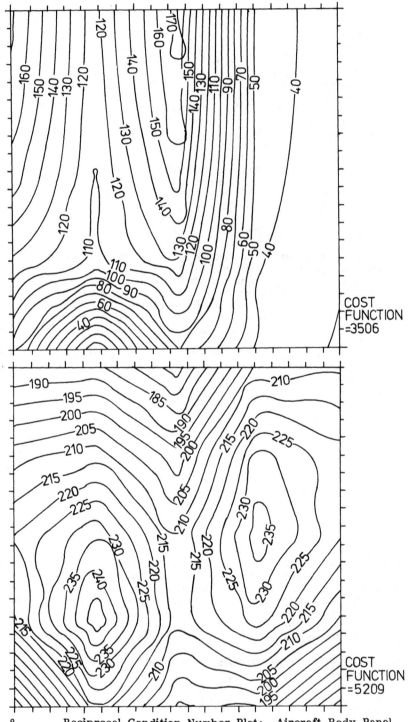

Fig. 8   Reciprocal Condition Number Plot: Aircraft Body Panel

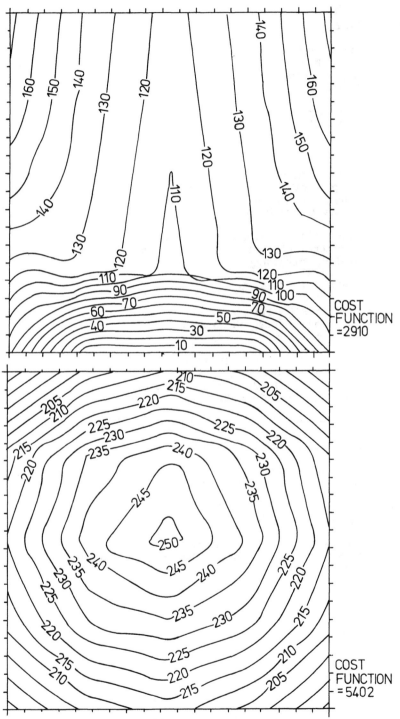

Fig. 9   Reciprocal Condition Number Plot: Square Plane

REFERENCES

1. Jared, G.E.M., 1983,
    'The BUILD Geometric Modeller', Proceedings of a joint
    Anglo-Hungarian Seminar on Computer Aided Geometric Design,
    Computer and Automation Institute, Budapest, Hungary.

2. Parkinson, A, 1984
    'Automatic NC Generation from a Geometric Model', Proceedings
    of the 15th CIRP International Seminar on Manufacturing
    Systems, Tokyo, Japan.

3. Jared, G.E.M., 1983
    'Shape Features in Geometric Modelling', Proceedings of the
    General Motors Research Labs. Symposium, 'Solid Modelling from
    Theory to Applications, Detroit, MI, USA.

4. Jared, G.E.M., Varady, T., 1984,
    'Synthesis of Volume Modelling and Sculptured Surfaces in
    BUILD', Proceedings of CAD 84, Brighton, England.

5. Craig, J.J., 1985, Robotics Today, 7, 45-47.

6. Cameron S.J., 1984,
    'Modelling Solids in Motion', PhD Thesis, University of
    Edinburgh.

7. Hunter, J.J., Kay, R.N. 1985,
    'Robot and Vision Developments for Industry', Proceedings of the
    8th Annual British Robot Association Conference, 'Robotic
    Trends', Birmingham, England.

8. Lozano-Perez, T., 1983, Proceedings of the IEEE, 71, 821-841.

9. Daniel, R., Cook, I.C., 1983,
    'The Relationship between Trajectory Sensitivity and
    Manoeuverability as a Design Aid for Revolute Fixed Arm
    Industrial Robots', Digest No. 1983/77, 4.1-4.7.

10. Salisbury, J.K., 1982, International Journal of Robotics
    Research, 1, 4-17.

11. Bazarra, M.S., Shetty, C.M., 1979,
    'Non-linear Programming, Theory and Algorithms', Wiley, New
    York, USA.

APPENDIX

Definition of the manipulation language.

1   Data Types

Real ⎱
Integer ⎰   as specified for the IBM370 ALGOL 68-C compiler

Setpoint: record of 6 real values to represent a cartesian position and tool orientation. The set point components may be selected using the field selectors x, y, z for the cartesian co-ordinates and r, p, y for roll pitch and yaw.

Boolean: variable taking the value TRUE or FALSE.

2   Arithmetic Operators

Real ⎱
Integer ⎰   +, -, *, / , : = (assignment)

Setpoint   +, -, : = (assignment)

3   Logical Operators

Real ⎱
Integer ⎰   <, >, / = , = , % =

Setpoint   / =, =, % =

(where % = means equal within a specified tolerance)

Boolean: =, /=, AND, OR, NOT

4   Subroutines

      Subroutines are defined in the form of functions: all functions return a value of type real, integer, setpoint or boolean.
      The general form of a function is:

      FUNCTION name = (TYPE Var 1 .....) TYPE:
      BEGIN
      <Body of function>
      END

      Type checking is done on all parameters at run time. Functions may not be defined inside other functions.

5   Control Structures

    i)   For i to n STEP m
                D <code> OD;

    ii)   IF <logical condition>
                THEN <code>
                ELSE <code>
                FI;

    iii)   GOTO label

6   Labels

Labels may be assigned to points in the program thus permitting transfer of control to that point. The label may be of alphanumeric characters and must be of the form:

    <label>:

7   Instrinsic Functions

The manipulator language has a small number of built in functions.

7.1   MOVEV (<setpoint increment>)
This moves the tool incrementally where the setpoint increment is supplied in the function call.

7.2   MOVEA (<absolute setpoint>)
This moves the tool to a position and orientation supplied in the function call. Both move functions return a value of type boolean indicating success or otherwise of the move command.

7.3   DELAY (<time>)
This causes a delay in the operation of the simulated robot.

7.4   GRASP
            RELEASE
These functions respectively close and open the robot's gripper.

The practical problems involved in off-line programming a robot from a C.A.D. system

M. Edkins and C.R.T. Smith

1.1   INTRODUCTION

As part of a joint project between UMIST and Leyland Vehicles Limited under the Teaching Company Scheme, the feasibility of using a robot to automate the asssembly of components for commercial vehicles was investigated. The company were very keen to integrate the C.A.D. and C.A.M. aspects of the project. The envisaged system has many similarities to the system developed elsewhere for the manufacture of electrical looms, described by Gibbons(1). However, because of the low-volumes and variety of looms the production time lost with conventional teach-and-repeat methods would be significant.

The recent availability of an off-line robot-programming package for the company's C.A.D. system meant that both of these aspects could be tackled. The use of off-line programming from a C.A.D. system is a relatively new concept with associated high technical risk, and so a development robot-assembly cell was set up to demonstrate that the technique was practicable.

Many problems were encountered but were overcome, and the project has been very successful in confirming the viability of automating this assembly process.

1.2   PROGRAMMING TECHNIQUES

The main techniques for programming a robot are briefly reviewed below to illustrate why accuracy becomes significant in off-line programming.

1.2.1   Teach-and-Repeat

The traditional way of programming robots for simple applications uses the teach-and-repeat method. This method does not require accurate alignment of axes and accurately known link lengths, since errors in these have little effect on the user. A remote-control teach-box or teach-gun is used to move the robot to required locations and these positions are recorded. The robot is effectively being used to digitise locations in space and the prime requirement of the robot's motion performance is that it is able to return to that location in space with good repeatability. Many of the smaller robots used for assembly operations have repeatability figure of ± 0.1mm or better.

1.2.2   Programming Languages

Recently there has been a shift to the provision of robot programming languages (Bonner and Shin(2), Gruver(3)) which simplify more complex applications. A classic example of this is the case of palletising where there is a regular pattern in a stack. Three nested loops can generate a series of rows by columns by layers, the stacking points being calculated rather than taught, thus saving a great deal of time and effort in teaching locations. For such a system to operate, a mathematical model of the robot kinematics is solved by the controller to produce the required joint angles. A particular robot's physical kinematics will deviate from the idealised model due to manufacturing tolerances, and these will lead to a reduction in the accuracy of the robot's positioning with respect to a cartesian space. Specifications for commercially available robots rarely state an accuracy figure and instead only the robot's repeatability is quoted.

1.2.3   Off-line Programming from a CAD System

A logical further extension of the concept of off-line programming is the use of a graphics or C.A.D. system to generate the program.

Several commercial packages have been developed (Anon(4), Bonney(5), Howie(6)) and these aid in the design of robot cell layouts by allowing different robots from a library to be tried for reach, collisions and for cycle times. They may also then be used to generate robot programs without removing the robot from production. The structure of the system used in this project is illustrated in Fig.1. Note that the design information of the product (i.e. its geometry) is integrated with the robot and workplace model. An example of a screen display is given in Fig.2.

As with robot programming languages, the question of robot accuracy becomes significant since the C.A.D. system will be based upon an idealised mathematical model of the robot. In addition there are problems in alignment of the physical and C.A.D. models of the workspace. This arises because on the C.A.D. system the idealised robot axes can easily be aligned with a tables axes, whereas when mounting the physical robot one is relying on the accuracy of construction of the workcell over a large volume.

1.3   REPEATABILITY AND ACCURACY TESTS

From the previous discussion it will be apparent that the accuracy-errors of the robot used would be significant and could be expected to be worse than the repeatability-errors. Since the assembly task investigated was not demanding in terms of precision (± 1mm) and the stated repeatability of the robot was ± 0.1mm it was anticipated that the accuracy might be sufficient for this application.

To measure the repeatability and accuracy of positioning and attitude of a 6-axis robot throughout its working envelope and under different working conditions is a complex problem. Much research is being undertaken in this area to develop techniques to do this (Langmoen et al(7), Podoloff et al(8), Martinez and Rex(9)). Fortunately, the nature of the application involved here allowed several simplifications of the problem:

i) accurate location was only required in a thin plane above the assembly table. The 3-D volumetric accuracy problem could be reduced to a 2-D one.

ii) the grippper was always required to be in a vertical orientation, with only its angle of twist (or roll) changing. This and the design of clamping and the compliance of the pipe made the assembly operations insensitive to gripper orientation errors, and these were therefore ignored.

iii) the weight of pipe being handled was comparatively low and so inaccuracies due to the compliance of the arm were low.

The robot was mounted upside down on a gantry over the table. The gantry was initially designed so that the robot would operate with its elbow up, but early accuracy tests showed large errors. The gantry was then raised to allow elbow-down operation, its more usual configuration, which gave better results.

A very simple technique was used to measure robot accuracy. A sheet of metal was accurately scribed wth a 100 mm square grid using a coordinate measuring machine. A pointer was mounted on the end of the robot wrist. A VAL-II program was written to move the robot above each intersection in turn, and pass control to the teach-box. The pointer was then manually positioned at the intersection and control handled back to the program. The VAL-II program recorded the robot's apparent location in space against the known grid position and recorded the errors in a file.

The results of a single test for errors in the X-Y plane are shown graphically in Fig.3. The worst errors found were around 8mm. The errors appear to have an underlying spherical characteristic, and this might be due to limb length errors, or in the joint angles as opposed to alignment errors in construction (i.e. non-parallel or non-orthogonal pairs of joint axes of rotation).

It should be noted that no general conclusions should be drawn from the results given, since many more tests would be required to obtain a statistically significant set of results.

Since the robot was going to be relocated, and it is intended to automate robot calibration at a later date, no further checks were made. Instead efforts were switched to developing a method for improving the robot's accuracy.

1.4   ERROR COMPENSATION

Since the robot's repeatability was good it was possible to consider the use of an error compensation technique. Both the C.A.D. system and the robot controller had the potential to perform this. It was decided to use the robot controller since this would allow several robots, each with its own error map, to interchange programs based on an idealised robot.

Since the error map appears to be systematic and well behaved it is possible to interpolate between grid points i.e. the error plot is approximated by piecewise linear segments. The procedure adopted was to use a separate VAL-2 program which performed the following steps:

i)   Every relevant robot location obtained from the VAL location file was scanned in turn.

ii)   The square in the X-Y grid containing the location was identified, along with the errors at the corners of the square from an array.

iii)   The X Y & Z error components for the particular location were evaluated by interpolation from the corner values. This process is illustrated in Fig.4.

iv)   The error values were subtracted from the location value to obtain a corrected value and this was used to create a renamed location.

The correction program thus only has to be run once when a new assembly program is downloaded to the robot. There is therefore no additional run-time overhead on the assembly operation itself. The effect of this simple correction improved the static positioning accuracy to better than 1mm distance, which was adequate for this application.

An additional advantage of this technique is that it also solves the problem of obtaining precise alignment between the physical workspace and the C.A.D. model. For example the plane of the assembly plane might be tilted with respect to the robot's XY plane, but this error would automatically be included in the error compensation.

1.5   IMPROVED CALIBRATION PROCEDURES

The semi-manual calibration technique adopted here would not be practical in a production environment. It is too slow, and is prone to error because of the level of concentration required from the operator. It is suspected that if the robot mechanics are dismantled and reassembled during maintenance, the robot error map may change.

Since, in the production system, there will be a pattern of holes to accommodate various clamp and guide layouts, it is planned to use a simple displacement probe to automatically probe these features and so collect accuracy data. This would speed up the process and allow much more frequent alignment checks. Not only would this reduce the chance of faulty operation, but it would also enable maintenance personnel to detect deterioration in the robot's performance.

1.6   CONCLUSIONS

A graphical robot-programming system provides benefits to the designer of robot applications and to the production engineer. The simulation facility allows various cell configurations and robots to be tried and checked for access and cycle times. This can save mistakes in cell design that are costly in time and money to rectify. When programming the robot, the production engineer is able to exploit the geometric detail that has already been captured on the C.A.D. system when the product was designed. This saves time and errors over effectively re-entering geometric information by teaching the robot. Experience has shown that productivity gains of up to 15:1 may be obtained from a skilled system user. An example output file from the system is illustrated in Fig.5.

Many other practical problems were encountered in this project (Smith and Edkins(10)), but it is clear that the most immediate problem preventing the wide use of off-line programming is limited robot accuracy. Unfortunately, it is unlikely that in the short term there will be any great improvement in general robot accuracy until a user-demand is created for it. This is regrettable since simple software compensation for limb-length and individual axis motion would be relatively straightforward to implement at little extra cost (Furuya and Makimo(11)). However, the error-correction technique used here, or variations of it, could be used in many applications where only limited areas of the working volume require accurate positioning.

A longer term problem will be the absence of standard robot languages, since the C.A.D. vendor will not be able to design his software to exploit all of their facilities. It is noticeable that the graphics sytems are naturally very strong at handling geometric data, but are relatively weak in handling input/output and program-flow language statements. An analogous situation exists in programming systems for numerically controlled (N.C.) machine tools. The machine-tool controllers frequently have very powerful canned-cycles (analogous to library subroutines) that are not exploited.

One has only to use a robot off-line programming package on a C.A.D. system for a short while to realise what a powerful tool it is, and how great its potential could be for coping with complex robot applications. The next few years should see a much larger acceptance of this technique.

1.7   ACKNOWLEDGEMENTS

The authors wish to thank both Computervision and Unimation (Europe) Ltd., for their co-operation in supplying pre-release versions of their new products for use in this project. They also wish to thank Leyland Vehicles Limited and the S.E.R.C. Teaching Company Scheme for providing the funding and support for this work.

REFERENCES

1. Gibbons, R.D., 1981
    'Automatic loom-making apparatus',
    4th British Robot Association Conf.,
    Proceedings, Brighton, England, 127-132.

2. Bonner, S. and Shin, K.E., 1982,
    'A comparative study of robot languages',
    Computer (Journal), IEEE, Dec. 82, 82-96.

3. Gruver, W.A., 1982, "Commercially available robot
    programming languages", Int. Conf. on Cybernetics and
    Society, Proceedings, IEEE, Seattle, U.S.A, 294-296.

4. Anon., 'CADDS 4X Robographix User Guide', 1984,
    Computervision, Basingstoke, England.

5. Bonney, M.C. 'Robot simulation and off-line programming
    using GRASP-outline notes', 1985, IEE colloq. on Robot
    Control in Practice (Digest No.12), London, England.

6. Howie, P., 'Graphic simulation for off-line robot
    programming', 1984, Robotics Today (USA), 6, 1, 63-66.

7. Langmoen, R., Lien, T.K. and Ramsli, E., 'Testing of
    industrial robots', 1984, 14th Int. Symp. on Industrial
    Robots, Proceedings, Gothenburg, Sweden, 221-230.

8. Podoloff, R.M., Seering, W.P. and Hunter, B., 'An
    accuracy test procedure for robotic manipulators
    utilising a vision-based, 3-D position sensing system',
    1984, American Control Conference, Proceedings, 19-22.

9. Martinez, P.L. and Rex, D.K., 'Video test tool for
    robotics accuracy repeatability', 1983, IBM Tech.
    Disclosure Bull. (USA), 26, 4, 1816-1817.

10. Smith, C.R.T. and Edkins, M., 'Off-line programming of
    an industrial robot from a CAD system', 1985,
    Computervision European Users Conf., Proceedings,
    Munich, Germany, 251-275.

11. Furuya, N. and Makino, H., 'Calibration of SCARA robot
    dimensions by teaching', 1983, J. Japan. Soc. Precis.
    Eng. (Japan), 49, 9, 1223-1228.

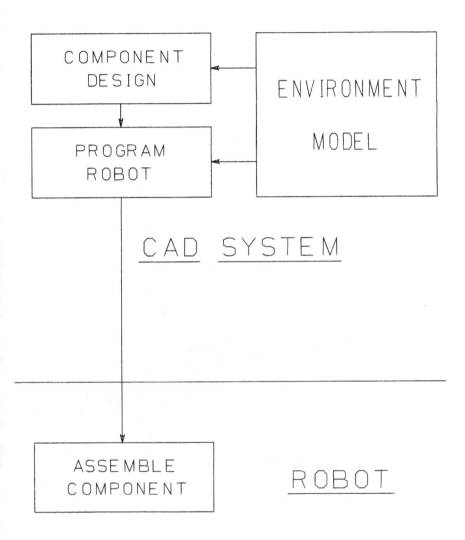

Fig. 1   Structure of robot programming system

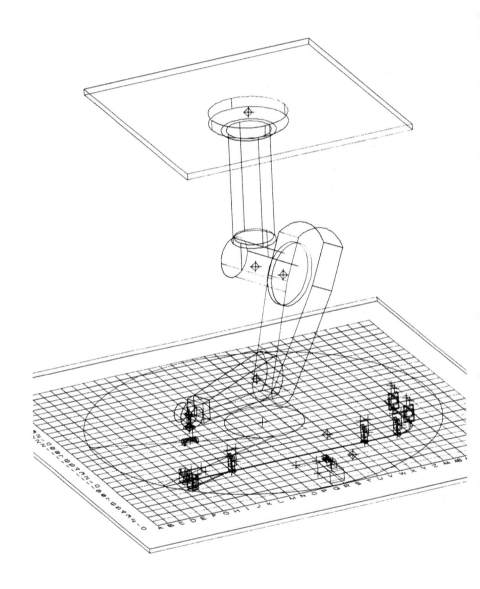

Fig. 2   Robographix generated model of the
         PUMA 560 robot over the assembly table

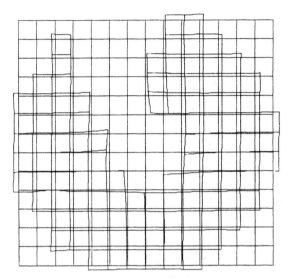

TOP VIEW: X & Y ERRORS, SCALED 5X,
ON 100MM GRID

SIDE VIEW: Z ERRORS, SCALED 5X

Fig. 3   Exaggerated graphical representation of
the X, Y and Z plane errors

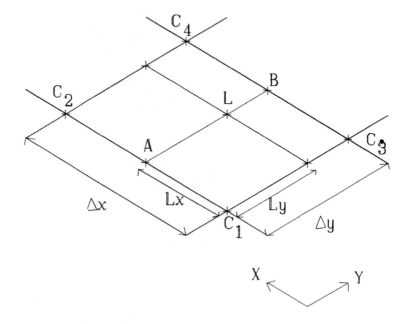

Let E(P) represent the error in x, y or z at point P. then:

$$E(A) = E(C_1) + ((E(C_2) - E(C_1)) \times Lx/\Delta x)$$

$$E(B) = E(C_3) + ((E(C_4) - E(C_3)) \times Lx/\Delta x)$$

$$\text{and} \quad E(L) = E(A) + ((E(B) - E(A)) \times Ly/\Delta y)$$

Fig. 4   The interpolation method used for error-compensation within a grid square

```
 1 ; *** COMPUTERVISION *** *** VAL2 PROCESSOR ***
 2 ;                *** VERSION DATE: XX-XX-XX ***
 3          BASE 0,0,0,0
 4          SPEED 25.4
 5          TOOL cad.0.tool [0]
 6 ;SET UP TO LAY PIPES
 7          TIMER 1 = 0
 8          WAIT TIMER(1) = 0              PO cad.0.tool [0]
 9          BREAK                          0,0,23.4,0
10          SIGNAL 2
11          OPENI
12          MOVES new.0.move [0]           PO cad.0.move [0]
13          MOVES new.0.move [1]           0,0,110.95,-180,-90,0
14 ; PICK UP PIPES
15          BREAK
16          CLOSEI                         PO cad.0.move [1]
17          MOVES new.0.move [2]           132.5,-619,361.01,-180,-90,0
18          MOVES new.0.move [3]
19          MOVES new.0.move [4]
20          MOVES new.0.move [5]           PO cad.0.move [2]
21          MOVES new.0.move [6]           135.7,-619,361.01,-180,-90,0
22          MOVES new.0.move [7]
23          MOVES new.0.move [8]
24 ; FEED THE PIPE                         PO cad.0.move [3]
25          BREAK                          142.6,-683.25,361.01,-180,-90,0
26          OPENI
27          MOVES new.0.move [9]
28          MOVES new.0.move [10]          PO cad.0.move [4]
29          BREAK                          142.6,-683.25,343,-180,90,0
30          SIGNAL 1
31          MOVES new.0.move [11]
32 ; END OF PROGRAM                        PO cad.0.move [5]
.END                                       142.6,-683.25,343,-180,0,0
```

PROCESS   TRANSFORMATION

Fig. 5   A VAL II program file and associate
         location file after correction

The application of rule based methods to inspection and quality control in a manufacturing process

D. Harrison, J. Billingsley, F. Naghdy

1. INTRODUCTION

The introduction of computer control has made possible the automation of many manufacturing operations. To fully automate a factory however requires more than computer control of fabrication, assembly, adjustment and testing. It is necessary to replace the information flow between these operations and the decision making process, together with control strategies based on this information. At present information flow often consists of linguistic communication between operators, supervisors and managers. Much of the decision making and control is based upon "rules of thumb" or heuristics also expressed in a linguistic form. Mamdani (1,2) has proposed rule based "fuzzy control" which provides a direct path between the type of loose linguistic statement of a control strategy often found in manufacturing and its implementation as a quantitative control algorithm. This paper describes how a controller based on similar concepts was used to control the adjustment of manufactured products and improve the quality of the final product.

Many techniques for the synthesis of automatic controllers will be found in the control literature, but most standard textbook approaches involve qualitative numeric calculations based on mathematical models. These controllers require highly specialised staff to build a model of the particular process. It is often the cost of this modelling, not that of computer hardware, which makes the automation of small scale manufacturing uneconomic. The rule based controller utilizes the existing linguistic model of the process for control purposes and could be tailored to a particular process by existing technical staff.

Assilian (3), Baaklini (4), King and Mamdani (5) and others have applied this type of control to various plants, from steam engines and sinter plants to a batch kettle reactor. The emergence of this control has done much to bridge the gap between human articulate reasoning (which governs many manufacturing processes at present), and system control. Thus it appeared well suited to the task of control in manufacturing. The dramatic fall in the cost of computer hardware made its application to small scale manufacturing a possibility.

2. THE MANUFACTURING PROCESS TO BE CONTROLLED

Earlier work at Portsmouth has already automated the adjustment and testing of energy regulators (6,7). The regulator consists of a number of mechanical components which are fabricated to fairly loose tolerances. The adjustment screw and backstop allows the variation in mechanical performance caused by the cumulative effect of these tolerances to be "tuned" out (see Figure 1). In the factory each unit is at present adjusted by hand to an intermediate mechanical specification. The performance criteria of interest to the product user is the mark-space ratio and the cycle time of the power supplied by the unit. This performance is a result of the combined mechanical and electro-mechanical properties of the unit, and is monitored at the final adjustment stage. At present when variation in and across batches causes a large number of units to fail the final test a supervisor is alerted. She consults a production manager and together they decide on a new intermediate specification to bring the output performance back within specified limits.

The aim was to replace this crude control with a system which constantly monitors the testing information and other measurements on unit parameters and uses this to update the adjustment specification, thus improving the quality of the final output. A number of "rules of thumb" which expressed the existing control strategy were embedded in the controller in the form of linguistic strings. These were then translated into a qualitative algorithm which could be executed by the controlling computers.

3. THE CONTROLLER

Initial experiments have been conducted with a simple rule based controller. The variables are grouped by thresholding into Positive Big, Positive Medium, Positive Small, Zero, Negative Small, Negative Medium, Negative Big, where the criteria form an exclusive set. The rule base is made up of ten rules of the form

If Ratio is PB and Closing Point is NB and Hysteresis is PS Then Make Change in Closing Point = PM

A more sophisticated controller using fuzzy logic, as described by Assilian in ref. 3, has been developed and tests with this controller on a computer model of the regulator appear promising.

4. INSPECTION THEORY

In addition to providing control over the specified adjustment angle rule-based control allows control of the inspection of failed regulators. The measurement of such parameters as winding resistance, switch pressure, and requests for visual inspection where necessary will be carried out under computer control. At present a certain proportion of components are inspected on arrival in the factory. In addition regular checks are carried out on samples to detect trends in component variation. The sample sizes of these checks are fixed at present, set by a rigid specification, and these checks have a significant fixed cost. With computer control of inspection the inspection can be directed to gather information only where appropriate, and thus reduce costs and improve control.

A typical sequence for inspection of units with faulty hysteresis is shown in Figure 7. The type of rule structure used to control the adjustment is being used to implement this inspection sequence.

5.   EXPERIMENTAL RESULTS

Large scale tests of regulator adjustment had established the incremental influence of adjustment angle on the mean mark space ratio. Comparing Figures 2 and 3 it can be seen that moving the specified angle from 267 in Data set 1 to 273 in Data set 2 produced a significant change in the mean ratio. It was therefore necessary to devise a feedback mechanism for modifying adjustment angle in the light of mark-space performance, bearing in mind the random variations of the latter from regulator to regulator. A number of smaller sample tests have been carried out with a simple rule based program. The regulators are tested in batches of four and the information from each batch is processed to influence the setting of the next by application of the following rules.

If Ratio is PB and Closing Point is NB and Hysteresis is PS Make Change in Closing Point = PM

If Ratio is PB and Closing Point is NB and Hysteresis is PM Make Change in Closing Point = PB

If Ratio is PM and Closing Point is NB Make Change in Closing Point = PM

If Ratio is PS Then Change in Closing Point = PS

If Ratio is Zero Then Change in Closing Point = Zero

If Ratio is NS and Closing Point is PS Then Make Change in Closing Point = PS

If Ratio is PM and Closing Point is NM Make Change in Closing Point = PM

If Ratio is PM and Closing Point is Zero or PS Make Change in Closing Point = PS

If Ratio is PB and Closing Point is NS Make Change in Closing Point = NM

If Ratio is NM and Closing Point is PM Make Change in Closing Point = PB

Regulators in a state not covered by one of these rules require rework or manual resetting of their backstops, and appropriate error messages are produced for each case. Figures 4 and 5 portray two separate tests which show the success in driving the adjustment angle to centralise the

spread of mark-space ratio. With only four regulators in
each batch, using simple averaging to derive a mean value of
ratio for the batch was prone to errors caused by rogue
points. By increasing the batch size to eight, stripping off
the top and bottom two, then taking the median value, it is
hoped this problem will be reduced. The adjustment of 2000
regulators also provided useful information on the applicability
of automated inspection. Of particular interest was a group of
17 regulators which had failed the ratio test. They passed
subsequent tests on their thermal parameters. However further
inspection of their mechanical behaviour revealed that although
they appeared to have been correctly adjusted when mounted
horizontally, when they were tipped into a vertical orientation their closing points shifted by an average of ten degrees,
(see distribution centred about 61 in Figure 6). This was a
result of large tolerances in the bimetal pivot support causing
such slack in the mechanism that it was significantly affected
by gravity. The discovery of this problem led to design
changes in the pivot mechanism. This was a good example of
how automated inspection could apply standard checks but
ultimately would require human expertise to identify
particularly "odd" failures.

6.   CONCLUSIONS

Variation of components from batch to batch causes
performance problems in many products. With the energy
regulator it is often necessary for the Quality Assurance
Department to extract samples from assembled batches before
they undergo testing (see Figure 8). These are then adjusted,
tested and analysed to determine a particular setting angle
for that batch. In this way the methods discussed in this
paper are already being applied in a cumbersome manual fashion
to achieve the quality of the final product. This paper
therefore describes the automation of this technique to permit
continuous adaptation and steady throughput.

7.   REFERENCES

(1)   Mamdani, E.H. (1974). Application of fuzzy algorithms
      for the control of a dynamic plant. I.E.E. Proceedings
      121(12), 1585-1588.

(2)   Mamdani, E.H. (1976). Application of fuzzy logic to
      controller design based on linguistic protocol.
      Workshop on Discrete Systems and Fuzzy Reasoning,
      Queen Mary College, London.

(3)   Assilian, S. (1974). Artificial Intelligence in the
      control of real dynamic systems. Ph.D. Thesis,
      London University.

(4)   Baaklini, N. (1976). Automatic learning control using
      fuzzy logic. Ph.D. Thesis, London University.

(5) King, P.J. and Mamdani, E.H. (1976). The Application
     of fuzzy control systems to industrial processes.
     Automatica, 13, 235-242.

(6) J. Billingsley, F. Naghdy and D. Harrison. The
     Craftsman Robot. Electronics and Power (Nov/Dec. 1983).

(7) F. Naghdy, J. Billingsley, and D. Harrison. Simulation
     of human judgement in a robot-based adjustment system.
     Robotica (1984), Vol.2, pp.209-214.

46

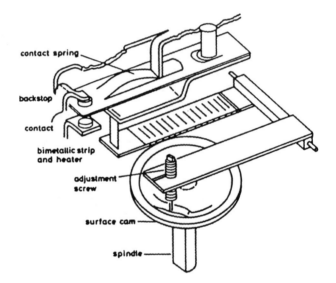

Mechanism of the regulator

Figure 1

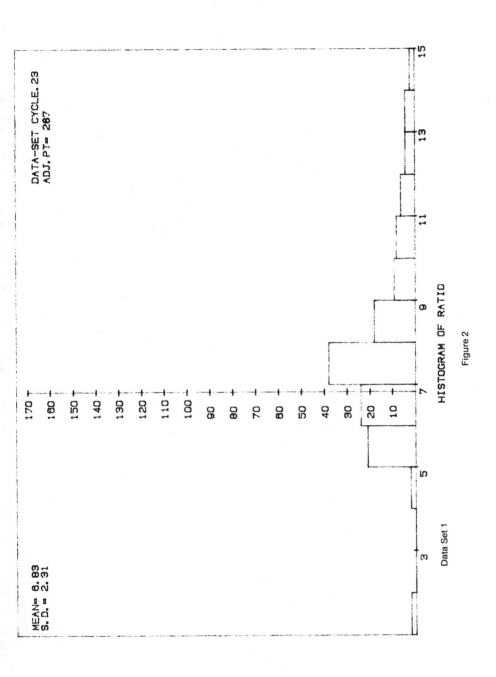

Figure 2

48

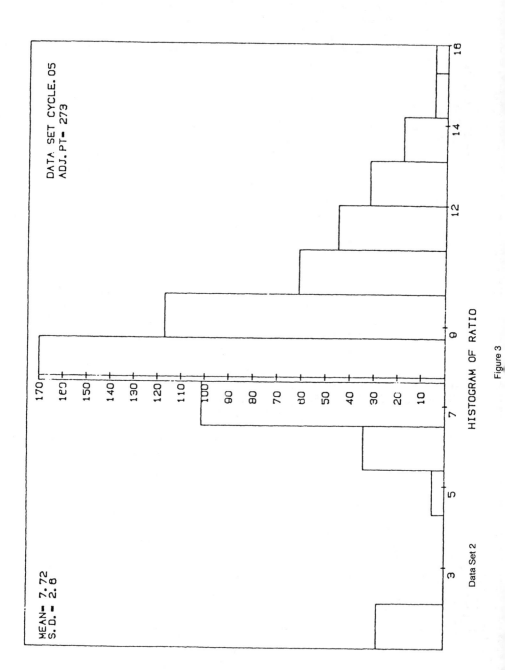

Figure 3

Data Set 2

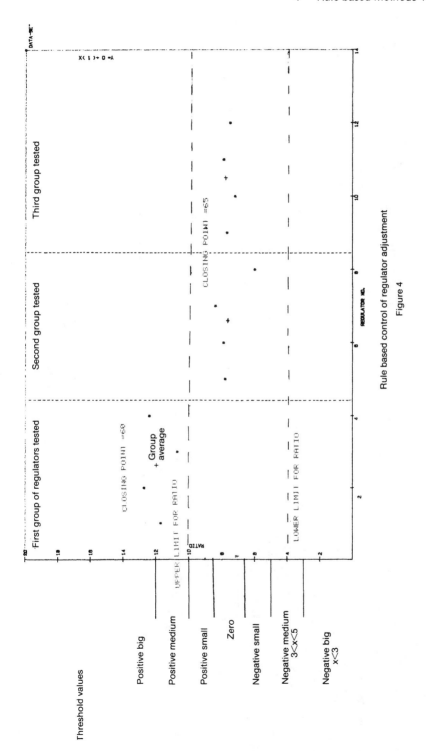

Rule based control of regulator adjustment

Figure 4

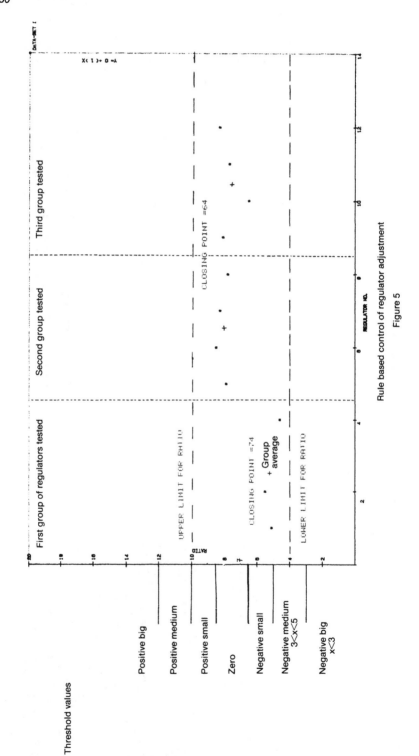

Rule based control of regulator adjustment

Figure 5

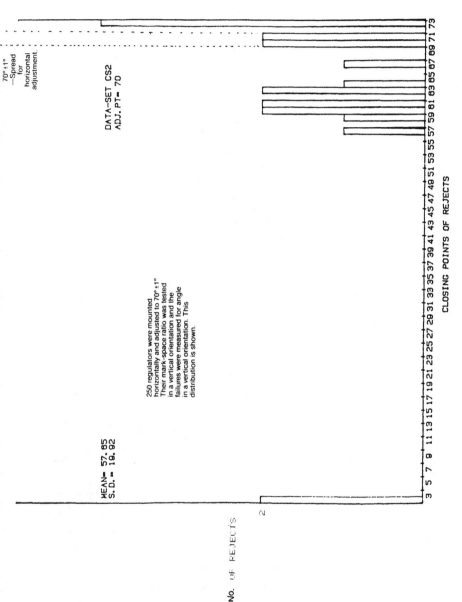

70°±1°
—Spread
for
horizontal
adjustment.

DATA-SET CS2
ADJ. PT= 70

250 regulators were mounted
horizontally and adjusted to 70°±1°
Their mark-space ratio was tested
in a vertical orientation and the
failures were measured for angle
in a vertical orientation. This
distribution is shown.

MEAN= 57. 85
S. D. = 19. 92

NO. OF REJECTS

CLOSING POINTS OF REJECTS

Figure 6

52

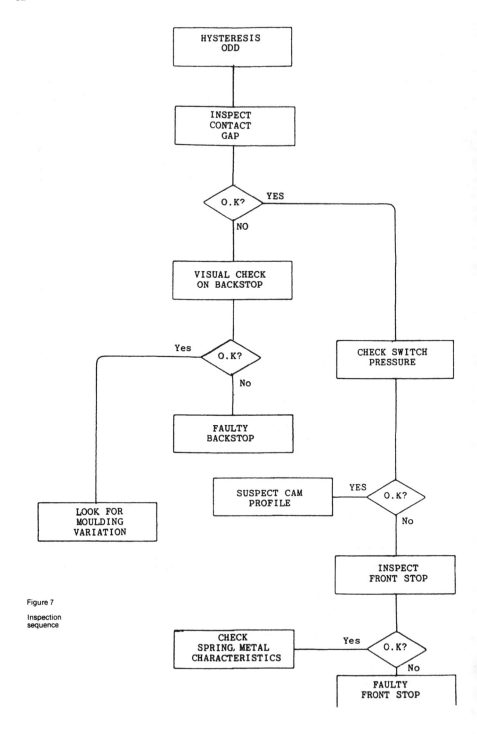

Figure 7

Inspection
sequence

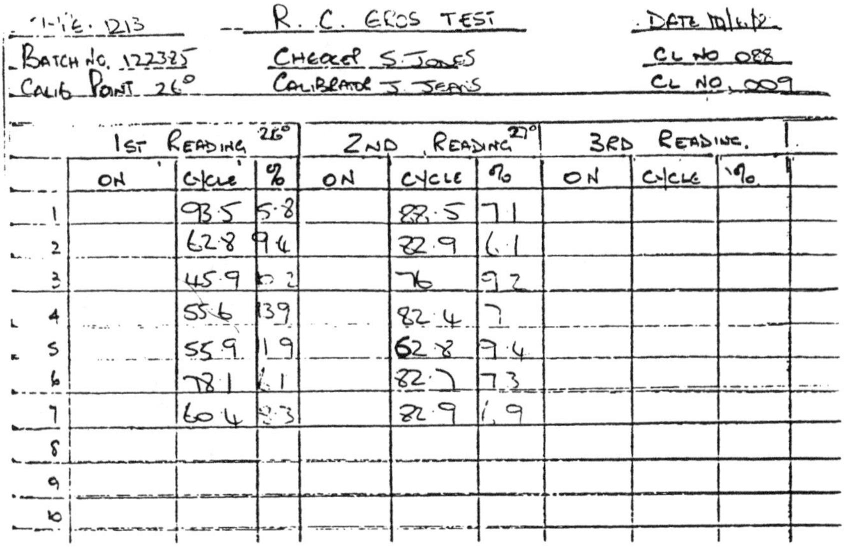

	TIE· 213	R. C. EROS TEST				DATE 10/6/12		
BATCH NO. 122325		CHECKED S. JONES				CL NO O88		
CALIB POINT 26°		CALIBRATED J JEANS				CL NO. 009		
	1ST READING 26°			2ND READING 27°			3RD READING.	
	ON	CYCLE	%	ON	CYCLE	%	ON	CYCLE	%
1		93.5	5.8		88.5	71			
2		62.8	94		82.9	6.1			
3		45.9	6.2		76	92			
4		55.6	139		82.4	7			
5		55.9	19		62.8	9.4			
6		78.1	6.1		82.7	73			
7		60.4	8.3		82.9	1.9			
8									
9									
10									

Data for manual quality control

Figure 8

Automatic location editing of assembly robot programs

M.A. Woollett

This paper examines the programming task required by a
robot workstation within a flexible assembly system. A
system is described which optimises programming by use of
off-line programming and automatic location editing.

The system uses a robot equipped with two simple two
state touch sensors which are manipulated to provide
redundant data capable of accurately describing various
types of 5 or 6 degree of freedom locations. Locations are
determined by automatic search routines and edit algorithms,
which enables automatic calibration of the sensors.

The implications of the results are discussed and it is
concluded that automatic location editing can increase both
productivity and flexibility of assembly workstation.

Keywords: Search Routine, Location Editing, Automatic
Predictive Programming, Flexible Assembly System.

1.   LOCATION EDITING FEASIBILITY AND USE

1.1   Present Methods of Robot Programming

Most robots are programmed using either manual teach or
leadthrough both of which require the robot to be programmed
on-line, i.e. the robot itself must be used. This requires
significant amounts of time and also the program may become
unreliable with use requiring either editing (if the
programming system allows) or re-programming, both of which
cause non-productive downtime.

Thus it is usually considered desirable to minimise on-
line set-up or programming time often by use of off-line
programming. Manual programming time is directly dependant
on the number of locations needed, the accuracy needed, the
number of degrees of freedom involved and the programming
and control facilities of the robot system.

1.2   Anatomy of a Robot Program

The robot program can be considered to have two

constituent parts, namely:

1. A sequence of instructions/operations for the robot to carry out, i.e. the algorithm.

2. A list containing precise details of the positions the robot arm will be required to move to. i.e. the location file.

The algorithm or logic of the program can be determined without the robot, it can be programmed automatically by a computer if the task to be performed can be broken down to a number of standard operations. The complete algorithm can then be built up from standard macros (1).

The location file almost invariably requires on-line programming usually by manual teach mode with consequent downtime. This is because the accuracy of locations programmed off-line is limited by the predictive accuracy of the off-line model of the robots workspace - often 10 - 100 times poorer than the robot repeatability.

1.3 Methods of Reducing On-Line Programming Time

There appears to be four ways of doing this:

1. Dynamic real time control of the robot using sensors, not just playback of pre-recorded programs.

2. Increase predictive accuracy by use of more complex off-line model.

3. Reduce accuracy required.

4. Automate the on-line programming.

The system described in this paper explores option 4 and is primarily concerned with location programming. The investigation was carried out using a Puma 560 robot system augmented with a touch probe sensing system and a micro-computer (for detailed location manipulation).

1.4 Automatic On-Line Programming

The system requires descriptions (or definitions) of the desired locations to be input, from which it can attempt to determine the location co-ordinate values to the desired accuracy.

The locations are defined by the requirements of the algorithm and the robot work-space. If an off-line model of the workspace is used then low accuracy location coordinates can be programmed off-line together with their descriptions. Alternatively the location coordinates can be roughly programmed on-line fairly rapidly using the established manual teach method, again using off-line location descriptions. The rough locations can then be automatically edited up to the accuracy required (Figure 1).

The location edit is an automatic control system and it is therefore necessary to:

1. detect and determine the error between the desired value and the actual value.

2. apply a correction to reduce the error.

The above desired value is a robot location transform with 5 or 6 degrees of freedom (DOF), not a single degree of freedom variable. Three of these degrees of freedom are the cartesian vectors x, y and z, the other three are the interdependant angular rotations. On the VAL system (2) these rotations are designated O, A and T.

1.5 Sensors for Automatic Location Programming/Editing

A sensing system is needed which is:

1. Capable of detecting errors in 6 DOF,

2. Sufficiently flexible to cope with a variety of assembly workpieces,

3. Accurate enough to enable successful assembly location editing.

It was decided to use tactile sensing but no 6 DOF sensor exists or was thought feasible. However by using the 6DOF robot to vary the measuring axis (axes) of the sensor 6 DOF data can be measured using a sensor with only 1 DOF. Furthermore the robot can be moved along the measuring axis and can use its own joint feedback sensors to determine its position, thus the sensor need only be a two state touch sensor (i.e. a touch trigger probe).

Methods similar to the above are widely used for measurement on CNC machine tools and on co-ordinate measuring machines using touch trigger probes (3). On machine tools touch trigger probes are used for single DOF location editing when compensating for variability in tool size.

1.6 Basic Search Program FIND1

This robot program is a single degree of freedom search along a line segment defined by two location variables for a point of contact which then defines a third location. The position of the first obstacle detected as the robot travels from AIR to TARGET is recorded in EDGE (see figure 2). The search is done in 3 or 4 stages at reducing speeds in alternate directions since a high initial speed is needed to minimise search time but a final very low speed search is required for accurate results.

FIND1 is used as a subroutine or macro and is called by the other search programs when a search is needed, the other program having set locations AIR and TARGET to appropriate values. The result of the search EDGE will usually then be recorded in the PUMA controller and subsequently be down-loaded to the master editing program, which can then determine the exact locations required by the user program (Fig. 3).

FIND1 was tested for repeatability, the results showed random errors with a standard deviation of 0.045mm. To reduce these it would be necessary to make several searches and then use the mean of their results (if a single DOF were involved). If more than one DOF were involved then extra searches could be done to make the results redundant and the method of least squares could then be used to obtain the data for location editing (eg to describe a plane at least 4 separate points should be used).

If systematic errors were present then an iterative series of searches could be used, continually updating the location (possibly using a moving average), ceasing the iteration when it became clear that optimum accuracy had been attained.

1.7 Operation of the Complete System

It was envisaged that on-line operation would proceed in the following four stages:

1. Initial calibration following installation

2. Automatic probe calibration

3. Automatic location edit of the user program

4. Production, operation of the user program.

While manual intervention is required in stage 1, this is only needed upon installation, once the system is set up only stages 2,3 and 4 need be performed and these can be fully automatic. The system would require off-line back up facilities, and while important, these are not a prime concern of this paper.

2. THE LOCATION EDIT TASK

Simple sensors cannot judge relative transforms between two locations, they can only determine the transform of one location with respect to a fixed location, usually the sensor casing (Fig. 4). More complex sensors (often visual) can judge relative transforms between two undefined locations (Fig. 5).

On robots, relative transforms are required between the

robot payload and a location in the robot workspace. An
example of this is the transform of a peg (P) with respect
to a hole (H).If this transform (P.H) is knownthen the robot
can vary the peg location to cause the peg to be inserted in
the hole.

If a relative location is required using a simple
sensor then two results must be used to determine the
relative transform (Figure 6).

Since the robot manipulator forms part of the sensor,
two sensor arrangements must be used (Figure 7), one with a
touch probe on the robot end effector used for determining
transforms of objects in the workspace (S1.H), and one with
a probe fixed in the workspace used to determine transforms
of objects held by the robot (P.S2). This effectively means
that two sensors are used and that the relative transform
between these must be known (S2.S1).

$$P.H = P.S2 : S2.S1 : S1.H$$

In keeping with the decision to build up the system
section by section no attempt was made to implement this
equation directly. Instead absolute locations (in robot
world co-ordinates) were determined and relative transforms
calculated when required. This implies that four locations
must be determined for the peg in hole assembly.

1.   The peg is replaced with a (mobile) probe which
      detects the hole position.

2.   The hole is replaced with a (fixed) probe which
      detects the peg position.

3.   The mobile probe is calibrated.

4.   The mobile probe is used to find the position of the
      fixed probe.

When a robot is being positioned in space using world
co-ordinates the controlled value is the position of the
robots tool point, on the Puma this is the centre of the
tool flange. The point of interest to the operator is the
position of the robot payload. Thus when the mobile probe
is in use the point of interest is the position of the
stylus tip. Using tool transforms it is possible to
directly control the location of any such position providing
its tool transform is known. Tool transforms are described
in reference 2.

2.1   Initial System Calibration

This is required to determine the transforms of the
fixed and mobile probes (FXPB and PBTX). Once this has
been accurately done the mobile probe can be used to
determine the value of static workspace locations, and

likewise the fixed probe may be used to determine the
transforms of objects held in the robot gripper.

Initially the position of the store for the mobile
probe must be accurately taught in order for the robot to be
able to pick up the mobile probe. This is the only location
which must be accurately taught and this need only be done
once.

2.2 Editing the Probe Tool Transforms, PBTX. TTX Programs.

PBTX is approximately determined (to 10 mm) using
program TTX and then accurately (0.1 mm) edited by TTX2 and
TTX3. These programs work by rotating the probe about the
transformed tool point and then measuring the change in x, y,
z position of the stylus (using search routine FIND1). An
appropriate correction proportional to the change in stylus
x, y, z position is then applied to PBTX.

Program TTX contains an erratic systematic error which
does not fall as PBTX becomes more accurately known.
However after three iterations of running TTX and applying
corrections PBTX is correct to 10 mm. TTX2 and TTX 3 use
180 degree rotations and can only safely be used once PBTX
is known to within 10 mm, unlike TTX which only uses 5
degree rotations and can cope with errors of 200 mm. TTX2
and TTX3 do contain systematic errors, however these fall to
zero as PBTX becomes more accurate. These errors can be
effectively eliminated by iterative correction of PBTX (4).

2.3 Editing FXPB. The Use of XSRCH, YSRCH and ZSRCH

Once PBTX is known and implemented as a tool transform
then the absolute location of the fixed probe (FXPB) can be
roughly taught. The external searches XSRCH, YSRCH and
ZSRCH can then be used to accurately edit FXPB.

XSRCH implements two searches towards the target
location from the positive and negative x directions as
defined by the target location. The two edges resulting
from these searches are recorded. Similarly YSRCH
implements searches in the y directions and ZSRCH is a
single search towards the set target from the negative z
direction.

The above describes the full calibration of the probes
from start up (stage 1). This is only necessary after
installation or after moving any of the system hardware, e.g.
if the fixed probe is moved.

The locations determined in stage 1 are recorded and
used as a basis for stage 2. Stage 2 is a fully automatic
accurate calibration carried out each time the mobile probe
is picked up. This was thought necessary because of
probable variation in the gripped position of the probe each
time the probe was picked up. FXPB should not change but

the check only takes about 30 seconds and a large error could indicate a major system fault.

2.4 Location Edit of User Program

The editing system has been developed to the point where the absolute value of certain location types can be determined. The system requires the following input:

1. The approximate location value.

2. A description of the location.

At present the location description is manually dealt with, however there is no known reason why this can not be automated.

The user program edit is stage 3 of the operating software. The edit begins with the robot holding the mobile probe and with both probes accurately calibrated (i.e. PBTX and FXPB accurately known).

Initially the mobile probe is used to edit all the static locations (i.e. locations of objects not being transported by the robot).

Then the mobile probe is put down (program PLPB). The robot then picks up its various payloads and determines their relative transforms using the fixed probe.

2.5 Types of Search Program

Three types of search program have been developed for three different tasks.

The External Searches XSRCH, YSRCH and ZSRCH. These are used to determine the positions of the centres of solid objects. This includes the fixed probe calibration (section 5.2).

The Internal Searches XHOLE and YHOLE. These are used to find the centre of a hollow surface e.g. a round hole in 2 DOF space. XHOLE implements two outward searches from CENT (the estimated location of the hole centre) in the positive and negative (transformed) x direction, and records the results. YHOLE implements similar searches in the y direction. The location CENT can then be edited (Fig 9).

The Surface Programs SURF1 and SURF2. These are used to determine the locations of 4 or 5 points a fixed distance (the probe **radius**) from a flat surface. This data is then processed to provide:

1. The x, y, z equation of the best fit plane for the data.

2. The transform needed to rotate the robot orientation so that the tool axis is perpendicular to the plane.

2.6 Edit Algorithm

This controls the edit process, selecting search programs and processing their results. The edit algorithm also decides whether a location is accurate, or whether further searches are needed. One of the most successful algorithms used to date (described in Fig 10) aquires data until the average of that data (and consequently the edited location) changes by such a small amount that it can be considered accurate. The edit algorithm should be optimised and there is scope for further development of edit algorithms.

2.7 Determination of the Gripper Transform, GPTX

This is the relative transform between the top centre of a gripped object and the robot tool point. When used as a tool transform it enables the robot to move to a location at which it may open or close the gripper around a object without moving the object or applying any net force or torque to it.

The gripper jaws are deliberately asymetric so that they can rigidly hold a variety of shapes. This asymetry causes GPTX to vary as the payload varies. This variation can be automatically calculated off line but at present it is dealt with by on line methods.

3.0 EXPERIEMENTAL RESULTS

3.1 Results used for System Development

It has been shown that the nominaly straight and square world co-ordinate system consists of curved lines. Thus a simple linear off line model of the robot workspace based on this system will contain significant errors and will thus be unsuitable for programming assembly tasks. Over short distances linear interpolation appears to be accurate enough for assembly location programming (Fig 8).

The robot braking test showed that there was a delay of about 40ms before input signals affected the robot. This delay (thinking time) was constant for all speeds and directions of operation. The actual deceleration (braking) time and distance were negligable at all normal operating speeds (0.5 m/s or less)

3.2 Repeatability Performance of FIND1

The repeatability of FIND1 (using a home made probe) was found to be +/- 0.16mm (+/-3σ limits). This is three times as large as the robot repeatability (+/-0.1mm) and spans some twenty minimum world increments. The resolution of the robot world x,y and z coordinates varies but is in the region of 0.03 to 0.04mm.

The absolute limit on the performance of FIND1 is the
resolution of the world system not the robot repeatability.
If the mean of several readings (EDGE s) is used then the
variation in the mean drops as the number of readings rises.
However the number of searches performed during a location
edit should be minimised and thus it may only be feasable to
increase the accuracy of FIND1 up to the robot repeatability.
It should be noted that when the user program of edited
locations is run there will be additional errors dependant
on the robots short range predictive accuracy.

During the fixed probe calibration process it was noted
that there was a systematic overshoot error present in FIND1.
This was found to be approximately 0.4mm. The minimum speed
of the Puma was determined as approx 2mm/s. With a 40ms
delay therefore an overshoot of 0.08mm can be expected.
This does not fully account for the actual overshoot so there
may be other causes of overshoot.

The source of the 40ms delay must be identified and if
possible eliminated. Its effects are similar to those of
the cycle time error problem described by Roe(Ref. 3). who
noted that it had been solved on co-ordinate measuring
machines. The proposed VAL2 language/operating system
provides asychronous parallel processing of robot motion and
program execution. This implies that in flight snapshots of
the robot position can be recorded.

In the present system the effects of the overshoot are
negated by the method of processing the FIND1 data. This
either uses the differnces between parallel direction
searches (as in TTX or SURF1) or the sum of opposite
direction searches (as in XSRCH or XHOLE) in both cases
thereby eliminating any fixed overshoot error.
If needed the error could be dealt with either by;
1 adding an appropriate correction to the EDGE results
or 2 modifying FIND1 to perform two increment speed searches
in opposite directions using the mean of their results as the
actual position of EDGE.

3.3 Performance of the Search Program

TTX2 and TTX3 appeared to be capable of determining PBTX
to within 0.3mm. The internal and external searches could
also achieve 0.3mm. Some improvements may be possible from
use of improved edit algorithms (which are off line) but the
performance of FIND1 is a key factor. The use of iterative
techniques involving a moving average may prove beneficial.

The orientation edit system was tested on a peg in hole
task and performed satisfactorily (section 3.4). The edit
algorithm which used a least squares program seemed to
perform perfectly with both SURF1 and SURF2.

3.4 Comparison of Programming Methods

The orientation edit was tested on the peg in hole
benchmark task using manual teach for the x,y and z co-
ordinates. Automatic editing of x,y and z was not tested on
the benchmark. Various manual teach methods were also tested.
The task involved insertion of a cylindrical steel peg

into a 25mm diameter hole with a 0.12mm diametral clearance.
The hole was inclined at roughly 15 degrees to the Puma's z
axis in an undefined direction (not in line with x or y axis)
The peg was to be inserted to a depth of 20mm and the hole
entry had a 45 degree chamfer.

Control/Programming Mode	On line robot time needed
Free = leadthrough	not tried
Joint Teach	after 15 min program tested, it failed
World Teach	6 Min
Tool Teach	4 Min
Tool Teach and tool transform	4 Min
Automatic edit for O.A.T to 50sec plus	
Tool Teach for x,y,z took 2min giving total <3 min	
Full auto edit	ESTIMATED AT <2 min

A world or tool control mode appears essential for
teach programming accurate general locations.
In tool mode roughly 2 minutes were spent teaching
orientation (O.A.T) followed by 2 minutes teaching position
(x,y,z). In world mode the teaching of orientation took 2
minutes as in tool mode but the teaching of position took
twice as long at 4 minutes.
Automatic location editing determined the orientation
in under half the time required to teach it. It is expected
that auto edit of position should take about 1 min thus the
on line time required for a full auto edit is estimated at
<2 min. It should be noted that this result is expected
with the present development system.
Since the speed of FIND1 can probably be quadrupled it
should be possible for a fully optimised system to edit 6
DOF locations at a rate of 1 per minute while achieving
accuracy approaching the robot repeatability.

4.0 DISCUSSION

It is assumed that initially automatic location editing
will only be used on an assembly task where parts from
different assemblies are universally interchangeable. It
should be possible to extend this principle such that
relative locations on workpieces are known to greater
accuracy than is required for the assembly operation (Thus
if a workpiece has two holes into which pegs are to be
inserted, then if these holes are 20mm apart on one workpiece they will be 20mm apart on all the other workpieces
of that batch). If this is so then once one 6 DOF location

on a workpiece has been determined (via location editing) then all other locations on that workpiece can be determined from the determined location and the workpiece specification using xompound transformations. World space curvature can be expected to reduce the accuracy of the locations if predictive programming is used over large distances.

Thus only a proportion of the locations needed by the user program would need on line editing. The remainder could be defined in terms of those which were on line edited and could then be calculated off line. This approach has been successfully implemented for stacking operations (1).

This approach means that locations defined by special features need not be directly edited. Instead the location of a standard feature (eg a hole) can be edited and the special location determined off line as described above.

It is suggested that the system be designed with edit macros. These would be capable of determining certain types of location information from certain workpiece features, which would be present on most workpieces. Such a system should be sufficiently flexible to be useful in a flexible assembly cell. If needed special edit programs could be written and used for special workpieces.

On some tasks certain locations may be determined by the part complete assembly and it may be necessary to partly assemble the workpieces so that these can be edited. The editing task could be kept simple if such locations could be determined off line using compound transforms as described above.

It is not certain that the workpiece itself requires a 6DOF assembly robot since many assembly tasks only require 3 DOF for vertical stack assemblies. It must be possible to align these 3 DOF with 3 of the robot DOF (such that the robot z axis is parallel with the workpiece assembly/ insertion axes) and this may to require a 5 or 6 DOF robot and editing system. A 4 DOF robot would require accurate initial alignment with its workspace and possibly regular re-calibration and re-alignment. The editing system described in this paper could be programmed to regularly (eg every hour or so) rapidly automatically check robot calibration, record the results and re-calibrate as required.

The possibility of using a multiple tool end effector should be considered, noting that one such tool could be a touch probe.

5.0 CONCLUSIONS

1. A fully automatic location editing system capable of editing 6 DOF locations at a rate of 1 a minute, accurate to within the robot repeatability, is technically feasible.

2. The touch trigger probes presently available commercially have a higher accuracy than is needed, and a higher cost than is desirable for the system described in 1 above.

(for most assembly robots, ie those with repeability =/- 0.1mm).

3. Automatic location editing has the potential to:

    1. reduce on line programming time and/or increase location accuracy.

    2. Provide a facility for rapid automatic checking and re-setting of robot - workspace calibration.

    3. Facilitate off line programming

4. Automatic location editing should be further investigated and evaluated.

5. A single stylus touch sensor (such as a touch trigger probe) is ideal for a flexible location edit system. It should be noted that considerable software (search routines, search programs and edit algorithms) is needed to use one of these sensors.

References

1. Adey, Holliday, o Gorman, Shanks, Shields & Watkins. The offline Programming report. CRAG Research Memorundum no.28, Cranfield Institute of Technology 1983

2. Unimation Inc. Users guide to VAL. Unimation UK Telford

3. Roe, In cycle gauging. 6th IFS Automated Inspection and Quality Control Conference 1982.

4. Woollett. Automatic Location Editing of Robot Programs MSc Thesis, Cranfield Institute of Technology 1984.

5. Williams. A User-Specific Approach to Robot programming using a Macro-Assembler Technique. Robotics & Automation Group, Cranfield Institute of Technology. Automan 85 Conference IFS 1985.

6. Wright. Accurate Robot Programming for Surface following Tasks using Automatic Location Editing. Robotics and Automation Group, Cranfield Institute of Technology. Automan 85 Conference IFS 1985.

7. Ambler Popplestone and Kempf. An experiment in the offline programming of Robots. DAI Research paper 170. University of Edinburgh 1982.

8. Williams, Walters, Ashton & Reay. A flexible Assembly cell. Thorn EMI Central Research Laboratories UK. 6th Assembly Automation Conference. IFS 1985

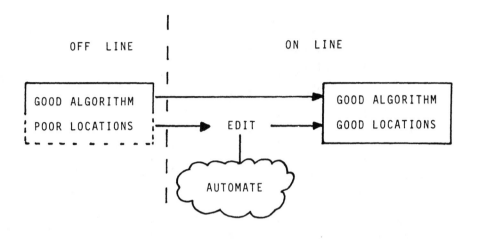

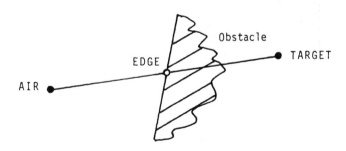

Fig. 1   Robot Program & System Concept

● Input to Search from
   Approx. Location
○ Output from Search = Input
   to Edit Algorithm

Fig. 2   Find 1 Search Routine

68

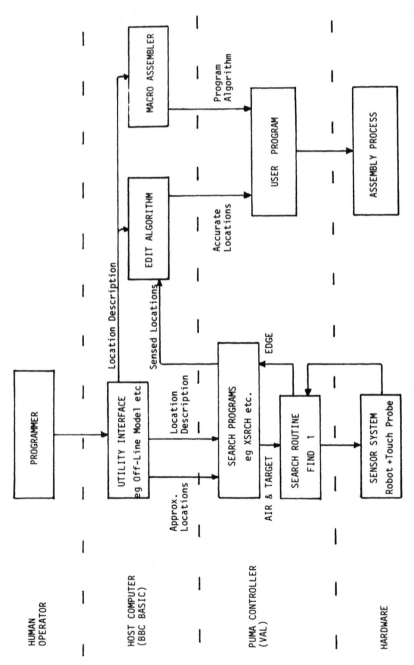

Fig. 3 AUTO EDIT SYSTEM HIERACHY

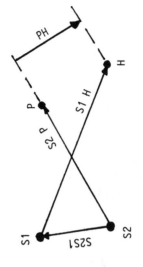

Fig. 5 Relative Sensor

PH = S2P^{-1}.S2S1.S1H

Fig. 7 Relative Transform using Two Simple Sensors

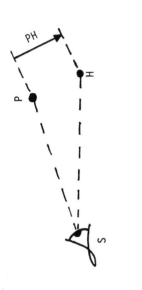

Fig. 4 Simple Sensor

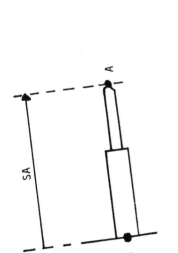

PH = SP^{-1}.SH

Fig. 6 Relative Transform using Simple Sensor

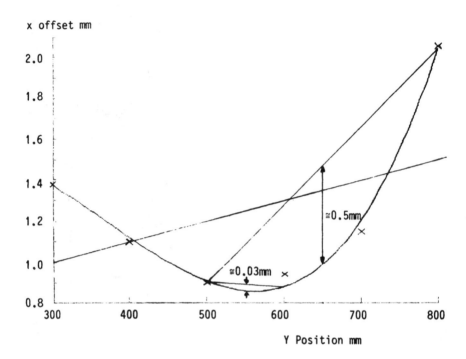

This diagram shows the variation in displacement (in the robots x direction) of the robot end effector from a reference straight edge as the robot moves in a straight line (the line x = -90, z = -100 in the Puma 560 robot's world) roughly parallel to the straight edge. The best fit linear and cubic lines have been plotted. The straight line has a correlation coefficient of 0.20 (1 would be perfect fit, 0 no correlation) and it is clear that a linear offline model of the robot will contain substantial errors. The cubic line has a correlation coefficient of 0.99 and is considered to represent accurately the shape of this part of the robot world. It can be seen that interpolating over large distances (e.g. 300mm) can give rise to substantial errors (e.g. 0.5mm) whereas over small distances (e.g. 100mm) it causes much smaller, more tolerable errors (under 0.03mm). Thus it is considered that while predictive location programming (using linear interpolation) cannot be accurate over large distances it is feasible over short distances.

FIG. 8   EXAMPLE OF ROBOT WORLD CURVATURE

Key to Figures 9 and 10

⊗   Input to search program, originally the rough location.

●   Input to search routine (FIND1, Fig. 2) derived from

○   Output from search routine, input to edit algorithm

⊙   Interim value of CENT calculated by edit algorithm, from edges E1-4

⊗   Output from edit, used for future edit or in user program

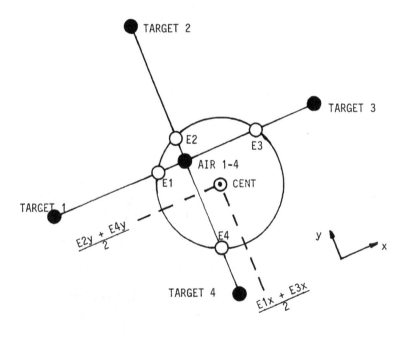

N.B. AIR 1-4 ● = CENT ⊗

FIG. 9 SEARCH PROGRAM EXAMPLE

72

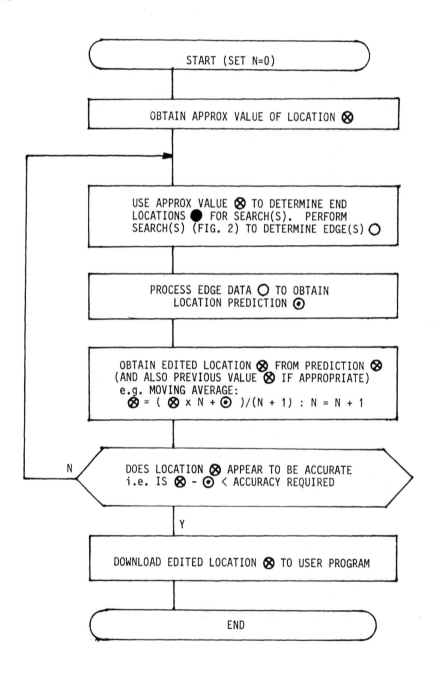

FIG. 10   EXAMPLE OF EDIT ALGORITHM

Dynamic analysis of industrial robots using decentralised control via kinematics

B.M.S. Zainol Anuar and P.D. Roberts

1. INTRODUCTION

The majority of industrial robots used in most practical applications move at a relatively slow speed. This means that the dynamics involved in the motion are affected dominantly by the gravitational and inertial components, where the centrifugal and Coriolis components are negligible. Under such a situation, if the precision of the tracking of any nominal or prescribed trajectory is not of major importance, but the accuracy at the terminal point is essential, the control of such robots may be performed adequately via kinematics, as shown by Vukobratovic and Stokic [1].With respect to this matter, it is necessary to have an insight on the dynamics of the robot,especially the contribution of gravitational and inertial components towards affecting the performance of the overall system.

There are several ways of analysing the dynamics of an industrial robot. Obviously, it would be more realistic if the analysis could be performed with respect to an application of a certain control algorithm. This would then provide the opportunity to analyse the dynamics, as well as the suitability and effectiveness of this control algorithm when the dynamics of the robot have to be taken into account. In order to achieve this proposition, the decentralized control algorithm in the form of local control (each subsystem has only information about its own state variables) is considered. This selection is made due to the fact that this algorithm may lead to a more reliable control of the system and is also simple to implement in practice. In order to apply the decentralized control structure the industrial robot is considered as a set of subsystems corresponding to separate mechanical degrees of freedom.

The theory of decentralized control has been highly developed and well discussed by various authors such as Davison [2] and Siljak [3]. The application of this concept in the field of robotics also has been dealt with extensively in papers such as those of Vukobratovic and Stokic [4] and [5]. In this case, the nominal programmed control (using the complete model of the robot) is applied so as to reduce the effects of interconnections at the perturbed stage where the decentralized control algorithm is then utilized. In references, Vukobratovic et al [6] and Stokic and Vukobratovic [7], a more robust decentralized control algorithm is used: where the nominal condition is based on an approximate model. This approach has been considered in order to cope with any variations of robot parameters.

However, with respect to the present work on dynamic analysis of industrial robots, only the basic decentralized control concept, without involving the nominal control, has been adopted. This implies that it is involved directly with the stage of perturbed dynamics where stabilization of the system has to be performed. The local controllers, which are designed by using the models of free subsystems, are then employed in controlling the complete system when the prescribed trajectory (in terms of kinematics), assumed as the nominal trajectory, is provided.

Since the controllers are designed by using the decoupled model, ignoring the interconnections (dynamics), there is a structural perturbation which affects the behaviour of the robot. However, for a relatively slow motion where the effect of the interconnections is not too strong, the local controllers are expected to be sufficient to stabilize the system around the nominal/prescribed trajectory [1].

The perturbation due to the initial condition and control input constraint are also ignored. In doing this, a wider scope is provided for analysing the dynamics with respect to variations of some of the parameters of the industrial robot and the controllers.

The organisation of the paper is as follows. In Section 2 the mathematical model of the system under scrutiny is briefly presented. In Section 3 the decentralized control algorithm and its implementation are described. Section 4 describes the prescribed trajectories, in terms of the system's kinematics, which are regarded as the nominal trajectories. In Section 5 the procedure for simulating the dynamics of the robotic system is described. In Section 6 the results of the dynamic simulation of an industrial robot with minimal configuration of revolute type are presented and discussed.

2. MATHEMATICAL MODEL OF THE SYSTEM

The industrial robot consists of two parts; namely, the mechanical part and actuators. Each degree of freedom of the system is powered by an actuator.

The model of the mechanical part may be formulated using the alternative (4x4) matrix approach given by Zainol Anuar and Roberts [8], and is expressed in the following compact general form,

$$A(\underline{\theta}) \ \ddot{\underline{\theta}} + \underline{H}(\underline{\theta},\dot{\underline{\theta}}) = \underline{T} \qquad (1)$$

where $\underline{\theta}(t)$ is the vector of internal joint coordinates, $A(\underline{\theta})$ represents the positive definite inertial coefficient matrix-valued-function, $\underline{H}(\underline{\theta},\dot{\underline{\theta}})$ is a vector function due to the effects of centrifugal, Coriolis and gravity, and $\underline{T}(t)$ represents a vector of the driving torques (forces).

Each actuator is represented by a linear time-invariant model in the following form [1],

$$\dot{\underline{x}}_i = C_i \underline{x}_i + \underline{d}_i u_i + \underline{f}_i T_i \quad (i=1,..,n) \tag{2}$$

where $\underline{x}_i$ is the vector of the state variables of the actuator, n_i is the order of the model, C_i is the constant system matrix, $\underline{d}_i$ is the input distribution vector, $\underline{f}_i$ is the load distribution vector, u_i is the input to the ith actuator and T_i is the driving torque acting on the ith actuator. For a direct drive case, T_i in (1) and (2) are equal. It is assumed that $\underline{x}_i$ has two variables which coincides with θ_i and $\dot{\theta}_i$.

Since the industrial robot has n degrees of freedom it is necessary to form a combined model for the whole actuating system. Hence, the model of the actuating system is

$$\dot{\underline{x}} = C \underline{x} + D \underline{u} + F \underline{T}, \quad \underline{x}(t_o) = \underline{x}_o \tag{3}$$

where, $\underline{x} = [\underline{x}_1{}^T, \underline{x}_2{}^T, ..\underline{x}_n{}^T]^T$, $\underline{u} = [u_1, u_2, ..u_n]^T$,
$C = \text{diag} [C_1, C_2, ...C_n]$, $D = \text{diag} [\underline{d}_1, \underline{d}_2, ...\underline{d}_n]$,
$F = \text{diag} [\underline{f}_1, \underline{f}_2, ...\underline{f}_n]$ and $\underline{T} = [T_1, T_2, ..., T_n]^T$.

Equations (1) and (3) may be combined by applying a transformation process [1] and through some mathematical manipulations to form an integrated model representing the overall system, giving (dropping (θ) and $(\theta, \dot{\theta})$)

$$\dot{\underline{x}} = \hat{C}(\underline{x}) + \hat{D}(\underline{x}) \underline{u} \tag{4}$$

where, $\hat{C}(\underline{x}) = C\underline{x} + F[I_n - ATF]^{-1} [ATC\underline{x} + \underline{H}]$,

$\hat{D}(\underline{x}) = D + F[I_n - ATF]^{-1} ATD$, I_n = unity matrix,

$T' = \text{diag} [T'_1, T'_2, ..., T'_n]$ obtained from $\ddot{\theta}_i = T'_i \cdot \dot{\underline{x}}_i$.

3. CONTROL ALGORITHM AND ITS IMPLEMENTATION

Since decentralized control is applied for stabilizing the system about the nominal trajectory, it is useful to consider the deviation models of the system's parts separately.

Based on the fact that each degree of freedom may be regarded as a subsystem, the overall robotic system consists of n interconnected subsystems as represented by the model of the actuator (2). The interconnection function is related to the model of the mechanical part and is non-linear in nature.

The deviation model of the subsystem about the nominal trajectory is given in the following form;

$$\delta \dot{\underline{x}}_i = C_i \delta \underline{x}_i + \underline{d}_i \delta u_i + \underline{f}_i \delta T_i, \quad \delta \underline{x}_i(0) \text{ is given} \tag{5}$$

where $\delta \underline{x}_i(t) = \underline{x}_i(t) - \underline{x}_i{}^o(t)$ is the deviation of the subsystem state vector from its nominal value, and $\delta u_i(t) = u_i(t) - u_i{}^o(t)$ is the deviation of the input from the nominal input. In the case when the nominal input is not introduced, $\delta u_i(t)$ becomes the input required by the actuator to realize the deviation in the state vector $\delta \underline{x}_i$.

In Equation (5), δT_i represents the deviation of the driving torques at the joint from the nominal torques and may be derived from the model of the mechanical part, as follows [1];

$$\delta \underline{T} = A(t, \delta\theta) \delta \ddot{\theta} + H(t, \delta\theta, \delta\dot{\theta}) \tag{6}$$
$$\delta\theta(0) \text{ and } \delta\dot{\theta}(0) \text{ are given.}$$

where $\delta \underline{T}(t) = \underline{T}(t) - \underline{T}^o(t)$, and $\delta\theta(t) = \theta(t) - \underline{\theta}^o(t)$ is the deviation of the joint coordinates from the nominal values. The superscript "o" signifies the nominal condition.

The separation of the overall system into Equations (5) and (6) may also be performed in an alternative manner. This is achieved by introducing the numbers $\tilde{A}_i$ and $\tilde{H}_i$ (which are related to the inertia and gravitational effects of the mechanical part, respectively) into the subsystem deviation model (5), as discussed in [1,5,6]. However, the matrix and vector coefficients of Equation (5) would then have to be modified.

3.1 Control algorithm

The control task is to stabilize the interconnected subsystems using the local controllers designed by employing the decoupled deviation model. The decoupled deviation model is obtained from (5) by ignoring the interconnection and is given in the following form,

$$\delta \dot{x}_i = C_i \, \delta x_i + d_i \, \delta u_i \tag{7}$$

Now the problem is to determine the control input δu_i which can stabilize the free subsystem about the nominal trajectory. Since the free subsystem is linear time-invariant, the stabilization of the free subsystem (7) can be performed by minimizing the standard quadratic criterion involving a prescribed degree of stability as follows, (Anderson and Moore [9]),

$$J_i(\delta x_i(0), \delta u_i(t)) = \int_0^\infty e^{2\Pi t}(\delta x_i{}^T Q_i \delta x_i + \delta u_i r_i \delta u_i) dt \tag{8}$$

where $\Pi > 0$ is the prescribed degree of the exponential stability, Q_i is a $n_i \times n_i$ symmetric non-negative definite matrix, and $r_i > 0$ is a positive number. It should be mentioned here that the criterion is considered over an infinite time interval even though the system is observed over a finite time interval.

As described in [9], under the assumption that the pair (C_i, d_i) is completely controllable and the input constraint is not taken into account, then there exists a unique optimal control law minimizing (8), with the constraint (7), in the following form:

$$\delta u_i(t) = - \, r_i \, d_i{}^T \, K_i \, \delta x_i(t) \tag{9}$$
$$= G \, \delta x_i(t)$$

where G is a $(1 \times n_i)$ vector of the controller gains, K_i is a $n_i \times n_i$ symmetric positive definite matrix given from the solution of the algebraic Riccati matrix equation

$$K_i(C_i + \Pi I_n) + (C_i{}^T + \Pi I_n)K_i - K_i d_i r_i^{-1} d_i{}^T K_i + Q_i = 0 \tag{10}$$

Under the assumption that Q_i can be expressed in the form of $Q_i = \bar{Q}_i \, \bar{Q}_i$, where $\bar{Q}_i$ is a $n_i \times n_i$ constant matrix, so that the pair $(C_i, \bar{Q}_i)$ is completely observable, then the subsystem (7) under a closed-loop condition will be exponentially stable around the nominal trajectory $x_i{}^o(t)$ with stability degree Π [9].

However, if the control (9) is to be used for controlling the interconnected subsystem, it is expected that the system will experience some deterioration in its performance, which depends on the strength of the interaction present between the subsystems. Although the

problem of the presence of strong interconnection between
subsystems may be overcome by using the concept of a
multilevel scheme, this idea has not been considered in this
case. For the time being it is assumed that the presence of
dynamics between the robotic subsystems, moving at a
relatively slow speed, may be overcome by increasing the
"strength" of the local controller. The strength of the
local controllers may be varied by respecifying the
prescribed degree of stability or weighting matrix Q_i or r_i.
The effect of these variations on the overall dynamics or
behaviour of the industrial robot will be illustrated and
discussed in the example.

3.2 Implementation of the control algorithm

The decentralized control algorithm described in the
preceding section is implemented as shown schematically in
Fig.1. In this case the prescribed or desired trajectory
(based on the kinematics of the system) is assumed to be the
nominal trajectory. Therefore the deviation of the state, as
defined by δx_i, necessary for determining the actuator
input, is obtained by finding the difference between the
nominal $(x_i{}^0)$ and the "actual" (x_i) state vectors. This
"actual" state vector is formed from variables obtained by
solving the overall integrated model representing the
complete robotic system, as defined by Equation (4).

Since the perturbation at the initial condition is
not considered, the initial state vector at t=0, necessary
for the integration of the overall model, must be the same
as that of the prescribed values at identical times.
However, if a variation at the initial condition is to be
included, the initial state variables for solving the
overall model depend on the position of the robot at t=0.

The dimension of x_i and $x_i{}^0$ depends on the order of
the actuator model used. For an actuator model of order 2,
x_i and $x_i{}^0$ become (2x1) vectors and functions of the angular
displacements and angular velocities for rotational joints,
and linear displacements and velocities for prismatic
joints. If a third order actuator model is adopted the
currents also have to be included in the state vectors. In
this condition, the nominal trajectories of the currents may
be determined from the actuator model with known angular
displacements and angular velocities as described in [7].

4. PRESCRIBED TRAJECTORIES/KINEMATICS

In many applications the motion of the end-effector
is normally defined. The desired or prescribed trajectory is
defined in terms of the position vector with respect to the
external (fixed) coordinate system. For a given vector of
the joint coordinates, the position vector is defined by the
following position (kinematic) model (Vukobratovic and
Stokic [1], Kircanski [10], Paul et al [11] and Lenarcic
[12]):

$$p_v(t) = \bar{f}(\theta(t)) \tag{11}$$
where $\bar{f}: R^n \to R^\ell$

In the case of an industrial robot with minimal configuration (three degrees of freedom), the dimension of $\underline{p}_v$ equals that of $\underline{\theta}$.

$$\underline{p}_v = (x,y,z)^T = \bar{f}(\underline{\theta}) = \bar{f}(\theta_1, \theta_2, \theta_3)$$
$$= [\bar{f}_1(\underline{\theta}), \bar{f}_2(\underline{\theta}), \bar{f}_3(\underline{\theta})]^T \qquad (12)$$

where $n = \ell = 3$

In general the movement of the tip along a prescribed trajectory may be represented in the following form (Vukobratovic and Stokic [1], Vukobratovic and Kircanski [13]):

$$\underline{p}_v = \underline{S}(\rho) \qquad (13)$$

where, $0 < \rho < 1$ is a scalar parameter and function of time, $\underline{S}:R \rightarrow R^3$ is a given vectorial function, $\underline{S}(0)=\underline{p}_v^I$ ($\underline{p}_v^I$ is the Initial position vector), $\underline{S}(1)=\underline{p}_v^F$ ($\underline{p}_v^F$ is the final position vector).

For a straight line trajectory, which will be considered in the example, Equation (13) becomes,

$$\underline{p}_v(t) = \underline{p}_v^I + \rho(t) (\underline{p}_v^F - \underline{p}_v^I) \qquad (14)$$

There are various velocity distribution functions along the straight line that can be adopted, but among the most common is the sinusoidal velocity distribution given by Zabala [14]. This is due to the fact that the optimal velocity distributions are similar to parabolic functions and that this profile should be applied in the practical control of an industrial robot in order to minimize the actuator power consumptions (Vukobratovic and Potkonjak [15]).

Since the desired values for servoing the robot have to be in joint coordinates, the inverse problem of Equation (11) has to be solved. This may be achieved efficiently by the use of a Newton-Raphson algorithm (Whitney [16]).

The linear velocities of the desired trajectory may be determined by taking the first derivative of Equation (11), leading to,

$$\dot{\underline{p}}_v(t) = J_c(\underline{\theta}).\dot{\underline{\theta}} \qquad (15)$$

where $J_c(\underline{\theta})$ is the Jacobian matrix.

Therefore, for any given linear velocities, the corresponding joint angular velocities can be determined by using the following expression,

$$\dot{\underline{\theta}} = [J_c(\underline{\theta})]^{-1} \dot{\underline{p}}_v \qquad (16)$$

In the case of a straight line motion the derivative of Equation (14) yields,

$$\dot{\underline{p}}_v(t) = \dot{\rho}(t) (\underline{p}_v^F - \underline{p}_v^I) \qquad (17)$$

Substituting Equation (17) into (16) results in

$$\dot{\underline{\theta}} = \dot{\rho}(t) [J_c(\underline{\theta})]^{-1} (\underline{p}_v^F - \underline{p}_v^I) \qquad (18)$$

where $\dot{\rho}(t)$ represents the velocity distribution along the straight line.

5. **SIMULATION PROCEDURE**

A computer program has been developed for simulating the dynamics of an industrial robot (minimal configuration) with respect to the application of decentralized control with prescribed kinematics as depicted in Fig.1. This program is designed to accomodate several kinds of variations in the parameters of both the robot and the control algorithm. Several of these aspects will be considered and illustrated in the example; namely, (i) effect of varying the prescribed degree of stability, (ii) effect of varying the travelling time, (iii) effect of varying the mass at the tip of the robot and (iv) effect of varying the weighting matrix Q_i in the performance criterion but at specified values of r_i and prescribed degree of stability.

The error in tracking may be considered as an indicator for describing the dynamical effect on the performance of the robot due to certain variations of parameters, as well as illustrating the efficiency of the applied control algorithm. This error is determined by calculating the Euclidean norm of the difference between the prescribed position vector $\underline{p}_v$ (14) and the "actual" position vector calculated using Equation (11) with the known vector of joint coordinates (obtained from the solution of Equation (4)). That is

$$\text{Error } E(t) = \left\| \underline{p}_v{}^a - \underline{p}_v{}^d \right\| \tag{19}$$

where $\underline{p}_v{}^a$ is the "actual" position vector and $\underline{p}_v{}^d$ is the desired position vector.

The overall model of the industrial robot (4) is solved by utilizing a numerical integration subroutine which has been developed in the NAG package. The particular routine used is based on the Runge Kutta - Merson method.

The generation of the prescribed trajectory has to be performed within the numerical integration subroutine so as to maintain the same time interval throughout the simulation.

Since the performance criterion (8) is observed over an infinite time interval, the local controller gains can be determined beforehand and kept constant throughout the simulation. As a result of using the same actuator at every joint the algebraic Riccati matrix equation (10) may be solved only once, instead of solving it for every subsystem.

5.1 Stability of the system

Since the local performance criterion involves the prescribed degree of stability the stability of the system is guaranteed by specifying a certain value of .. It should be borne in mind that a stable condition of the system does not necessarily result in good tracking of the desired trajectory. This is due to the effect of robot dynamics, represented by the interconnection function, acting as a disturbance or perturbation. The effect of these dynamics on the tracking of the trajectory, as represented by the error response, may be scrutinized with respect to the variation of the prescribed degree of stability.

Obviously, as the prescribed degree of stability is increased, the systems will become more stable. Thus it is expected that the tracking of the trajectory will be improved. This will be illustrated and discussed in the example.

5.2 Robustness of the control algorithm

In this case the robustness or the sensitivity of the applied decentralized control algorithm is checked by varying a lumped mass with its centre of gravity coinciding with the tip of the industrial robot with minimal configuration. In the real situation this lumped mass may represent the integrated mass of the end-effector plus the load at a fixed orientation.

The variation of this mass changes the dynamics of the robot, which then affects the effectiveness of the applied control algorithm. For this purpose, the integrated model of the industrial robot has to be modified in order to incorporate the variation of the mass at the tip. This involves modifying the expression for the moment of inertia of the last moving segment. The extra torque developed due to the addition of the lumped mass will be transmitted accordingly to all the joints of the industrial robot.

The robustness of the control algorithm may be studied by observing the error response with respect to the variation of the amount of mass at the robot tip. This will also provide an indication concerning the limit of load that the robot can carry at a particular set of gains of the local controllers and within a certain amount of error limit especially at the terminal point.

6. EXAMPLE

In this example an industrial robot with minimal configuration is considered (Fig. 2). The detailed model of the mechanical part may be obtained from Zainol Anuar and Roberts [17] and the inertial parameters are given in Table 1.

Each joint is powered by the same permanent magnet direct current motor for which a second order model ($n_i = 2$, $i=1,2,3$) is adopted. The state vector becomes $\underline{x}_i = [\theta_i, \dot{\theta}_i]^T$ where θ_i is the angular displacement and $\dot{\theta}_i$ is the angular velocity. As a result, the matrix and vectors of the model represented by (2) are in the following form,

$$C_i = \begin{bmatrix} 0 & 1 \\ 0 & \alpha_i \end{bmatrix} \qquad \underline{d}_i = \begin{bmatrix} 0 \\ \gamma_i \end{bmatrix} \qquad \underline{f}_i = \begin{bmatrix} 0 \\ \beta_i \end{bmatrix} \tag{20}$$

where $\alpha_i = -\left[\dfrac{Bv_i}{Jr_i} + \dfrac{(Kti)_i (Kvs)_i}{Jr_i Rr_i} \right]$

$\beta_i = -\dfrac{1}{Jr_i} \qquad \gamma_i = \dfrac{(Kti)_i}{Jr_i Rr_i}$

Bv_i is the viscous friction coefficient, Rr_i is the rotor resistance, Jr_i is the rotor moment of inertia, $(Kti)_i$ is the torque constant and $(Kvs)_i$ is the back emf constant. The values of these parameters are given in Table 2.

The desired straight line trajectory of the robot tip is defined by the initial and final position vector $\underline{p}_v^I$ and $\underline{p}_v^F$ as $(0.0, 0.3, 0.3)^T$ [m.] and $(-0.4, 0.5, 0.6)^T$ [m.], respectively. The initial and final velocities of the tip are zero and the velocity distribution along the prescribed trajectory is given by

$$\dot{\rho}(t) = \frac{\pi}{2T} \sin\left(\frac{\pi}{T} t\right) \tag{21}$$

where T is the travelling time between the initial and final points. T = 1.25 seconds (unless it is stated otherwise).

The prescribed velocity distribution, desired linear velocities with respect to x, y and z axes and the desired angular displacements of the arm corresponding to each joint are shown in Figs. 3a, 3b and 3c, respectively.

Consider the case when Q_i and r_i are specified as follows:

$$Q_i = \begin{bmatrix} 10.0 & 0 \\ 0 & 10.0 \end{bmatrix} \quad r_i = 0.5$$

If the prescribed degree of stability is taken as zero, meaning that each local performance criterion is of normal quadratic type, the resulting response of the overall system is found to be very poor. The resulting controller gains, (-4.47,-1.36), are rather small. Thus, these "weak" controllers are unable to overcome the dynamic effects of the interconnections which have not been considered in establishing the control law. Presumably the dynamics have been dominated by the inertial and gravitational components. The difference in angular displacements between the desired and "actual" values (Fig. 4a) seems to increase with respect to time. As a result, the tracking of the desired tip trajectory becomes very sluggish (Figs. 4b and 4c).

However, as the prescribed degree of stability is introduced at an increasing value, the gains of the controllers are also increased (Table 3). This implies that the "strength" of the local controllers has also been increased. Under such a situation it is expected that the controllers would be able to suppress the effect of interconnection dynamics (as represented by the interconnection function) in a better manner.

As the value of Π is increased the eigenvalues of the subsystem are moved further away from the origin, thus increasing the degree of exponential stability of each subsystem. Consequently, if a sufficiently large value of Π is chosen, meaning larger values of controller gains, then the local controllers may stabilize the overall system. Obviously the tracking of the desired/nominal tip trajectory is also improved. Fig. 5a shows that the minimum error is produced when $\Pi = 30$, occuring at controller gains of (-576.63,-11.10). Due to the improved tracking the error at the terminal point is also reduced (Fig. 5b).

Since the movement of the robot is considered as relatively slow the dynamics present are dominated by the inertial and gravity terms. In order to have some insight of this fact consider Fig.6, which shows the comparison between the inertial, centrifugal, Coriolis and gravitational terms for the case of $\Pi = 25$. The dynamics which affect the first controller (Fig. 6a) are found to be dominated only by the inertial term. In this case there is no gravitational effect because the axis of rotation of the segment is vertical. The Coriolis effect which is present seems to be negligible. The dynamics of the second segment which affect the second controller are dominated by the gravity and inertial terms. The contribution from the centrifugal and Coriolis terms are found to be negligible (Fig. 6b). Although the presence of the centrifugal term is quite distinct for the third segment (Fig. 6c) its magnitude is still small when compared to the gravity term. Since the nominal control has been ignored the input to each actuator is represented by δu (Fig. 7).

At constant controller gains the dynamic performance of the industrial robot is also affected by the variation of the travelling time. It is important to note that the efficiency in tracking the prescribed trajectory deterioriates, to some extent, when the travelling time is altered from 1.25 seconds to 1.0 second and 1.5 seconds. This phenomenon is illustrated by the tip error responses at $\Pi = 25$ (Fig. 8). When the travelling time is reduced to 1.0 second, a faster motion is experienced by the robot, occuring at a larger magnitude of acceleration. Thus this has led to the generation of greater inertial torques but without affecting the gravitational terms. Consequently the interconnections between the subsystems are also increased. Since the controller gains are kept constant this dynamical effect has given rise to an increase in the error (curve A). The comparison between the contributing terms in the dynamics is shown in Fig. 9 .

Although the travelling time is increased to 1.5 seconds, implying that the robot is moving at a slower speed, the error in tracking still increases (Curve C). As a result of slowing down the motion, resulting in a lower acceleration, the inertial effects in all segments are reduced. Since the gravitational effect/torque remains the same a reduction in the inertial torque will increase the resultant torque which is exerted by each actuator, as given by the following expression,

$$\underset{\text{torque}}{\text{inertial}} + \underset{\text{torque}}{\text{centrifugal}} + \underset{\text{torque}}{\text{Coriolis}} - \underset{\text{torque}}{\text{gravitational}} = \underset{\substack{\text{tion} \\ \text{torque}}}{\text{interac-}}$$

Hence, the increase in interaction torque (interconnections) has affected the tracking by increasing the error. Fig.10 shows the comparison between the contributing terms in the dynamics when the travelling time equals 1.5 seconds.

The dynamical effect on the performance of the industrial robot due to a variation of mass at the tip may be scrutinized by considering the error responses.

Since the controller gains are maintained at the same values (when $\Pi = 25$), the addition of the mass to the robot tip has the tendency of slowing down the movement. As a result it is presumed that the variation in the robot dynamics is affected considerably by the gravitational terms instead of inertial, centrifugal or Coriolis components. Obviously a greater gravitational torque due to the addition of the mass would increase the interconnection effect between the subsystems. Thus this leads to a larger error in the tracking of the prescribed trajectory (Fig. 11). Therefore in order to cope with an extra load at the tip, a larger value of Π (hence, larger controller gains) has to be adopted. The resulting error at the terminal point seems to increase linearly with respect to an increase of mass at the tip (Fig. 12). Consequently for a particular set of controller gains the maximum limit of the mass has to be considered such that the error at the terminal point falls within an allowable tolerance. This fact, in some manner, describes the robustness or sensitivity of the applied control algorithm.

Apart from increasing the prescribed degree of stability the strength of the local controllers may also be increased by varying the weighting matrix Q_i in the performance criterion, with r_i kept constant. If Q_i is expressed as

$$Q_i = \alpha_i \begin{bmatrix} 1.0 & 0 \\ 0 & 1.0 \end{bmatrix} \quad \text{and fixing} \quad \Pi = 15,$$

the effect of varying α_i on the dynamic performance of the system may be shown by the error response in Fig. 13. It is noticed that the higher value of α_i, corresponding to higher controller gains, improves the performance by reducing the error in tracking.

7. **CONCLUSION**

Observation of the dynamic responses of the industrial robot leads to the fact that, in this particular situation at least, the design of the decentralized control, by ignoring the interconnection dynamics of the system, is insufficient. The dynamics of the system as represented by the interconnection between the subsystems has to be taken into account in order to achieve good control of the system. Nevertheless, the dynamical effect may be overcome, to some extent, by increasing the strength of the local controllers; that is by increasing the gains. This may be achieved by means of increasing the prescribed degree of stability or respecifying other parameters such as the weighting matrix Q_i or r_i in the performance criterion. However, the values of gains may only be increased to the practical limit of each local controller.

For a particular set of controller gains varying the travelling time also affects the dynamic performance of the industrial robot. This leads to a deterioration of the performance of the robot in tracking the prescribed trajectory. From these results it may be concluded that there may be an optimal travelling time which is suitable for a certain distance of travel with a particular set of controller gains.

Finally the analysis of the results has given further insight into the dynamic behaviour of a robot in the task space. It has also provided an opportunity to understand the concept and application of decentralized control in the field of robotics.

REFERENCES

1. Vukobratovic, M., and Stokic, D, 1982, "Control of
   Manipulation Robots : Theory and Application",
   Springer-Verlag, Berlin.

2. Davison, E.J., 1976, IEEE Trans., AC-21, 14-24.

3. Siljak , D.D., 1978, "Large-Scale Dynamic Systems:
   Stability and Structure", North-Holland, New York.

4. Vukobratovic, M., and Stokic, D., 1980, Automatica, 16
   9-21.

5. Vukobratovic, M., and Stokic, D., 1981, Large Scale
   Systems, 2, 159-170.

6. Vukobratovic, M., Stokic,D., and Kircanski, N., 1984,
   IEEE Trans., AC-29, 841-844.

7. Stokic, D., and Vukobratovic, M., 1984., Automatica,
   20, 353-358

8. Zainol Anuar, B.M.S. and Roberts, P.D., 1985,
   Modelling, Simulation and Control, B, AMSE Press, 3,
   41-64.

9. Anderson, B.D.O., and Moore, J.B., 1971, "Linear
   Optimal Control", Prentice-Hall, Englewood Cliffs,
   N.J.

10. Kircanski, M., 1980, Proceedings of Int. Conf. on
    Systems Engineering (Coventry Lanchester Poly),
    114-122

11. Paul, R.P., Shimano , B., and Mayer, G.E., 1981,
    IEEE Trans., SMC - 11, 449-445

12. Lenarcic, J. 1985, Robotica, 3, 21-26.

13. Vukobratovic, M., and Kircanski, M.,1982, J. Dyn. Syst
    Measurement and Control, 104, 188-193.

14. Zabala, J. 1978, "Control of Robot Manipulators Based
    on Dynamic Modelling", Thesis, Toulouse.

15. Vukobratovic, M., and Potkonjak, V., 1982, "Dynamics of
    Manipulation Robots: Theory and Application", Springer-
    Verlag, Berlin.

16. Whitney, D.E., 1969, IEEE Trans, AC-14, 572-574

17. Zainol Anuar, B.M.S., and Roberts,P.D.,1984,"Modelling
    and Simulation of an Industrial Robot with minimal con-
    figuration" Memorandum CEC/ZABMS/PDR-16, Control
    Engineering Centre, The City Univ., London.

TABLE 1. Parameters for the mechanical part of the robot
cg* is the centre of gravity

	Segments number			
	0	1	2	3
Mass (kg)	---	---	2.0	1.5
Length, L (m)	0.3	0.5	0.6	0.5
Position of the centre of gravity ℓ (m)	---	---	0.3	0.25
Moment of inertia with respect to, (kg m^2)				
x-axis at cg*	---	---	0.06	0.03
y-axis at cg*	---	---	0.008	0.0025
z-axis at cg*	---	0.04	0.06	0.03

TABLE 2. Parameters of the actuators.

$$Jr_1 = Jr_2 = Jr_3 = 1.52 \text{ kg } m^2$$
$$Rr_1 = Rr_2 = Rr_3 = 2.45 \text{ ohms}$$
$$Bv_1 = Bv_2 = Bv_3 = 1.5 \text{ Nm/rad/s}$$
$$(Kti)_1 = (Kti)_2 = (Kti)_3 = 4.31 \text{ Nm/A}$$
$$(Kvs)_1 = (Kvs)_2 = (Kvs)_3 = 7.0 \text{ V/rad/s}$$

TABLE 3. Controller gains at varying Π .

Π	Controller gains	
	G_{11} (V/rad)	G_{12} (V/rad/s)
0.0	-4.47	-1.36
3.0	-51.4	-2.22
5.0	-85.73	-2.85
10.0	-173.89	-4.43
15.0	-265.76	-6.03
20.0	-362.30	-7.67
25.0	-465.31	-9.36
30.0	-576.53	-11.10

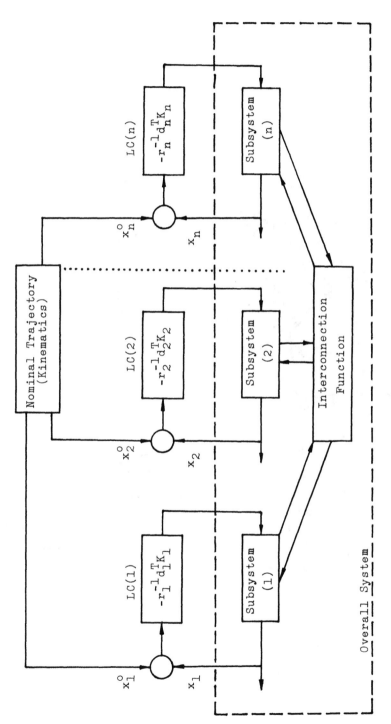

Fig.1. Decentralized control of an industrial robot via kinematics

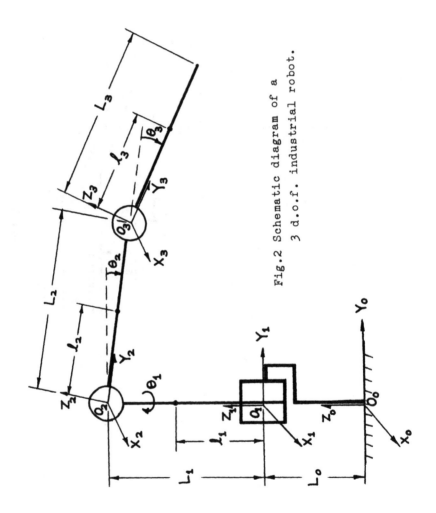

Fig.2 Schematic diagram of a
3 d.o.f. industrial robot.

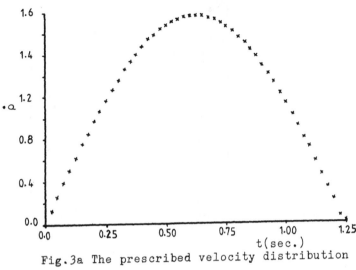

Fig.3a The prescribed velocity distribution

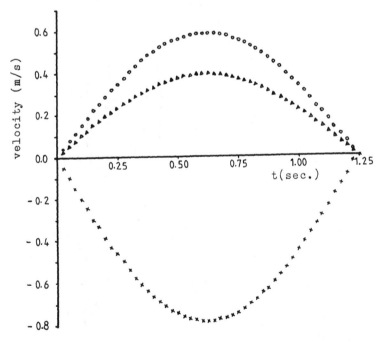

Fig.3b The prescribed linear velocities of
       the robot tip

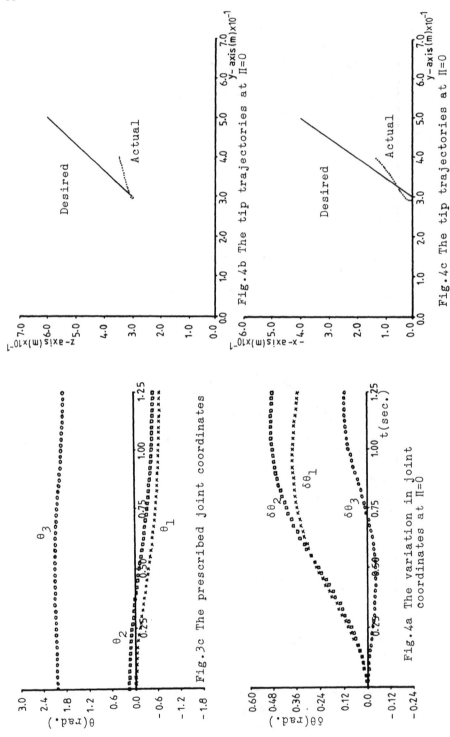

Fig.3c The prescribed joint coordinates

Fig.4a The variation in joint
coordinates at Π=0

Fig.4b The tip trajectories at Π=0

Fig.4c The tip trajectories at Π=0

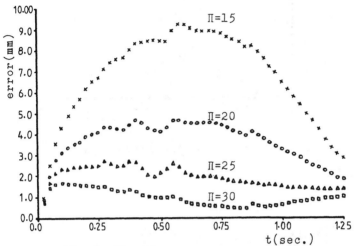

Fig.5a The tip error responses at
different Π

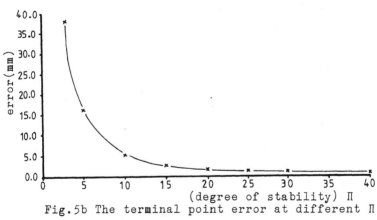

Fig.5b The terminal point error at different Π

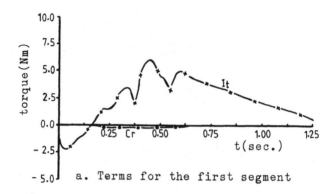

a. Terms for the first segment

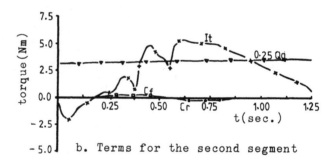

b. Terms for the second segment

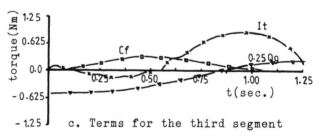

c. Terms for the third segment

Fig.6 Comparison of terms, T=1.25 sec.
(Inertial It,Centrifugal Cf,
Coriolis Cr,Gravitational Qg)

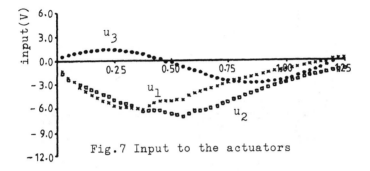

Fig.7 Input to the actuators

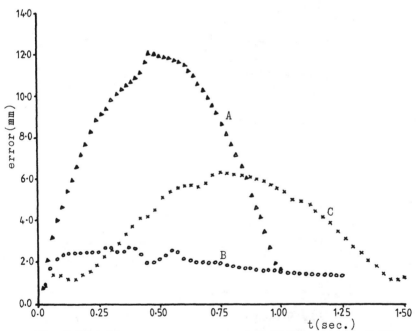

Fig.8 The tip error responses at different T, Π=25
Curve A for T=1.0 sec.,Curve B for T=1.25 sec.,
and Curve C for T=1.5 sec..

94

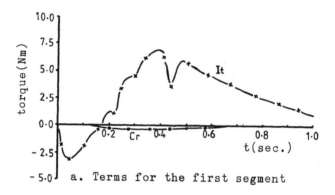

a. Terms for the first segment

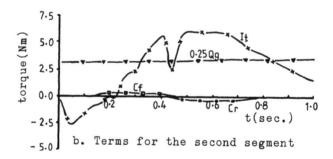

b. Terms for the second segment

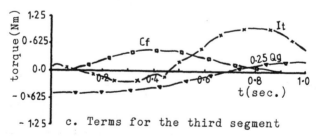

c. Terms for the third segment

Fig.9 Comparison of terms, T=1.0 sec.
(Inertial It,Centrifugal Cf,
Coriolis Cr,Gravitational Qg)

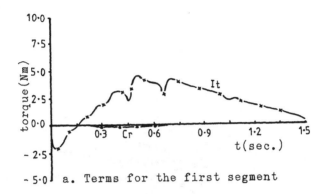

a. Terms for the first segment

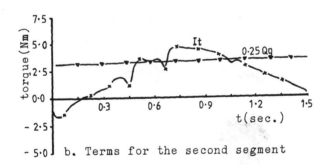

b. Terms for the second segment

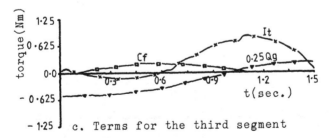

c. Terms for the third segment

Fig.10 Comparison of terms, T=1.5 sec.
(Inertial It, Centrifugal Cf,
Coriolis Cr, Gravitational Qg)

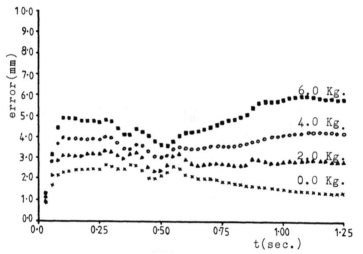

Fig.11 The tip error responses with
different mass at the tip, Π=25

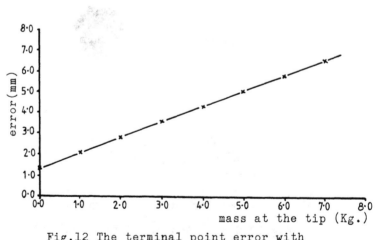

Fig.12 The terminal point error with
different mass at the tip, Π=25

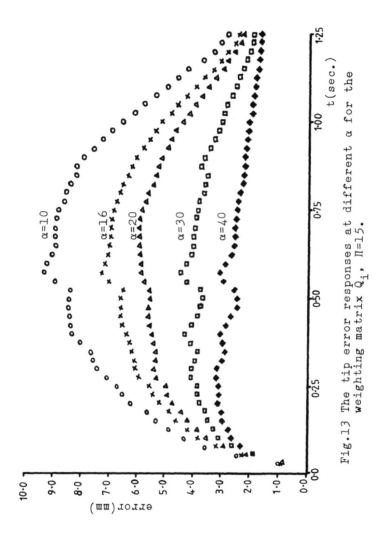

Fig.13 The tip error responses at different α for the weighting matrix Q_i, $\Pi=15$.

Collision avoidance between robots operating in the same cell

D.J. Rhodes, E.H. Stenson and P. Blanchfield

1 ROBOT CELLS AND FLEXIBLE OPERATIONS

Manufacturers are increasingly forced by competitive pressure to supply products which to some degree are customised. This means a wider range of parts and smaller batch sizes, both of which dictate a need for efficient, flexible methods of manufacture. If automation is to be of any use it must thus address this need. However, current robot cells are set up to deal with a limited number of types of part and usually operate as fixed rather than flexible automation, performing identical tasks for very long periods. One reason for this is that hitherto robots have been used for repetitive, simple jobs. Another is that the predominant method of programming is "teach-by-show" which requires the robot to be withdrawn from production prior to a new task or product and hence reduces productivity. The need to locate the part-presentation devices very accurately so that the robot may find the parts also requires withdrawal and lost production.

For these reasons, off-line programming systems, such as RAPT (1) and GRASP (2), and sensors, particularly vision systems (3,4), are under current development. They should eventually reduce both the sensitivity of robots to inaccuracies in part location and the need to take robots out of service when changing product. They should enable manufacturers to flexibly handle the less-simple tasks to a high and consistent standard with reasonable efficiency.

A further possibility in helping to achieve flexibility is to introduce several robots to the same cell. This would help the cell to deal with several parts at the same time and despite seeming a trifle ambitious, is nevertheless worth considering since there are circumstances in which the manufacturing process requires the cooperation of more than one robot anyway. An example, currently under development at NEL involves three robots, two of which share part of the work space. Furthermore, the process times of the tasks at various workstations vary with the condition of the material and stage of final assembly, and the arrival of parts (material) at the cell is irregular. In the circumstances sequence control does not ensure maximum productivity and is

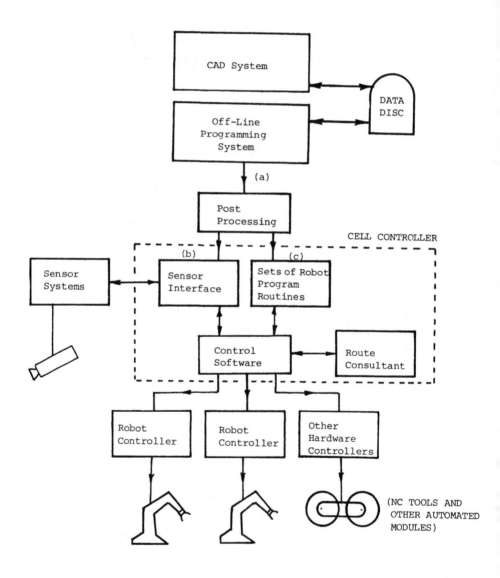

Fig.1.   The Components of a Robotic, Flexible
         Manufacturing Cell with Multi-Robot Control

only practical for the simplest arrangement of workstations and assembly points. Other examples occur wherever several robots share an assembly task and in circumstances where, due to some unexpected occurrence such as the build-up of swarf, damaged or dropped parts, one robot must intervene in the work envelope of another. Though not identical in form, there is also the related problem of robots cooperating on a task (5). The robots may not be working in the same envelope at all but their control is interactive. The control of multi-robot cells is therefore of some concern and non-trivial.

Fig.1 illustrates current thinking on the general approach to single-robot cell control. In principle, some CAD and off-line programming system will produce a specification for the manufacturing requirements (a). The post-processor converts these into a form suitable for driving the cell and sensors, (b) and (c). Alternatively the cell (b) and sensor (c) information can be supplied directly (as now). The cell-controller operates the cell by sending instructions to the individual (local) hardware and robot controllers, the latter as a sequence of subroutines (VAL for example). Such is the form of control system for non-interacting and single robots. For multi-robot cells the Route Consultant is an additional requirement. Its function is to interpret the overall status of the cell and thence to advise the cell controller accordingly.

2 ROUTE CONSULTANT (RC) SPECIFICATION

The status of the manufacturing cell is determined by the current position of each robot, by the locations at the workstations of product parts and by the progress of any process or assembly task. From a knowledge of this status the route consultant (RC) is intended to "select from the set of possible safe moves (routes) for each robot that which is probably going to be the most productive".

To achieve this requires some kind of expert system, and although it is possible to write a single system for a particular manufacturing cell and product, our view is that a truly flexible system will contain four distinct parts as shown schematically in Fig.2. The four parts are clearly distinguished by type and purpose and their separation is a critical factor in the formulation of the simple, user-friendly procedures which are so important in fostering the use of innovative, practical systems.

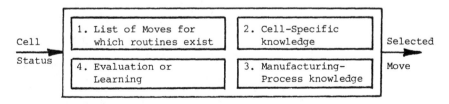

Fig.2. The Four Parts of the Route Consultant

Part 1

Lists of the robot dependent subroutines (eg. VAL) which instruct the robots for individual moves. These subroutines are either written for a specific move or are derived via an off-line system. The subroutines may contain inner loops and even jump statements provided they are all internal to the routine and do not cause the robot concerned to move outside the assumed trajectory. Some routines will be primarily for process tasks at a workstation, others with movements between workstations. The latter are referred to here as routes and are the prime concern in regard to interactions between robots.

Part 2

This covers the cell-specific knowledge such as the range of possible moves of all robots separately, the locations of workstations and either a direct tabulation of prohibited combinations of robot moves or an algorithm to detect such combinations.

Part 3

This includes details of the manufacturing process involved for each product, covering the parts, and any machining and assembly sequences. These details are largely independent of the cell-specific knowledge in part 2 with the exception that the locations for parts, machining and assembly points must be designated consistent with the locations in part 2.

Part 4

The selection of the most productive moves from alternative available moves, requires either an evaluation procedure or the ability to record experience and learn from it, since productivity derives not from a single move but from the cumulative effect of several. (This part has not yet been implemented in our model system).

Additionally, the RC must be able to operate under the cell controller and be subordinate to over-rides from sensors installed to detect abnormal conditions. Error recovery can, however, be incorporated in principle, as another kind of move.

Part 1 is covered in this application by the existing robot subroutines held in the cell controller. Part 4 has received little attention though the concepts of evaluation and learning have already been established (6,7). Parts 2 and 3 are the basis for this paper and involve both matters of principle and scale. They have already been implemented in a computer simulation and are currently being implemented in a demonstration unit comprising two small, 4-axis robots. They are best explained in relation to an example, against which background the more general scope and limitations can be discussed. The example is shown in Fig.3(a) and consists

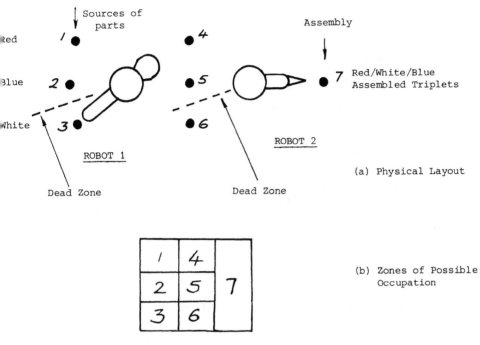

(a) Physical Layout

(b) Zones of Possible Occupation

Fig.3.   Two-Robot Configuration with Common
Envelopes at Zones 4, 5 and 6

of two robots engaged in the assembly of a ´product´ made up
of three discs in the sequence RED-WHITE-BLUE.   One robot
carries out the assembly, the other collects the parts.
Intermediate workstations facilitate transfer of parts from
source to assembly and it is inherent that, although all
pick-up and movement times are of a similar order, the times
taken for pick-up or assembly may vary over a range of
several 100% of the notional time allowed.   Pick-up and
placements could be extended to include machining etc. so
the example is really a stylised version of the practical
application being addressed.   The following sections explain
how the cell-specific knowledge and manufacturing process
are formalised to support parts 2 and 3.

3   CELL-SPECIFIC KNOWLEDGE

Practically acceptable moves are determined from the
cell-specific knowledge which ideally starts with a
consideration of all possible robot moves.   These are
defined by subdividing the volume of the cell into sub-
elements or zones as shown schematically in Fig.3(b); one
hundred elements being more typical of a real cell.   These
zones will not normally be of equal volume, they reflect the
distribution of robots, workstations, tooling and tasks
around the cell and are resolved more finely in areas where
the robots could potentially collide.   Zones above and below
the workstations are useful for providing moves which offer
collision-free routes for one or other robot to pass above
the other.

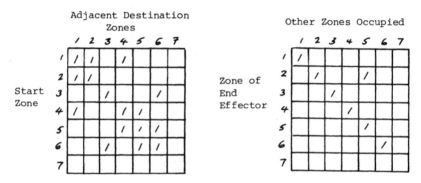

(a) Possible Moves to Adjacent Zones for Robot 1 alone

(b) Zones Occupied for a given position of the Robot 1, End Effector

Fiq.4. Robot Occupation of Zones

Fig.4 contains the interpretation of this zone information in relation to the adjacent moves and space occupancy of one robot. Fig.4(a) indicates which zones the effector end of the robot must pass in moving between all practical combinations of start and destination zones. Note the effect of the dead zones. Fig.4(b) indicates which other zones are necessarily occupied for the given position of the end effector. Note the overhang at position 2 which causes occupation of zone 5. Given this information, a request to this part of the route consultant to advise on a move from say 2 to 6 for this robot would be treated as follows :

- count the number of entries in the start and destination rows of the array in Fig.4(a), 2 for row 2 and 3 for row 6. Choose the row with the smallest number because this reduces the initial options without loss of opportunity.

- build up a tree of alternatives as shown in Fig.5. Note the 2, 1 and 4 alternatives in this tree can be discarded as they simply indicate the backwards option of the previous move.

- note all the zones occupied for each alternative route between the start and finish zones using the array in Fig.4(b).

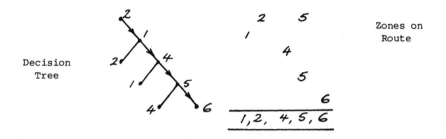

Fig.5.   Possible Moves for a Route between
         Zones 2 and 6

The same procedure is followed for the second and
further robots.   The zones occupied by each are next
compared and provided they contain no common zones, they are
authorised to move as requested.   If not, then one or other
or a no-move, move is authorised according to zone overlap.
In the case of a "no move", part 3 of the system will
request alternative action since it deals with decisions
about what should be done to carry out the overall
manufacturing process.   A request to this part can of course
either be referred to the algorithm above or to a table of
predetermined values.   The latter is a useful means of
assessing which routes to program in the first place.   This
part of the system has an off-line role in both single- and
multi-robot cells to determine the robot moves that it is
most desirable to program (Part 1) before programing is
undertaken.   Where a route is not program med, even though
there is no physical impediment, the array in Fig.4(a) would
not include the move.
     In the case of more than one move being available
evaluation is necessary to determine which one to accept.
(The shortest or quickest may be expedient but the
evaluation of several moves ahead is really necessary and is
a planned part of future development).

4   THE MANUFACTURING PROCESS

     The manufacturing knowledge is held as a set of
CONDITION-ACTION production rules (8) (an AI term, not a
manufacturing term).   The function of these rules is to act
as an interface between the cell-specific information and

the product. Whilst the complexity of the cell-specific rule base is dependent upon the arrangement of the cell (the number of robots, workstations, etc.) the number of rules necessary to specify a manufacturing task is totally independent of any particular robot configuration. This separation of the knowledge bases permits modular definitions of parts 2 and 3 of the ROUTE CONSULTANT. The cell-specific detail can be supplied quite separately from the data on the manufacturing task and each can be altered without fear of unwanted side-effects on the other.

The manufacturing knowledge concerns what has to be done, with which parts and, if important, in what order. Its connection with the cell-specific knowledge is thus primarily location-type information such as the zone locations for the parts, which robots serve which zones, and which zones are workstations used for processing, assembly or waiting. Facts such as "served-by(loc1,robot1)" simply state that location 1 is served by robot 1. These facts specify the operating positions of each robot and can be altered to assess the implications of rearranging the cell layout by moving or adding workstations.

A typical production rule (drafted in no specific language) is shown below. Lower case words represent variables.

```
IF robot SERVES ASSEMBLY POINT
AND robot IS NOT HOLDING ANYTHING
AND robot CAN REACH next part AT x
AND robot IS NOT AT x

THEN MOVE robot TO x
```

The rule will fire whenever the assembly robot finds that it happens to have the next part it requires within its reach. More complex rules cover circumstances when the assembly robot cannot reach the necessary part and the second robot must fetch the part, place it at a mutually acceptable workstation and move out of the area to allow the assembly robot to collect the part. Yet more complex rules govern extreme situations such as when a robot initially finds itself holding a part for which it has no immediate use (possibly complicated further by the absence of a free workstation at which the part can be temporarily left).

Production rules are also used to specify of the manufacturing sequence. For the RED-WHITE-BLUE prototype of Fig.3 the rules are :

```
IF blank THEN ADD RED
IF red   THEN ADD WHITE
IF white THEN ADD BLUE
IF blue  THEN STOP
```

More powerful representations are presently under consideration since it is recognised that this formalism has certain inherent weaknesses. For example, some manufacturing processes feature options of the form "following part A add part B or part C"

which cannot readily be expressed in production rule format. Also, the rules cannot directly cope with a requirement for a RED-WHITE-WHITE-BLUE sandwich. Solution by defining WHITE-1 and WHITE-2 to come from the same source is not really satisfactory since the distinction must be made throughout the cell when, in reality, all white counters are identical. The assembly robot will occasionally reject a WHITE-1 counter within reach in favour of a WHITE-2 counter it cannot reach although a human observer would realise the error. Future versions of Route Consultant will probably employ a heirarchical list structure to overcome these weaknesses.

5 PRACTICAL VIABILITY

A prototype system has been implemented in Prolog to test and debug the productions and has been found to select and authorise collision-free moves. It can be demonstrated for any arbitrary, initial condition of cell status for the system in Fig.3.

Work continues with the objective of controlling a particular industrial cell for which the complete manufacturing and performance specification is available. This is likely to be achieved. However, one of the foremost requirements of the Route Consultant is that the system must accept details of the cell and manufacturing sequence in an uncomplicated form via a reasonably user-friendly interface.

Fig.6 indicates the way in which a specification would be formulated for a wide range of applications. It concerns the assembly of a casting and three similar sub-assemblies each of four parts. The spring is assembled to a specified tension as part of the assembly process and the times for doing this vary.

The specification process is intended to convert the specific details of the product and its processes into a general form for which rules and a control strategy exist. Each separate part is first identified, numbered and listed in order of assembly. Where options are appropriate they are simply expressed as a tree as in Fig.6(b). For example, once the assembly has reached the stage of having parts 1, 2 and 3 then 4, 6 or 10 are possible alternatives.

All workstations should next be identified by the numbers assigned to pre-defined spacial positions. (Earlier in this paper the two are defined together but in general there will be more positions in space than occupied by workstations). Those which are sources of particular parts (with the part numbers) should be indicated as sources. Assembly and sub-assembly points should also be defined with an indication of which robot serves them. The allocation of workstations and the process activities that occur must be consistent with the availability of robot routines to both move to each workstation and to carry out the relevant processes. The moves between workstations are likely to be

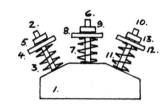

(a) Parts and Assembly   (b) Assembly sequence

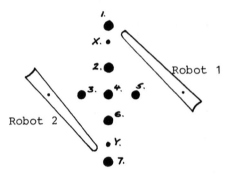

(c) Spacial Assignments

Spacial Point	Use	Robot
1	Assembly	both
1a	Decision	"
2	Source part (1)	"
2a	Decision	"
3	Source part (2)	"
X	Decision	"
etc	---	"

Fig.6. Practical Control Specification

general purpose for the cell. The product-specific routines
will currently require teaching mode or off-line development. So far as the user is concerned they are expected to
be available in Part 1 of the controller. From the parts
and assembly data and this workstation information, Part 3
of the RC can automatically provide the relevant robots with
valid manufacturing moves though not necessarily safe ones.

If no new routines have been added to Part 1, Part 2
will however select safe moves and avoid collisions.

If new routines are added to Part 1 then the
information in Part 2 must be augmented. This may mean
referring to spacial positions not previously used. This
effectively increases the number of points at which
decisions are made (move authorisation). The natural
choices of such points apart from workstations, are where
the envelopes of independent robot activity intersect. The
set of all probable positions is an important
characteristics of a cell.

In Fig.6 points with a suffix are physically above
those without. A robot at a suffixed point is clear of one
in a non-suffixed point and vice versa. Points X and Y are
introduced because assembly activity by either robot in that
region precludes assembly activity by the other. Points X
and Y thus feature in both Part 1 and Part 2 of the
controller and should be included in the prior set of
spacial positions.

The user is thus primarily concerned with the allocation of robots to workstations, workstations to spacial positions, and parts and process subroutines to workstations. The procedures for doing this are simple.

If flexibility is to be achieved with very small down times, there should be no requirement for an experienced programmer to maintain the knowledge bases following changes in specification. Rather, the system should be re-programmable by users at shop-floor level. The issue appears to be, not whether one can control a multi-robot cell flexibly (at any price) but whether the formalisms being explored can evoke the kind of specification that leads to easily assimilated, economic flexibility. An ideal specification is one which enables the user to enter common sense, user type insights at a level of aggregation appropriate to the cell's tasks and allows the cell controller, or route consultant in particular, to detect and exploit opportunities in an efficient productive manner. It is certainly the case with this robot problem, and probably with robots generally, that the relationships between manufacturing process specification, programs, control parameters and hardware configuration require much more attention.

6 REFERENCES

1. Ambler, A. P., 1982, RAPT: An Object Level Robot Programming Language, I.E.E. Seminar, Savoy Place, London.

2. Yong, Y. F. et al., 1985, "Off Line Programming of Robots", Chapter 4.5, Industrial Handbook of Robotics, John Wiley and Sons.

3. Whitehead, D. G. et al., 1984, "Industrial Robotic Assembly: A Multi-Sensor Approach", International Conference on the Development of Flexible Automation Systems, I.E.E., Savoy Place, London, pp.38-41.

4. Aleksander, I. (editor), 1983, "Artificial Vision for Robots", Kogan Page.

5. Morecki, A., Bianchi, G. and Kedzior, K., 1985, "The Theory and Practice of Robots and Manipulators", Kogan Page, pp.163-167.

6. Levy, D., 1983, "Computer Gamesmanship", Century Publishing.

7. Michie, D., Chambers, R. A., 1968, "BOXES: an experiment in adaptive control" in "Machine Intelligence 2", Oliver and Boyd.

8. O'Shea, T. and Eisenstadt, M., 1984, "Artificial Intelligence", Harper and Row.

9.   Spur, G., Mertins, K. and Albrecht, R., 1984, "Interactive Shop Floor Control for Flexible Automated Systems", International Conference on the Development of Flexible Automation Systems, I.E.E., Savoy Place, London, pp.38-41.

10.   Alty, J. L. and Coombes, M. J., 1984, "Expert Systems, Concepts and Examples", NCC Publications.

Flexible automated assembly systems

J.D. Lock, I. McLeod and G.J. Smith

This paper describes the research work being carried out at Napier College with the aim of creating a pallet based Flexible Automated Assembly System capable of building a range of assemblies on demand without the need for human intervention.

The paper outlines the programme of research being undertaken to establish the requirements for such a system and the criteria against which it may be evaluated in the following areas:

1. Investigation into flexible assembly requirements and criteria;
2. Possible configurations for a flexible assembly system; these would include both mechanical structures and control requirements;
3. Systems for interchangeable grippers in the system;
4. Systems for interchanging pallet-based tooling (fixturing); these would include both passive (location) and active (clamping) elements;
5. Indentification of parts for assembly, transportation of parts into the cell, methods of temporarily storing parts within the cell and methods of presentation of parts, in a known orientation, to the robot. Included in this section would be an investigation of parts design changes in order to facilitate robot handling;
6. Strategies for integrating the Flexible Assembly Cell with the FMS.

1.   INTRODUCTION

It is intended that the proposed Flexible Assembly System will interface with the Napier College Flexible Manufacturing System (FMS).

A modular approach has been adopted for the existing FMS whereby flexible manufacturing cells are linked by an automated materials handling system which is capable of transporting work pallets within the modular configuration according to a work specific routing plan. To fit in logically with the existing FMS design, the Flexible Assembly System is being treated as an additional cell within the modular structure, so allowing for future replication by making use of the existing FMS materials handling system and executive computer.

One of the reasons for the slow development of automated assembly systems in industry generally has been the dedicated nature of much of the equipment available. So much so that in order to justify automated assembly it has been necessary to have a product life of about 3 million units with about 1 million requirement per annum.

In recent years the reduction in cost and the increase in speed and capacity of control computers has led to the production of programmable arms (robots) at a price which brings them into the serious financial consideration of most companies. In particular, robots are widely used in the car industry principally for welding, painting and assembly tasks. The use of robots in the car production lines has led to a degree of flexibility in that various body shells can be passed down the same production line and the robot control programs altered to suit.

An obvious advantage of flexible automated assembly emerges from the above example, that of producing smaller batches of different models down the production line hence keeping up a steady volume. The development of this idea to its ultimate conclusion is a batch size of one, leads to the philosophy of Flexible Automated Assembly as we see it at Napier College.

2.   SPECIFICATION OF FLEXIBLE ASSEMBLY SYSTEMS

In order to build a Flexible Automated Assembly System (FAAS) which would accommodate very small batch sizes various requirements become apparent:-

(1)   Provision of a suitable robot - the specification varies greatly depending upon the assembly task but generally the machine will have six degrees of freedom and a repeatability of at least ± 0.1mm. It will also have a robust user friendly programming system.
(2)   As the assembly system is flexible, the component parts it handles will vary both in size and shape depending upon the assembly.
(3)   The system will require some form of input/output arrangement for component parts and assembled products.
(4)   An essentially metamorphic assembly fixturing system is required such that it can be changed to suit the assembly passing through. A sensing system is also required to ensure that parts are properly in place before final fixing is applied.
(5)   The system must allow optimum assembly times for assembled products and at the same time show a reasonable utilisation efficiency for the robot.
(6)   The FAAS should be capable of integrating with other automated production facilities, eg shape creation and sub-assembly production system.
(7)   Programming for the FAAS should be offline and down loaded when required to the local control computer and robot controller and there is a requirement for an overall executive computer to take care of scheduling and overall management.

From the above seven design criteria it is possible to construct a schematic diagram showing on some further detail the requirements for design of a FAAS. Figures 1A and 1B illustrate this approach.

3. FLEXIBLE ASSEMBLY SYSTEM CONFIGURATION

There are many ways of integrating the above criteria into a design but to make the best logical use of the existing Napier College Flexible Manufacturing System and taking particular account of items 3 and 6 above, a pallet based system seemed to be the best approach and an assembly cell construction around the robot offered good utilisation of the robot's operational envelope in terms of both space and time.

Figures 2 and 3 show a possible configuration of a Flexible Assembly Cell. This cell has been designed to integrate with the Napier College Flexible Manufacturing System, in particular with regards to the pallet for work transfer and processing, pallet transfer system and the overall control of the system.

At present, a Unimate, PUMA 560 robot is used, in conjunction with four assembly stations, in the cell. Pallets, which will arrive on a transfer track surrounding the cell, enter the principle assembly station via an input buffer. This allows a queue of pallets to be loaded sequentially to the principle station. The remaining three assembly stations may be loaded directly from the track and are used for magazines and tool storage.

Some component parts, tooling and grippers are held within the cell structure. Additional components parts, tooling and grippers can be fed into the cell on pallets.

On arrival at an assembly station, a pallet is located accurately by driving it against locating pins and then clamping it in position. A low profile clamping system has been devised to leave the maximum possible task space for the robot arm. Although the size and shape of the pallets conform to those in the Napier FMS, the type and pattern of holes is different. A grid pattern of equispaced through holes was chosen initially. This allows tooling to be added to the pallet before assembly and also provides a basis for evaluating the performance characteristics of the robot. Experiments were carried out on the effect of programming tolerance errors on positioning parts relative to the pallet holes.

The advantage of a cell construction for a FAAS is that many such cells could be arranged in parallel, serviced either by conveyors or automated guided vehicles, and could be used to produce sub-assemblies of large products or alternatively, many different sizes of assemblies in a company's product range.

4. RESEARCH PROGRAMME

Research work to achieve the Flexible Assembly System is being supported by the Science and Engineering Research Council.

The initial research work was directed towards obtaining a more accurate picture of the robot's capabilities than that given in the manufacturer's specifications. This involved studying accuracy relative to known points in space. The results of this work was a known task space and a set of programming rules that would provide a known positional accuracy for assembly.

Another series of experiments was carried out to study the magazing and assembly of variously-shaped small parts. The idea was to represent typical assembly and assembly tooling items by using fundamental shapes.

Initially, parts providing very little compliance were used as a worst-case test for assembly. Basic shapes such as plates, pins, blocks and sleeves were then used to gain experience in magazine presentation of parts and in programming for assembly. Vertical magazining with single-point parts pick-up proved very successful and provided full use of the cell's work space. the use of compliant fasteners also suggested a sensible path to follow in designing parts for assembly.

Computer-aided engineering methods have been used to facilitate research into studies of cell designs for the FAAS. This has enabled extensive work to be carried out in configuration, spatial and interference areas. It has also led to work for the development of a computer-aided system for off line planning and programming of the cell's operation. It is clear from the research work undertaken in this area, that the off line planning and programming facilities will be a critical component in the economic viability of FAAS, particularly for companies with short production runs.

The other research work being undertaken to ensure the successful development of an FAAS is as follows:

(1) Development of a modular assembly tooling subsystem, which would allow the system to automatically construct tooling for particular assemblies.

(2) Development of gripper technology; this would enable the robot to automatically change manipulators to meet requirements for assembly. This work is linked to the design of grippers for particular functions.

(3) Development of control systems; this work is based on the present FMS control system which uses a microcomputer to supervise the assembly processes within the system and a communications microcomputer to control the real time processes such as routing of material in the system.

(4) Simulation studies for operation procedures; this work enables the researchers to consider the dynamic aspects of materials management as emphasised in terms of availability, work in progress and time in progress. It also aids the development of communications structures within the system and between the system and the outside world.

(5) Economic and performance evaluation studies. This work helps establish the criteria against which the performance of the Flexible Assembly System should be gauged. The economic viability of the system is also considered relative to other assembly processes.

5.   CONCLUSIONS

The results of the research to date have been encouraging and work is now underway to construct a prototype of the Flexible Assembly System. Once a flexible assembly cell has been constructed, much of the research detailed above will contribute to the development of a true Flexible Automated Assembly System.

It is clear that the automation of the assembly process for small batch production is a complex and challenging activity. The successful outcome of this work will depend on many factors, not least of which will be product design. The computer and robot offer many potential advantages, but until the technical requirements of automation are absorbed by the prudent designer, little of the possible economic gain for user and consumer will be realised.

REFERENCES

1.   Lock et al, 1982, "A Flexible Manufacturing System for Prismatic and Cylindrical Components", Conference Proceedings 3rd Polytechnic Symposium on Manufacture, Wolverhampton, England.

2.   Young et al, 1984, "Performance and Evaluation of Flexible Manufacturing Systems", Conference Proceedings 4th Polytechnic Symposium on Manufacture, Birmingham, England.

3.   Young and Drake, 1982, "Integration of CAD/CAM with FMS", Conference Proceedings 2nd International Conference on FMS, Brighton, England.

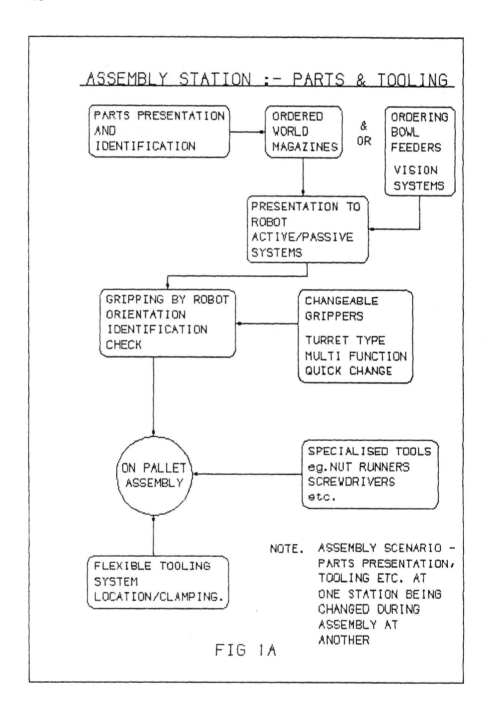

ASSEMBLY STATION :- PARTS & TOOLING

PARTS PRESENTATION AND IDENTIFICATION

ORDERED WORLD MAGAZINES

& OR

ORDERING BOWL FEEDERS

VISION SYSTEMS

PRESENTATION TO ROBOT ACTIVE/PASSIVE SYSTEMS

GRIPPING BY ROBOT ORIENTATION IDENTIFICATION CHECK

CHANGEABLE GRIPPERS

TURRET TYPE MULTI FUNCTION QUICK CHANGE

ON PALLET ASSEMBLY

SPECIALISED TOOLS eg. NUT RUNNERS SCREWDRIVERS etc.

FLEXIBLE TOOLING SYSTEM LOCATION/CLAMPING.

NOTE. ASSEMBLY SCENARIO – PARTS PRESENTATION, TOOLING ETC. AT ONE STATION BEING CHANGED DURING ASSEMBLY AT ANOTHER

FIG 1A

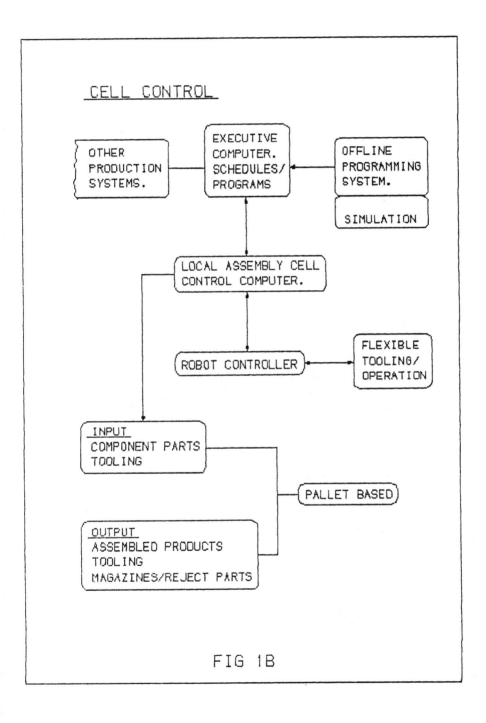

FIG 1B

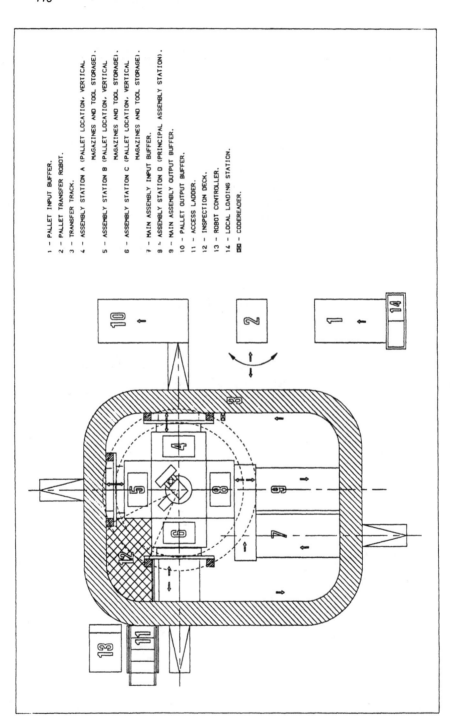

1 - PALLET INPUT BUFFER.

2 - PALLET TRANSFER ROBOT.

3 - TRANSFER TRACK.

4 - ASSEMBLY STATION A (PALLET LOCATION, VERTICAL
MAGAZINES AND TOOL STORAGE).

5 - ASSEMBLY STATION B (PALLET LOCATION, VERTICAL
MAGAZINES AND TOOL STORAGE).

6 - ASSEMBLY STATION C (PALLET LOCATION, VERTICAL
MAGAZINES AND TOOL STORAGE).

7 - MAIN ASSEMBLY INPUT BUFFER.

8 - ASSEMBLY STATION D (PRINCIPAL ASSEMBLY STATION).

9 - MAIN ASSEMBLY OUTPUT BUFFER.

10 - PALLET OUTPUT BUFFER.

11 - ACCESS LADDER.

12 - INSPECTION DECK.

13 - ROBOT CONTROLLER.

14 - LOCAL LOADING STATION.

▨ - CODEREADER.

FIGURE 2

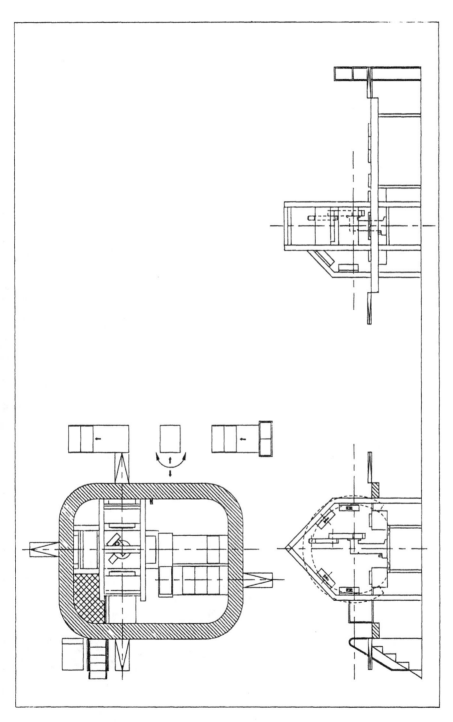

FIGURE 3

Autonomous assembly devices and related control strategies

P.J. Drazan, S.H. Hopkins and C.J. Bland

1. INTRODUCTION

Within many industries e.g. car manufacturing, assembly requires a large proportion of the labour force. In a bid to reduce the labour requirement, and increase productivity, automatic assembly machines have been developed. These classical machines may be considered to be a series of work stations each adding a single component or performing a single operation. In addition to the basic machines complex conveying systems are required to transfer assemblies from station to station. This type of system is very expensive to install and hence is rarely economically viable at production volumes below one million assemblies per year. Another disadvantage is that being custom built for a particular assembly the machinery cannot be easily adapted to keep up with future developments in product design.

Industrial robots have recently been introduced to the assembly environment with the aims of doing the work of several work stations. This results in a cheaper assembly system that is economically viable at lower production volumes and can be modified as product developments are introduced. There are still some problems that are common to both classical and robot enhanced assembly systems. One of the most important being that any misalignment between parts will prevent successful assembly. Hence, high accuracy is needed not only in the assembly machine and any jigs and fixtures but also in the components themselves.

The need to overcome such problems in automatic assembly is illustrated by the considerable amount of research work currently being performed throughout the world. This paper will concentrate on the research work being carried out at the University of Wales Institute of Science and Technology (U.W.I.S.T), sponsored by the Science and Engineering Research Council, concerned with the development of "intelligent" robotic devices to enhance automatic assembly systems. Two such devices and the associated hardware and software are described. In particular the use of search routines to complement a vectoring technique are explained.

2. PASSIVE AND ACTIVE ACCOMMODATION.

Basically there are two systems which will compensate for part misalignment. Passive units were initially developed by researchers (1) (2) (3) (4). These mechanical units operate by elastically deforming about a centre of compliance to zero rotational and lateral misalignments. Passive units are carefully designed structures and have a preset stiffness and centre of compliance. They have the advantage of being simple and cheap and have been successfully used in a variety of applications. Their limitations are firstly, being mechanical structures of relatively low stiffness they have a low natural frequency and are susceptible to vibrations. Also, each unit is designed for one particular "peg" length and in addition one of the parts must have a chamfer greater than the largest possible lateral misalignment.

To reduce these limitations active accommodation units are being used where the contact forces during assembly are monitored by transducers and the signals used to compensate for the misalignment. Two force sensors were developed at the Charles Stark Draper Laboratory (5) one is an instrumented version of their passive unit, the other a six degree of freedom sensor employing strain gauges. Piller (6) also developed a six degree of freedom sensor suitable for assembly. Kasai et al (7) have developed an active sensory table comprising sensory and actuation components.

There are problems associated by mounting the active unit on the manipulator several of which are:-

(i) Six degree of freedom sensors are expensive and generally involve complex signal processing before they can be used to control the manipulator.

(ii) Due to the manipulator's high inertial mass it is difficult to compensate for small misalignments.

(iii) It is not always possible to adaptively control the manipulator since relatively few robot languages have the facility for sensor controlled movement.

The approach at U.W.I.S.T. and elsewhere (8) (9) (10) is to develop autonomous units comprising both sensing and local actuation.

The team at U.W.I.S.T. have reduced the complexity of the unit by using only three degrees of freedom in sensing and two degrees of freedom in actuation. The loss in sensory information is compensated for by intelligent signal processing in a local computer.

3. AUTONOMOUS UNITS DEVELOPED AT U.W.I.S.T.

Two electric units have been developed at U.W.I.S.T. Both have three degree of freedom sensing, but one utilises strain gauges and the other photo-electric transducers. The sensor outputs are processed on a micro-computer which then controls a stepper motor driven X-Y table. The insertion is performed by a vertical stepper motor drive, see Fig.3.1.

3.1 Computer Interface and Stepper Motor Drives.

The computer is connected via a dual eight bit digital input/output card containing a PIA to an Eurocard rack system, see Fig. 3.2. One half of the PIA is used to select the card in the Eurocard system and the other half to provide eight bits of data transfer to and from the computer. The rack will carry a variety of cards for driving motors and sensing from a variety of different transducers.

3.1.1. Strain Gauge and Photo-electric Devices.
The sensor consists of a flat plate in the shape of a cross, see Fig. 3.1. Onto each arm of the cross are mounted four strain gauges connected into a bridge. The output of each bridge is connected to a card in the rack system where the appropriate signal conditioning and eight bit A/D conversion is performed.

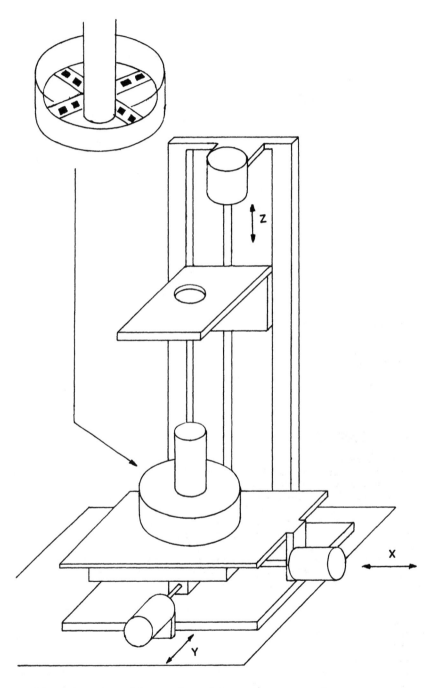

Fig.3.1.   Assembly station with strain gauge unit.

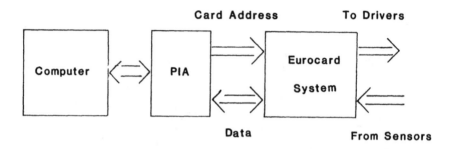

Fig.3.2.   Computer Interface

The photo-electric unit incorporates a recently developed position sensitive detector employing lateral photo effect on a semi-conductor light detecting surface. By combination of this sensor with a carefully designed mechanical structure a three degree of freedom unit has been developed. The electronics for this unit were developed in two sections. The basic signal conditioning circuitry is within the body of the unit itself whilst the A/D convertors required for each axis are on a card in the rack system.

4.   CONTROL STRATEGIES

The programs developed for active accommodation control have been written using a combination of BASIC and 6502 machine code. The machine code subroutines were written to obtain maximum speed when driving the stepper motors. Parameters between BASIC and the machine code sub-routines are transferred by the BASIC command POKE.

By monitoring the vertical load on initial contact it can be estimated whether or not the chamfer is within the hole, see Fig.4.1. The flow chart for assembly of a chamfered peg into a hole is shown in Fig.4.2. Assuming that the chamfer on the peg touches the hole then a lateral force vector is produced, its direction being towards the centre of the hold, see Fig.4.3. The X-Y table is then driven in the direction of this vector.

If the chamfer on the peg does not make contact with the side of the hole then a search routine is executed until the axial load reduces. When the axial load has reduced the chamfered insertion routine is executed.

Two search patterns have been implemented, a spiral search for situations where the positional peg-hole relationship is unknown, and a wave pattern for situations when the peg is offset to one side of the hole, see Fig.4.4.

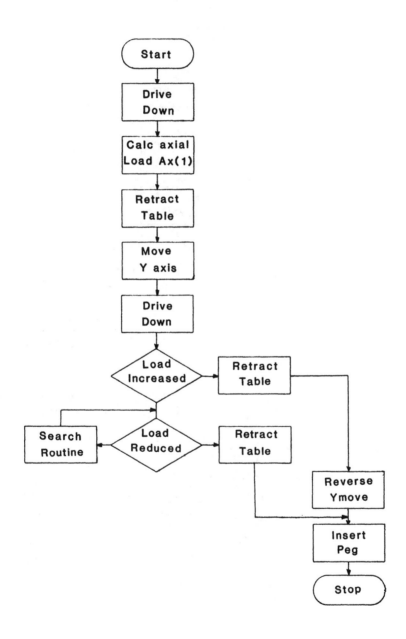

Fig.4.1.   Flow chart for insertion process.

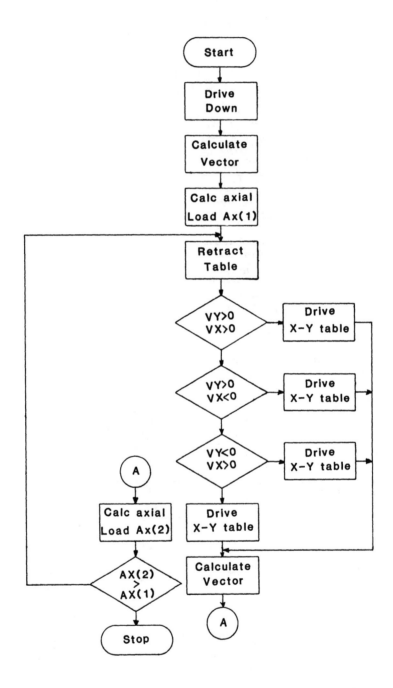

Fig.4.2.   Flow chart for chamfered assembly.

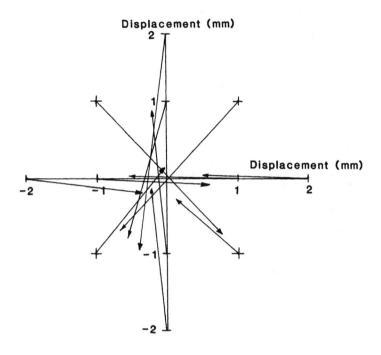

Fig.4.3. Typical vectors produced.

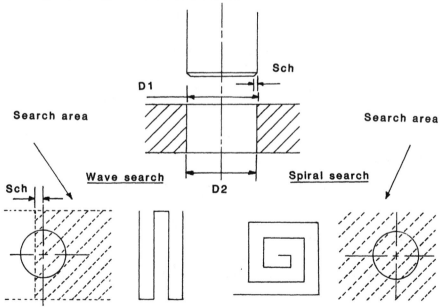

Fig.4.4. Spiral and wave search patterns.

Further developments are now being investigated to increase the execution speed of the programs by rewriting the software in FORTH. Also, strategies which will enable components without a chamfer to be assembled autonomously are being developed. Again methods of combining search routines with a vectored insertion are being studied. There are also plans to implement the signal processing and control strategies on a single board which will comprise all the computing and interfacing requirement.

5. CONCLUSIONS

The authors work on adaptable handling systems shows that relatively simple devices can be manufactured to accommodate misalignments between parts. By the use of appropriate control strategies chamfered pegs can be assembled whether or not initial contact occurs on the chamfer. Work is now proceeding to decrease the assembly time and also to enable unchamfered components to be assembled.

6. REFERENCES

1. Havlik, S., 1983, "A new elastic structure for a compliant robot wrist", Robotica, 1, 95-102.

2. Romiti, A., Belforte, G., and D'Alfio, D., 1982, "A self-adaptive guided assembler (SAGA)", Robots VI Conference Proceedings, 2, Paper No. MS82-225.

3. McCallion, H., Alexander, K.V., and Pham, D.T., 1980, "Aids for automatic assembly", Proceedings, 1st Conf. on Assembly Automation, Brighton, England, 313-323.

4. Watson, P.C., 1978, "Remote center compliance system", U.S. Patent No. 4098001.

5. Charles Stark Draper Laboratory, 1983, "The model 4 IRCC", Report No. C-5601.

6. Piller, G., 1982, "A compact six degree of freedom force sensor for assembly robot", 12th Int.Sym. on Industrial Robots/6th Int. Conf. on Industrial Robot Technology, 121-129.

7. Kasai, M., Takeyasu, K., Uno, M., and Muraoku, K., 1981, "Trainable assembly system with an active sensory table possessing six axes", 11th Int.Sym. on Industrial Robots, 393-404.

8. Drazan, P.J., and Hopkins, S.H., 1984, "Semi-Autonomous systems for automatic assembly", Annals of CIRP, 33/1.

9. Brussel, H.V., and Simons, J., 1981, "Adaptive assembly", 4th BRA Proceedings, England.

10. Goto, T., Inoyama, T., and Takeyasu, K., 1974, "Precise insert operation by tactile controlled robot", The Industrial Robot, 1, No. 5, 225-228.

An aid to effective off-line programming of assembly robots

A.H. Redford

1.   INTRODUCTION

With the expected increased utilisation of assembly robots and once the current 'appraisal' exercises have been completed, potential users will need more substantial evidence of the economic viability of the equipment and, if this is to be given, a key factor will be utilisation. As is the case for any high capital cost equipment, it will be essential that assembly robots are operated on at least a two, and preferably a three shift basis and, perhaps more significantly, it is essential that they are kept running. To achieve this, three factors will emerge as being important and these are:-

(i)   Interactive user-friendly programming languages which will allow fast <u>and</u> accurate program generation

(ii)   Effective off-line programming such that valuable robot application time is not lost as a result of having to carry out tedious and time consuming 'self-teach' exercises.

(iii)   The development of recovery strategies which can minimise robot down-time which results from faulty handling and insertion operations

2.   INTERACTIVE USER-FRIENDLY LANGUAGE

As was the case with early NC there are many different robot programming languages each with its own characteristics and syntax and even though it can be expected that some rationalisation will take place, it can also be expected that, as with NC, user-friendly languages will be developed, and these will be structured in a similar way to those for NC, i.e. geometry and motion processors will be generated which will be task dependent , and this information will then be post-processed to suit the requirements of specific robot controllers. With suitable algorithms, complex geometric functions can be quickly and accurately derived and transcribed to be syntactically correct. Unfortunately, unlike NC, two important features necessary for robotic applications would not be accommodated by this facility and, as a consequence, the benefits of this limited exercise would only be to speed up programming and reduce the possibility of both syntactical and <u>basic</u> geometric errors which would result in reduced lead time and slightly improved robot utilisation.

The missing features which are vital for effective robot utilisation are off-line programming and effective recovery strategies; a possible means of achieving the former will now be described.

3.   OFF-LINE PROGRAMMING

Off-line programming can perhaps best be defined as having the ability to write programs for machine controllers which can be loaded and which will carry out the required tasks without any further actions. For NC this is accepted as routine but for assembly robots the principal problem associated with off-line programming results from the conflicting requirements of assembly robots. Because of the nature of the task involved, it is generally accepted that the equipment should be capable of high maximum velocities and high accelerations. For limited power availability this implies a light structure. Conversely, many mating operations require precise alignment of the robot end effector which requires a rigid (heavy) structure between the point at which the robot's position is measured and the point at which the insertion operation is taking place.

The general tendency has been to concentrate on increasing robot speed and, consequently, robot structures tend to be light and lacking in stiffness. Thus, whilst assembly robots are usually adequate regarding repeatability (the ability to return to a pre-taught point), they are often lacking in accuracy (the ability to go to a pre-determined point). This feature of assembly robots is recognised and work on off-line programming reflects this basic characteristic. Research into off-line programming to date, therefore, has attempted to tackle the problem in one of three ways. These are:-

(i)   Program to nominal co-ordinates and then use either passive or active compliance.

(ii)   Monitor the position of the end effector as opposed to the position of the individual axis drives.

(iii)   Map the 'real' position of the end effector for a wide variety of nominal positions and use this data base to compute real co-ordinates.

Passive compliance is achieved by devices such as the remote centre compliance in conjunction with good product design, where the use of chamfers and tapers facilitate insertion operations for parts which are incorrectly aligned. Alternatively, active compliance using either tactile or vision sensing with appropriate feed-back to modify the end effector position can be used such that satisfactory insertion operations can be performed.

The problems associated with compliance are that passive compliance is reliant on the designer being able to incorporate the necessary features. It is relatively expensive and in its usual form where the compliant device is fitted to the end effector, it is relatively heavy. Active compliance is currently very expensive and slow, and although it can be expected that the cost will reduce considerably, it is likely that considerable time penalties will always be accrued when using this method of compensation.

Sensing the position of the end effector appears to be a logical method of overcoming mis-alignment problems but this also has several disadvantages. It is expensive to install the tracking equipment, the equipment is both bulky and 'delicate' and there are many operating conditions where the end effector is 'hidden' from the tracking equipment. It is thought that end effector tracking will be shown to be effective in the near future but its costs and reliability will be subject to some doubt.

Mapping is both expensive and slow. Further, with time (and wear) with breakdown of either robot arms or controller, or with the requirement to re-set, the data base would need to be regularly refreshed.

A preliminary study has been carried out at Salford to look at the possibility of using compliance but only as a means of considerably reducing self-teach time. In the study, consideration has been given to using passive compliance, in the form of a compliant fixture (XY plane), a template with suitably located holes which represent all the insertion locations and modified 'parts' having good insertion features which are to locate in the holes.

4.   SPECIFICATION FOR THE COMPLIANT FIXTURE

Whilst conducting insertion tests on a Pragma A3000 assembly robot, using round chamfered end pins, the following phenomena were observed.

(i)   Pins of 3mm diameter or less with clearances of 0.005 mm inserted reliably with an offset not exceeding the size of their chamfer.

(ii)   Chamfered pins greater than 3 mm diameter with clearances of 0.005 mm inserted reliably until an offset defined by their kinematic insertion envelope is reached.

(iii)   The percentage clearance between the pin and hole bore no relationship to the allowable offset.

Clearly if the above is true then, assuming the work is rigidly mounted, either the robot arm is deflected (in a lateral plane) or the carriage forced along the track. In either case a force must be exerted on the robot arm and it is implicit that a corresponding force is experienced on the fixture.

It is suggested that this be used as the basis for the design of a compliant device which could be utilised in one of the ways outlined below:

A calibration curve of force experienced at the end effector against known lateral deflections could be obtained. Thus, by measuring the lateral force exerted on the work by the appropriate insertion operations, it would be possible to evaluate the error in alignment prior to insertion and to then calculate the correction required.

For each insertion operation, the presence of any force on the work fixture may be sensed. Subsequent traversing of the fixture, until this force is eliminated, would yield a position where the axis of the mating parts would be in alignment. A measurement of the distance traversed would therefore correspond to the error in alignment prior to insertion.

For each insertion operation, rather than traverse the fixture as above, the robot arm could be driven such that the lateral insertion force is nullified.

The first method suggested suffers from three distinct disadvantages:

Firstly, any evaluation made relies upon the accuracy and consistency of the force measurement. Secondly, the deflection characteristics of the end effector will be dependent on its position within the working envelope (this may result in the requirement of a vast quantity of data to be of practical use). Lastly, this data would need to be acquired for each individual robot since the deflection characteristics cannot be guaranteed to be identical, even for robots of similar type. It was felt that these disadvantages were sufficient to discount the method.

In a practical situation it is considered undesirable to move a work fixture once placed, since it would subsequently have to be returned to its datum position. Further, the cost of the appropriate drives and measuring equipment would be high. For these reasons the second method was rejected. Both the second and the third methods offer the advantage that no actual magnitude of force need be measured, merely its presence detected. However the means of error elimination selected in the third method shows excellent potential and it was decided that this should be developed.

By causing the robot arm to provide the necessary movements it should prove possible to make use of the robot's own positioning, measuring and processing facilities. The subsequent assessment and interpretation of 'error' information and correction of co-ordinate data will therefore be much simplified. It will be noted that although only the presence of the force need be detected its direction will also be required. By basing a design on the principles described above, two major potential problems are eliminated; the lack of stiffness of the robot arm is now of no consequence and only very restricted compliant movement is necessary.

Based on the above discussion a specification for the compliance device has been compiled and this is detailed below:

(i)   It should be capable of measuring a 5 N lateral force in two orthogonal directions i.e. it shall be aligned with the X and Y axes of the robot arm.

(ii)   It shall remain in place throughout the normal assembly process; the implications of this are covered by the following specifications

(iii)   Parts to be assembled shall be temporarily replaced by purpose designed substitute components for initial calibration procedures.

(iv)   It should adopt a stable datum position.

(v)   It must withstand the rigours of everyday workshop use and in addition must be reliable and have a useful working lifespan.

(vi)   It should be manufactured from materials which are compatible with (iv) and (v) above.

(vii)   Consideration must be given to the incorporation or fitment of assembly jigs.

(viii)   It should be simple to use and should require little operator intervention.

(ix)   Because it would be general purpose equipment, quality or performance to reduce cost should not be compromised but obviously cost is important.

5. DESIGN OF THE DEVICE

For the specification outlined above, the basic requirement is for a method of measuring small forces with small deflections in two orthogonal directions. Many types of device would meet this specification but after examining a variety of these, it was decided that one based on the principles of a metal cutting dynamometer would be most suitable. Clearly, a metal cutting dynamometer is designed to measure and withstand very large forces with very little deflection and the former characteristic is not ideal for the limited compliance fixture specified. However, it was thought that a device of this type could be used to obtain an indication of the likely effectiveness of the concept.

The dynamometer chosen, because of its configuration, its very low orthogonal interaction and its availability, was one designed by Boothroyd and this is shown in Fig. 1. A 'tool-holder' was designed for this which had a 10 mm drilled and reamed hole in an appropriate position into which the robot was to insert a ground steel peg with 0.01 mm clearance. The signal which results from an inbalance of the strain bridges of the dynamometer was monitored on a digital voltmeter capable of reading 1 microvolt.

6. EXPERIMENTAL TESTS

A series of insertion tests were carried out using one arm of the PRAGMA twin arm assembly robot shown in Fig. 2. The peg was held in one of the robot grippers (Fig. 1) and the arm was then manually driven into an XY position where the peg would enter the hole in the dynamometer tooling area (Fig. 1) without any resistance to motion. The co-ordinates of the robot arm could then be read directly using a command at the robot controller VDU. The peg was then withdrawn from the hole, the robot arm was moved 0.6 mm in the X direction and the peg was re-inserted in the hole. The 0.6 mm offset was the limiting value possible because of the size of the chamfer on the peg but this is a much larger value than should be encountered in practice as the difference between nominally programmed co-ordinates and actual co-ordinates.

Because the dynamometer is very stiff, the 0.6 mm offset between the peg and the hole is accommodated by the natural compliance of the robot arm which deflects to allow the insertion to take place and which results in a force on the dynamometer and a reading on the digital voltmeter. The robot arm was then driven manually in the X direction until the reading on the voltmeter was the same as that for the dynamometer unloaded. The XY co-ordinates of the robot arms were then read at the robot controller VDU and were compared with the known position of the hole.

One problem which arose was a function of the very small voltage generated by the strain gauges. As could be expected, a device which is capable of measuring forces up to 10,000 N is not really suitable for measuring forces of less than 50 N and the main problem was voltage drift which was typically 2 microvolts per minute on a signal of 25 microvolts per mm deflection. However it should be recognised that with a compliant fixture based on the dynamometer principal but designed specifically for assembly, the signal per unit deflection would be significantly larger (say 250 microvolts per mm) and that for this and a time to move the robot arm from nominal insertion position to true insertion position of 12 seconds, the error resulting from drift would be 0.0016 mm; this is a figure which is considerably better than the resolution of the robot system.

Many tests were carried out for displacements in X, Y and in both X and Y and the various extensions of the robot arm. Table 1 shows a typical set of results where a 0.6 m offset was used and the test was repeated 23 times. It can be seen that the maximum difference between the X co-ordinates of the hole and the X co-ordinate of the peg resulting from attempting to null the force on the dynamometer was 0.16 mm. For insertion operations in assembly, it is recognised that assembly of 'good' parts will generally occur if misalignment is less than 0.05 mm and will invariably occur if the misalignment is less than 0.025 mm. The values obtained from the tests were considerably in excess of the requirement but with an increase in sensitivity as desired above, it is likely that the requirement can easily be met.

7.   CONCLUSIONS

A device has been successfully tested which can be used to obtain accurate insertion co-ordinates from off-line programming an assembly robot. In the form outlined it is only suitable for insertion operations and equivalent devices for handling (picking) operations would be more expensive and more difficult to develop. However, since insertion is far more critical than handling it could be expected that, particularly with good design of the picking location, nominal positioning of the gripper will be sufficient.

TEST NUMBER	ERROR, mm
1	−0.05
2	+0.08
3	+0.04
4	+0.125
5	+0.125
6	−0.13
7	+0.16
8	+0.02
9	−0.08
10	0
11	−0.05
12	+0.04
13	+0.025
14	+0.03
15	0
16	+0.04
17	+0.12
18	+0.04
19	+0.08
20	+0.01
21	−0.02
22	+0.03
23	+0.005

Table 1   Error between true insertion co-ordinate and that determined using the compliant fixture.

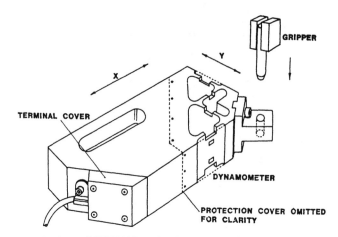

FIG. 1 CUTTING FORCE DYNAMOMETER

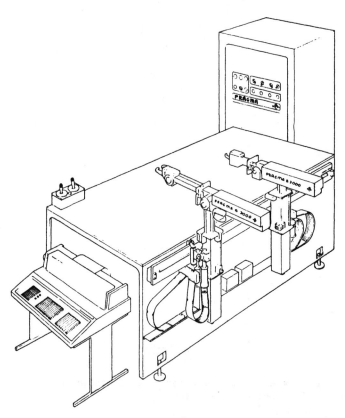

FIG. 2 PRAGMA A3000 – TWO ARM CONFIGURATION

Robot force sensing using stochastic monitoring of the actuator torque

F. Naghdy, J. Lidbury, J. Billingsley

1.   INTRODUCTION

To integrate a robot into the manufacturing system will usually require the use of sensors of one sort or another, particularly for complex tasks such as adjustment and testing (Naghdy et al (1)). Force sensing can be of great importance for overcoming position tolerance in assembly (Harrison et al (2)) and for obstacle detection and avoidance (Dothy et al (3)).

Force sensing has been reported by many authors, some using an added sensor such as a strain-gauge mounted at the joint (James (4)) while others monitor the applied torque (3,2). The latter method is adopted in this paper, adding the new approach of stochastic modelling and time series analysis of the currents driving the actuators. Feasibility of the method is shown to be established, by experiments using a Puma 600 robot for obstacle detection and avoidance.

The results of the feasibility study are discussed, and comparisons are made against the merits of other methods.

2.   ACTUATORS AS FORCE FEEDBACK SENSORS

Force information is conventionally captured from certain gauges or other specific force sensors· Although the addition of such sensors may carry a substantial cost, their effectiveness may often be limited. A force sensor's signal will often be restricted to sensing in a local area concerning a single joint. What is required is a system covering all sections of the robot, to give effective protection to the robot, to the operator and to the environment.

An alternative method is to monitor the actuator torque. The motor current of a d.c. motor varies as the square of the motor torque. Therefore to monitor the torque of the joint, it suffices to monitor the current flowing in the armature. During the insertion of a workpiece in a jig, if the workpiece is not positioned and aligned accurately external forces will be applied to the object from the jig, opposing its motion (Paul et al (5)). These forces are transmitted to the different joints of the manipulator and the actuators try to overcome these forces by driving the manipulator in the opposite direction. Since the jig is not movable, the amount of torque will increase in the manipulator. Thus abnormal changes in the torque can be detected by monitoring the currents of the actuators. An obstruction on the way of movement of the robot will produce a similar effect.

This method has the advantage of requiring no additional hardware and can be applied to any robot without any serious modification or reconstruction of the joints.

3. STOCHASTIC MONITORING OF THE ACTUATORS TORQUE

To monitor the current in a d.c. motor for the purpose of force sensing is not itself a novel idea. Rolling machine engineers have used it since the fifties. However it has only been used when relatively large forces have been involved (Colvile (6)).

For various reasons, the analysis and monitoring of the actuator current is not simple and straightforward. The oscillations in the servo loops and amplifiers introduce a high level of noise in the armature current. The start up current, always present during the change of direction, also adds significantly to the variation of the current in the joint. Figures 1, 2 and 3 are good examples of current wave-forms. They are obtained from a PUMA 600 robot in motion and illustrate the current variation in the first three joints of the robot. The signals are in fact the voltages read across the output resistor of each servo amplifier. The robot was programmed by simple "MOVE" instruction to move between a number of taught locations.

In an earlier work (2) some features of the motor current/time profile for a satisfactory insertion of a work-piece into a jig was taught to a computer running in parallel with the robot controller. The features comprised level and gradient of two consecutive samples. During the robot operation, the computer could then apply a simple template matching to determine whether the profile has diverged from that of a satisfactory insertion. The technique was of course restricted to the defined task and could not cope with noise and transients in the wave-form.

Existence of a strong stochastic component in the current waveforms suggested that a statistical approach might produce a more satisfactory result. In this method the actuator current is converted into a discrete-time, stochastic, sampled data model suitable for attack by modern control methods or time series analysis. Time series analysis techniques have been used for the monitoring of other engineering systems especially in the field of patient monitoring (Naghdy et al (7)). The results produced have been very encouraging.

A time series consists of a set of observations on a variable taken at equally spaced intervals in time. In engineering terms, it is the same as the discrete or sampled data model of the process represented by the variable. A stochastic model built on a time series is illustrated in general form as

$$\phi(B)Z(i) = \theta(B)a(i)$$

where

$$\phi(B) = 1 - \phi_1 B - \ldots - \phi_p B^p$$

is the autoregressive term and

$$\Theta(B) = 1 + \Theta_1 B + \ldots + \Theta_q B^q$$

the moving average term. $a(i)$ is a series of shocks generated by a white noise process for which $a(i)$ are independently and normally distributed with zero mean and constant variance.

4.   CASE STUDY

4.1   Application of Stochastic Force Sensing to PUMA

Before launching an overall research project, a feasibility study was essential to establish the credibility of stochastic force sensing technique.

A system based on a Unimation Puma robot and a 16-bit Apricot micro computer was set up (Figure 4). The voltage across the output resistor of each servo amplifier was digitized through a 6-channel A/D converter and fed into the Apricot. The Apricot was interfaced to the robot controller via serial terminal port and I/O module of the robot.

A series of experiments were conducted to identify the behaviour and the model of each actuator. The emphasis was put initially on three actuators driving waist (joint 1), shoulder (joint 2) and elbow (joint 3) of the robot. It was logical first to look at the idling current wave-forms, i.e. when the motors are turned on but the robot is not moving. They are shown in Figures 5 - 7. Then while the robot was at standstill, each joint was pushed or pulled in a variety of directions. Figures 8 - 9 illustrate the results for joints 1 and 2. Steady state movement of the robot, i.e. motion at a constant speed and no change of direction produced wave-forms similar to the idling state.

Afterwards an experimental task was defined for the robot and an objective was set. The task was for the robot to collect an object from an inlet tray, insert it into a jig and then place it on the outlet tray. The object is an energy regulator (1) which has a spindle shaft held downwards. During insertion this spindle is pushed into the jig. The inlet and outlet trays have holes through which retain the regulator spindle. If during insertion of the spindle into the jig or outlet tray the previous workpiece is not removed, the result can be catastrophic. The declared aim was set to detect the obstruction of the jig and outlet tray by the stochastic force sensing technique.

4.2   Noise Reduction of the Signals

First of all some effort was made to filter the signals and reduce the noise level. Two different forms of low pass filters were applied to the currents:

(a)   Simple RC filter with a break point frequency of 10 Hz

(b)   Sallen and Key Active Second order low pass filter
with a measured cut off frequency of 9.1 Hz. It
was applied only to joint 2 and no improvement in
the signal was observed. The reason is still not
clear.

The results of the filters in terms of standard
deviation of the signals are shown in table 1. There is
still much more work required to determine the optimum
sampling time of the currents and an effective method of
filtering the signals.

TABLE 1

Joint	Standard Deviations		
	1	2	3
No Filter	15.94	22.54	27.60
RC Filter	7.63	5.00	15.5
Active		8.29	

4.3   The Model of the Wave-forms

The auto-correlation function of each current time
series was calculated to determine the stochastic model of
the joint. The results for joints 2 and 3 are shown in
figures 9 and 10. Since there is no peak in the auto-
correlation functions, the model identified for all three
joints is ARIMA (0,1,0) (Autoregressive Integrated Moving
Average). This represents a random walk where the next value
in the time series is equal to the previous value plus noise.
The corresponding forecast algorithm is:

$$\hat{Z}_t(k) = Z_t \quad k = 1,2,\ldots\ldots$$

Z_t is the current observation and $\hat{Z}_t(k)$ is the k step ahead
forecast for $Z_{(t+k)}$ made at time t.

4.4   Control Strategy

The actuator currents are monitored by a double sided
CUSUM scheme to detect any large deviation of the current
from the forecast value in any direction. The scheme has two
moving parameters which fluctuate about a constant value when
the process is in control. The scheme uses a minimum amount
of computer storage. The CUSUM algorithms are:

$$D_t = MIN \ (D_{(t-1)}, \ L_0) + C-e_t \qquad D_0 = L_0$$

$$F_t = MAX \ (F_{(t-1)}, \ -L_0) - C-e_t \qquad F_0 = L_0$$

Max and Min functions determine the maximum and minimum
values between the two parameters in the bracket. D_t is the
moving parameter for the lower limit at time, t and F_t is the

upper limit at time t.

C and L_0 determine the sensitivity of the algorithm. $L_0 = w1$, $C = w2$ is the standard deviation of the time series, w1 and w2 should be optimised for the system monitored.

e_t is the one step ahead forecast at time t, i.e. $e_t = Z_{(t+1)} - \hat{Z}_t(1)$.

"Out of control" is signalled when either $F_t > 0$ or $D_t < 0$. The CUSUM is an iterative process. Once out of control has occurred, the scheme must reset to its initial conditions.

4.5   Interfacing the Control Algorithm to Puma

The "out of control" produced by the computer must be passed to the robot. Initially, the message was passed via the robot's command terminal. However due to the restrictions imposed by the operating system and the "VAL" language of the robot, the communication and reaction of the robot to the out of control message proved to be slow and inefficient. Therefore it was decided to use the I/O module which permits eight state line to the robot. As soon as an out of control is detected on one of the joints, an input line of the robot is pulled high. The robot reacts to the flag only when it is approaching to one of the insertion points. On receiving the "out of control" signal, the robot branches to a recovery subroutine and takes necessary action to overcome the obstacle. The "REACTI" is ignored if the robot has arrived at the insertion point. In this experiment the information produced by joints 2 and 3 is sufficient for detection of the obstacle. Therefore only these two joints are monitored by the computer. All the software in the Apricot is written in the programming language C. Hence execution of the code is reasonably fast and an obstacle is detected before any damage is incurred by the robot or the jig. The sampling time is around 0.05 second. The sampling rate is not optimum and some work is needed to find an improvement. The significant portions of the VAL routines are as follows:

MAIN PROGRAM

```
        .
        .
        .
        .
APPRO   JIG, 150.00
SPEED   50.00
REACTI  1, OBSTACLE ALWAYS
APPRO   TESTP, 00.00
IGNORE  1 ALWAYS
        .
        .
        .
```

OBSTACLE SUBROUTINE

```
MOVE        DEPOSIT
OPENI       0.00
DELAY       1.00
APPRO       TESTP,150.00
SPEED       50.00
MOVES       JIG
CLOSEI      0.00
DELAY       1.00
DEPARTS     150.00
RETURN      1
```

5.   RESULTS AND CONCLUSIONS

The experimental system designed and developed for the feasibility study of the stochastic force sensing technique worked successfully. The first full trial of the system was a nail-biting occasion. The robot had to be allowed to hit an obstacle and nobody was sure of the outcome. However, on reaching the obstacle, the robot reacted promptly and switched to the recovery routine.

The encouraging results of this study shows that stochastic force sensing can be a useful, inexpensive and straightforward technique for integrating force monitoring in a robot. More study is needed to transfer this technique to an efficient and universal force sensing methodology, suitable for all robot control applications.

REFERENCES

1.   Naghdy,F, Billingsley, J., Harrison, D., "Simulation of human judgement in a robot based adjustment system", Robotica, 1984, Volume 2, pp.209-214.

2.   Harrison, D., Billingsley, J., Naghdy, F., "Simple force sensing with the Unimation Puma", Symposium: Robotics-tactile sensing and force feedback, London, 25th October, 1984.

3.   Dothy, K., Govindaraj, S., "Robot obstacle detection and avoidance determined by actuator torques and joint positions", CH1749-1/82/0000-047, 1982, IEEE.

4.   James, L. Nevins et al, "Adaptive control, learning and cost effective sensor systems for robotic or advanced automation systems", Robotic Research, The First International Symposium, 1984, MIT Press.

5.   Paul R., Shimano, B., "Compliance and Control", Joint Automatic Control Conference, San Francisco, 1976.

6.   Colvile, R.F., digest of "Symposium : Robotics-tactile sensing and force sensing" in Measurement & Control" Journal of Institute of Measurement and Control, Vol.18, No.4, May, 1984.

7.   Naghdy, F., Stoodley, K.D.C., Henry, R.M. and Crew,
       A.D., "Development of a microprocessor based
       monitoring system for post-surgical cardiac patients",
       Journal of Microcomputer Applications, 1984 7, 41-49.

8.   Luh, J.Y.S., Fisher, W.D., Paul, R.C., "Joint torque
       control by a direct feedback for industrial robots",
       Proceedings of 20th IEEE Conference on Decision and
       Control, 1981.

9.   Whitney, D.E., Junkel, E.F., "Applying stochastic
       control theory to robot sensory, teaching and long term
       control", 12th International Symposium on Industrial
       Robots, June, 1982, Paris.

10.   Railbert, M.H., Craig, J.J., "Hybrid position/force
       control of manipulators", ASME Journal of Dynamic
       Systems, Measurement and Control, June, 1981.

146

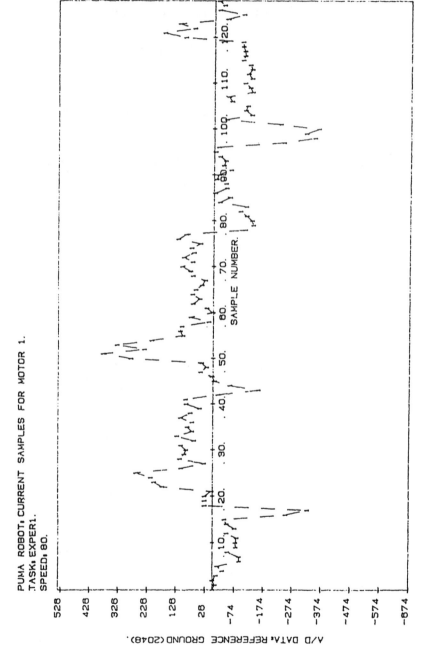

Figure 1

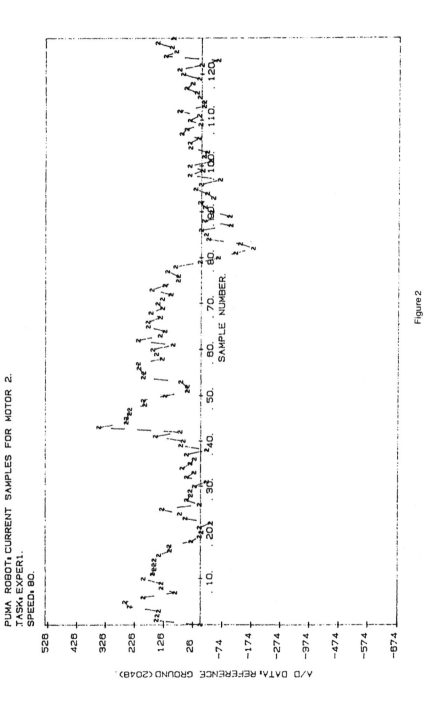

Figure 2

148

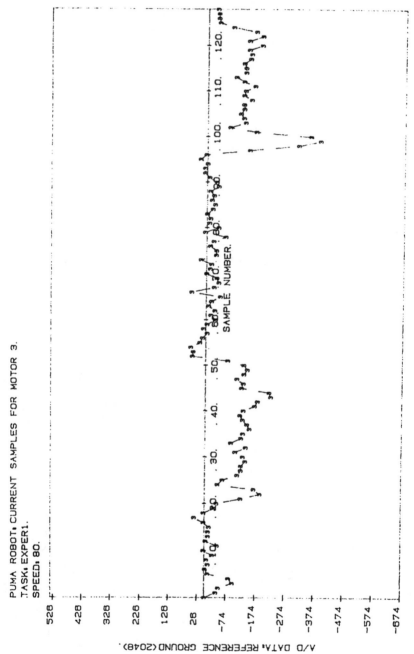

Figure 3

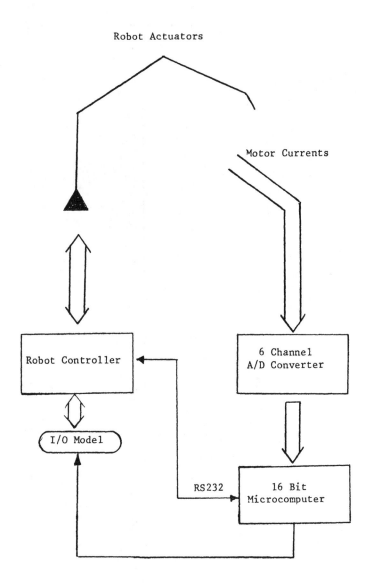

Figure 4

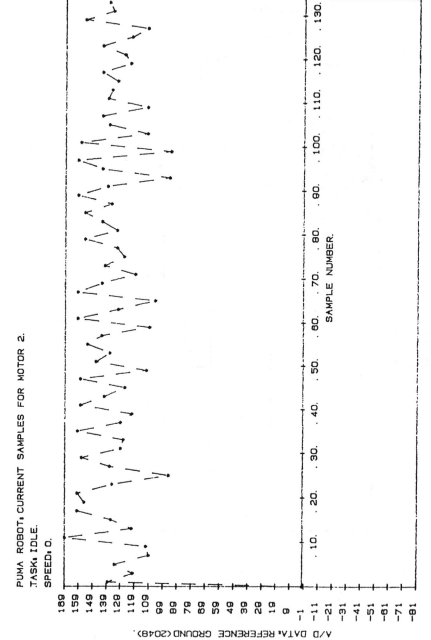

Figure 6

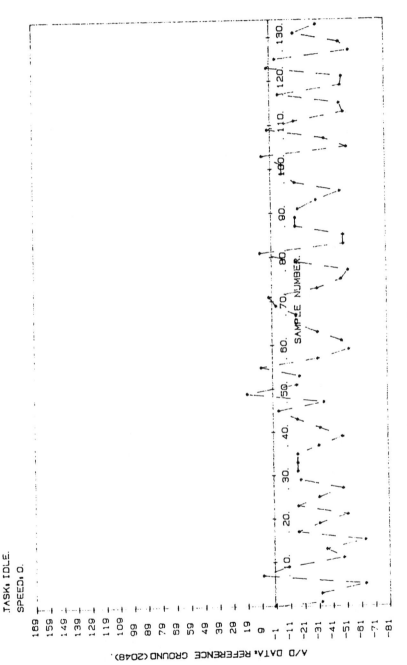

Figure 5

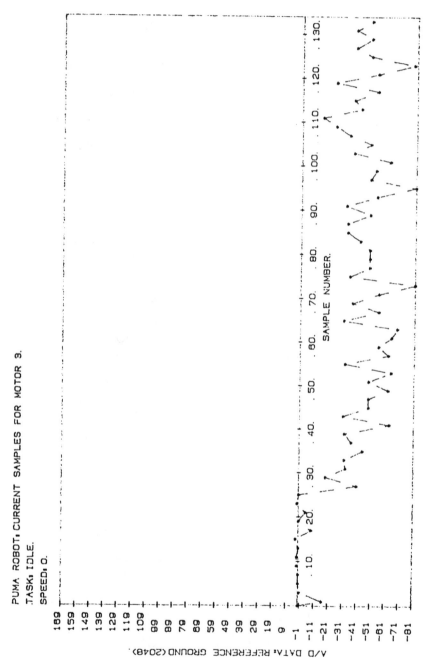

PUMA ROBOT, CURRENT SAMPLES FOR MOTOR 3.
TASK, IDLE.
SPEED, 0.

Figure 7

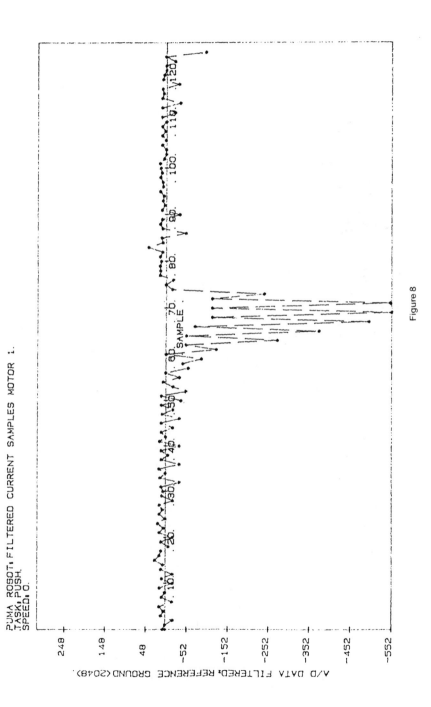

Figure 8

154

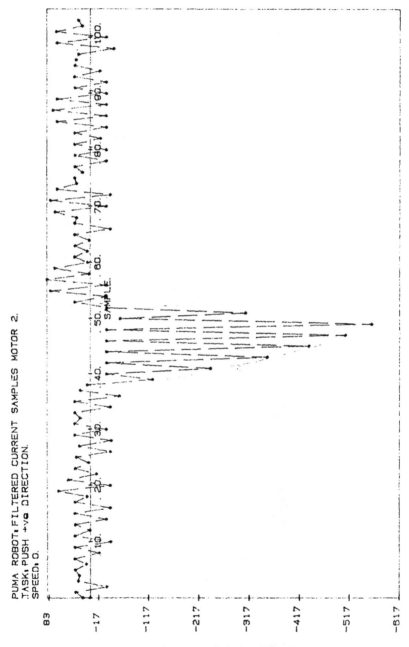

PUMA ROBOT: FILTERED CURRENT SAMPLES MOTOR 2.
TASK: PUSH +ve DIRECTION.
SPEED: 0.

A/D DATA FILTERED: REFERENCE GROUND (2048).

Figure 9

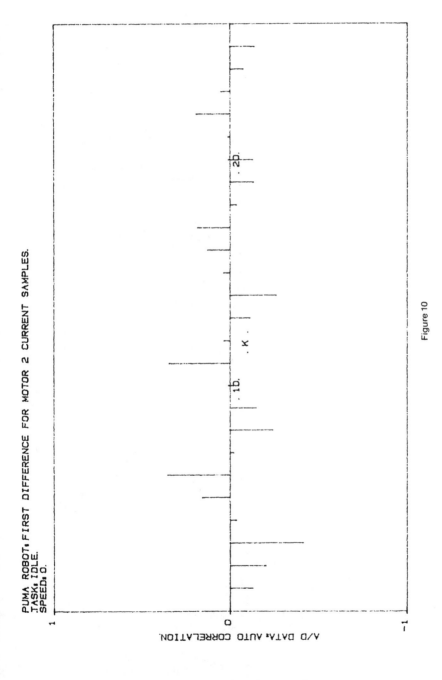

Figure 10

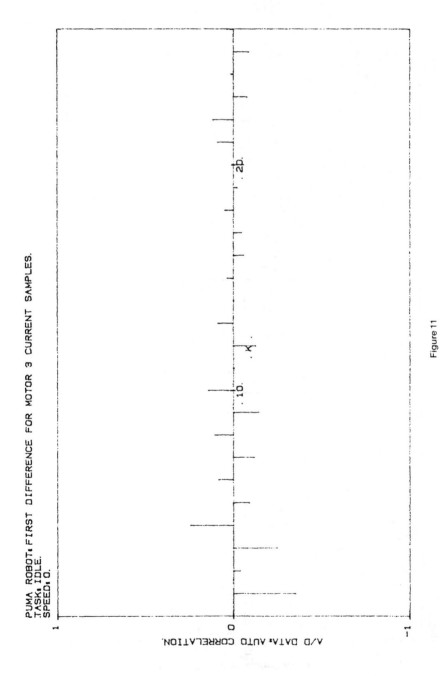

Figure 11

Precise measurement of radial dimensions in automatic visual inspection and quality control— a new approach

E.R. Davies

Automatic visual inspection of manufactured goods is commonplace in many parts of industry: an important aspect of inspection is checking that the sizes of products lie within predetermined limits. This paper analyses how circular products may rapidly and conveniently be measured with high precision. A new approach is developed which is valuable when product sizes do not vary by more than a few percent.

1. INTRODUCTION

Computer vision systems have by now become vitally important in many sectors of industry. A primary area of application is that of automatic inspection for quality control: another is the automatic assembly of complex products. Typical products for assembly include brake housings, alternators, carburettors and car frames - to name but a few from the automotive industry. If assembly tasks are to be carried out successfully, it will be necessary to check individual parts for defects such as burr, mis-shapen holes, etc, or they may not fit together properly and an expensive product may be ruined. Thus, automatic inspection will usually be a necessary part of automatic assembly, and the two processes should not be considered in isolation.

The process of visual inspection is valuable in three main ways: (1) to lead to the rejection of inadequately made parts; (2) to provide feedback to manufacturing plant of specific parameters such as product dimensions; and (3) to provide management with information and statistics on the state of manufacturing output. Item 1 will be involved particularly with defects such as broken, cracked, mis-shapen, mis-orientated or badly positioned parts, or (in the case of food products) inadequate chocolate, jam or batter cover. In addition, the checking of specific product features such as holes, flanges, spindles, etc will be important. With item 2, the emphasis changes from the identification of defects to the measurement of particular product (or sub-product) dimensions. Item 3 will normally be concerned just as much with product dimensions as with the rates at which particular defects occur.

The design of vision systems for industrial automation

is complicated by the enormous amount of processing required
to analyse even the simplest of images, and the cost of
hardware to implement vision algorithms in real time. Many
methods have been devised for simplifying processing in
order to enable it to be carried out rapidly and
economically: however, the result often reduces
significantly the accuracy with which measurements can be
made (1). It has been shown that these problems can be
tackled with a high degree of rigour for an important set of
manufactured products - those which are circular or which
possess circular features such as holes (2). Though this
might at first appear to imply an enormous restriction on
the possibilities for use of vision in industrial
applications, two points should be noted:

1. Circular objects constitute a large proportion of
    common manufactured goods, and include items ranging
    from nuts, bolts, washers, ball-bearings and pistons to
    pharmaceuticals and many foodstuffs - all of which are
    of considerable importance to life and to technology.
2. Obtaining general solutions for circular object
    inspection does not preclude finding similar solutions
    in other instances, and may in fact accelerate progress
    in this area (as indeed has occurred with the location
    of objects of arbitrary shape (3)).

It will be clear from the above discussion that it is
of immense importance to manufacturing industry not only to
have robust methods for inspecting circular objects for
faults, but also to have precise means for measuring
characteristic radii of these objects. This paper will be
concerned particularly with the latter objective, and will
follow on from previous publications (2,4) in which
inspection of circular objects with the aid of radial
intensity histograms has been discussed in some detail.

Section 2 of this paper will review the problems of
inspecting circular products; Section 3 will examine the
radial histogram approach to circular object inspection, and
will summarise the problems of obtaining accurate values of
radial dimensions. Section 4 will then develop a new
approach to the measurement of radii based on the concept of
the matched filter, and will go on to investigate the
accuracy and general usefulness of the new approach.

2.   INSPECTION OF CIRCULAR PRODUCTS

Industrial inspection tasks can normally be separated
into two sub-tasks - those of object location and object
scrutiny. Each of these sub-tasks is made easier in
industrial applications by the fact that the range of
objects to be considered at any one time is rather
restricted. This has the effect of making software run
inherently more rapidly, thereby permitting the use of less
expensive processing hardware.

Object location is frequently carried out by simple
image segmentation operations (5,6) involving the
thresholding of the initial digitised grey-scale image.
Such methods can be valuable for locating silhouetted parts,

but are prone to failure in other situations - particularly when facets of three-dimensional objects are to be examined. In these cases segmentation is usually carried out with the aid of edge detection techniques (5-7). After an edge-enhanced image has been obtained, complete object boundaries are extracted, and the existence of an object and its orientation are deduced.

In our work we have found that the Hough transform technique offers a particularly efficient means of locating circular objects (3,4,8). The edge orientation for each edge pixel (i.e. one whose intensity gradient exceeds a certain threshold value) is noted, and a possible position for the centre of a circle on which it could lie is recorded in a separate image space. When all edge pixels have been examined, each cluster of candidate circle centre points is averaged to obtain accurate estimates for circle centres. It is found that this method of locating the centres of circular objects is particularly robust and requires a minimum of computation. It is therefore ideally adapted to the current purpose. Further details of this method and its application to the inspection of foodproducts are described in (4).

It should be remarked that the accuracy of location of any circle centre by the Hough technique is limited in part by the precision of the product. On the other hand, for precision machined products, the method is clearly limited by the available resolution. However, averaging is an inherent part of the technique - a factor that compensates to some extent for lack of spatial resolution: the result is that circle centres can be located to within one pixel, provided that the image was originally digitised to at least 6-bit grey-scale resolution.

One of the most basic of all methods of scrutinising objects is that of template matching. However, this method is highly computation intensive when there are many degrees of freedom, and it is seldom practical to use it (9). For objects possessing circular symmetry there is one less degree of freedom and template matching becomes significantly more practical once the object has been located. In order to reduce computation further, we have studied the use of radial intensity histograms and applied them to the scrutiny of foodproducts (2,4). This approach to the inspection of circular objects seems very powerful, provided certain problems are dealt with. These are discussed in the following section.

3. RADIAL HISTOGRAMS

The underlying justification for the radial intensity histogram approach to product scrutiny is that it is a convenient and rapid means for reducing significantly the amount of redundant information in an image, and retaining only the information that is relevant to analysis of the highly symmetric object being examined. The result is that computational effort and storage requirements are limited to what are necessary for the task in hand.

This analysis of the value of the radial histogram approach implies that the method will have little value if the object being scrutinised is not axially symmetric. Such a conclusion would not really be valid, and the method can still be useful for inspecting objects such as buttons with four holes symmetrically placed around the centre (2).

In order to compute the radial histogram of a circular object, the radial step size has to be taken as an adjustable parameter. If the step size is small, then few pixels will contribute to the corresponding column of the histogram. Increasing the radial step size in order to obtain a histogram with a statistically significant set of ordinates will result in loss of radial resolution, and a compromise will have to be arrived at.

Another problem is that if the radial step size is constant, the histogram ordinates will (for an object of uniform luminance) vary more or less linearly with radial displacement r, because the number of pixels in a band of radius r and given step size is essentially proportional to r. Davies (2) has shown that this inconvenient situation may be tackled by plotting the histogram as a function of some parameter u(r) which will in principle make the distribution uniform: in fact the required parameter is $u(r)=r^2$. Unfortunately, the distribution will still not be uniform since it is affected by the statistics of pixel placement within the azimuthal bands. Thus it is preferable in addition to fully normalise the histogram by multiplying individual ordinates by suitable numerical factors.

Davies has given a number of practical applications of the use of normalised radial histograms for the detection of various types of defect in nominally circular objects (2). The types of defect that were detected include: badly coated foodstuffs; products that are broken, in contact or overlapping; and missing holes in products such as washers. We now extend this work to the measurement of radii.

3.1 Measurement of Radius - the 'Obvious' Approach

It will be clear from the above discussion that there is sufficient information in the radial histogram of a circular object to permit its radius to be estimated with considerable accuracy. Indeed, it will often be possible for the radius to be estimated within ~0.2 pixels: the reason that measurements might be made to sub-pixel precision is that estimates from a large number of pixels can be averaged to produce an accuracy of this order.

It is now necessary to consider quite how the radial histogram should be analysed in order to attain these high levels of precision. The most obvious approach is perhaps to differentiate the radial histogram, thereby generating a function that peaks at a value corresponding to the boundary of the object being measured, and to estimate this value as closely as possible. Such a series of operations is by no means trivial. First, differentiating a function is prone to accentuate noise. Second, it is not clear how the function should optimally be differentiated since it is a discrete function, and the required differencing operation will introduce an additional parameter into the situation. Third, finding the peak of the differentiated function is difficult, as its height varies only in second order near the crucial value of r. Fourth, minor sub-peaks and troughs due to noise (already accentuated by differentiation) may have to be negotiated. Thus, it is clear that for the highest accuracy the region near the peak will have to be modelled, and the position of the peak then computed by some reasonably sophisticated method.

All this means that this 'obvious' method of measuring radius is bound to involve considerable computation, and has to be based on finding suitable models of relevant functions. These considerations prompted the search for other methods of estimating radius from the intensity histograms. An ideal method would have the following characteristics:
1. it would be highly precise;
2. it would have optimal sensitivity;
3. it would be simple to apply;
4. it would involve minimal computational effort;
5. it would be able to cope with a variety of histogram profiles;
6. it would suppress noise automatically.

Such a set of requirements is clearly highly exacting, and might of course be impossible. Fortunately, the theory of matched filters in the field of signal detection leads to the possibility of completely solving item 2 above, and as a result helping with the solution of item 1. A new approach to the problem, described in Section 4, shows that the matched filter viewpoint permits a sound attempt to be made to attain all the ideal characteristics listed above.

4. A NEW APPROACH TO THE SENSITIVE MEASUREMENT OF RADIUS

4.1 The Basis of the Method

The work to be described here is based on the concept of a matched filter, and makes use of the fact that a given type of signal is detected with optimal sensitivity by a filter that is 'matched' to it (10,11). Signal detection theory shows that, under conditions of white noise, the design of a matched filter is simple: it merely has to have the same profile as the signal it is to detect. To detect a signal, its matched filter is convolved with the input waveform, and a significant peak indicates the presence of a signal.

The adaptation of this idea to the current application - that of accurate measurement of radius - is extremely simple. If we imagine that two objects are to be distinguished that are very close together in size, then applying two matched filters to detect these objects and comparing the results of matching will immediately show which of the two objects is present. Comparing the results in fact means subtracting the two correlation coefficients that result. This amounts to applying a single difference filter equivalent to the difference in shape between the two almost identical matched filters. Clearly, as the two objects are made to match each other ever more closely in size, the difference filter becomes ever more closely the derivative of the matched filter, and this is identical to the derivative of the original shape.

This idea can perhaps be proved more directly as follows. The signal that is to be detected when trying to distinguish between two objects of identical shape and similar size is the difference signal, which approaches the derivative of the original shape as the difference between the radii approaches zero. To detect this derivative signal with greatest sensitivity, the optimal method must be to use a filter that is matched to the shape: so the required matched filter is identical in form to the derivative of the original shape.

The result of this analysis is that the original radial histogram need only be correlated with a function identical in form to the derivative of the assumed shape in order to obtain a signal proportional to the difference in radius from that of the assumed shape. While the signal is basically proportional to the difference in radius, for sufficiently large changes in radius this will no longer be valid since higher order derivatives of the assumed radial profile (ignored here) become increasingly relevant. The result of this will usually be a levelling off of the difference signal for large changes in radius.

Intuitively, it is clear from Fig. 1 why the result will only be linear over a restricted range of radii. It is also clear why the slope will be data-dependent. However, it is interesting that the basic method is independent of data, so that it remains valid whatever the shape of the radial histogram. Thus the approach described here may well, under certain circumstances, be the only viable one.

One complication arises when the radial histogram profile is not zero outside a certain limited range of radii: the complication arises since detection of a signal of this shape is not covered by normal matched filter theory. In such a case, the derivative of the assumed radial profile does not give a correlation coefficient that is zero at the assumed value of radius: the reasons for this are apparent from Fig. 2. This can be thought of as a calibration problem and eliminated by subtracting a suitable value from the correlation coefficient. However, in order to reduce the effects of small changes in the level of illumination, we have found it best to adjust the height of the correlation template (the derivative of the assumed radial profile) so that it has a zero mean. This procedure completely eliminates errors in estimating radius that might have arisen from a constant height being added to the radial shape profile. Since changes in the level of illumination may also affect the constant of proportionality for measurement of radius, it will be preferable to check this constant periodically by a separate calibration procedure: calibration will in any case be necessary during the initial setup phase.

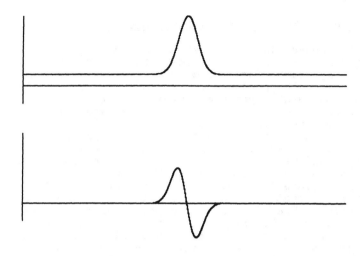

a - radial histogram
b - derivative template

Fig. 1.   Derivative template for a simple form of
          radial histogram

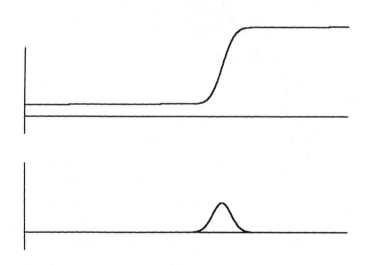

a - radial histogram
b - derivative template

Fig. 2.   Derivative template for a radial histogram
          approximating a step function

4.2 Further Theory and Practical Details

The above discussion has shown that there should be no insuperable theoretical problems to using the derivative filter approach for estimating radius from the radial histogram.   The present sub-section aims to show that there are no insuperable practical problems either.

Correlation of a pre-formed template with the radial histogram immediately gives a measure of the radius of an object: the process presents very little difficulty apart from that of performing sufficiently rapidly the necessary multiplications and additions. Hence the major practical problem to be considered is that of forming the required template. This is achieved in two stages: (1) forming an 'assumed' radial histogram, and (2) computing its derivative. The first of these processes may be carried out by averaging the radial histograms of a large number of sample objects of the same nominal size.   The second process, that of differentiation, is more problematic for the reasons outlined in Section 3.1: however, noise is less likely to be a problem in this case because the assumed radial histogram is the average of many signals and should therefore be essentially noise-free. In fact, we have found very little problem with either of these processes, and have used the simplest possible method of differentiating - namely taking the differences between adjacent discrete histogram values and then smoothing the derivative function slightly to eliminate remanent noise.   Smoothing was implemented by a one-dimensional convolution with a near gaussian (1:3:3:1 binomial) profile. The results obtained using this approach are described in the next sub-section.

4.3 Experimental Results

With a new measurement technique it is important to obtain an overall assessment of its capabilities, and to show how well it matches relevant theoretical predictions. Accordingly, it was decided to test the technique on three objects having substantially different radial histogram profiles, and to examine carefully the response curves obtained. In particular, it would be useful to find the range over which radial measurements could reasonably be made, and also the range over which measurements could be expected to be linear.

Fig. 3 shows the images of three objects used for this purpose: these include a flat black metal disc, a biscuit with rounded and slightly browned edges, and a knob (for electronic equipment) whose body happens to have a similar luminance to the background.   The radial intensity histograms of these objects are shown in Fig. 4: it will be seen that they are quite distinctive and different in shape near the positions corresponding to the outermost radii.

For this test, each histogram was differentiated and smoothed slightly, and a zero-mean template derived as indicated in Section 4.1 (see Fig. 5). Then the correlation coefficient R between the template and the original radial

Fig. 3.   Images of three objects with different radial profiles

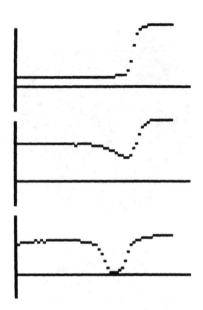

Fig. 4.   Radial histograms of the three objects shown
in Fig. 3

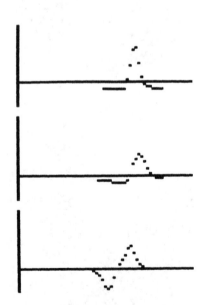

Fig. 5.   Zero-mean templates for the three objects
shown in Fig. 3

histogram was computed as a function of lateral displacement of the histogram. Since the zero-mean template has quite restricted range, the resulting response curve is very similar to what would be obtained if the objects in question appeared in a range of sizes (Fig. 6). This makes the test particularly useful, since it generates a realistic approximation to the response curve with a minimum of input data. Thus the consequences of the assumptions made in formulating the new approach become very clear.

It is remarkable that the response curves for the three dissimilar objects are actually so similar in form and in scale (Fig. 6). The symmetry about the central position, and the linearity near this position essentially corroborate the theory of Section 4.1. Following the need to determine the range of linearity, the central regions of the response curves were modelled in terms of a cubic approximation. Since the quadratic component is negligible, we take

$$R = ax + bx^3 \qquad (1)$$

where $u = kr^2$ and x is the deviation between u and the central value u_o. We can now proceed to estimate the deviation from linearity. This is

$$R - R_o = (ax + bx^3) - ax = bx^3 \qquad (2)$$

and the fractional non-linearity may be interpreted as

$$Q = \pm bx^3/2ax = \pm(b/2a)x^2 \qquad (3)$$

Taking $\pm 10\%$ non-linearity as a practical limit, this determines the effective linear range of x as bounded by the values $\pm \sqrt{|a/5b|}$.

Since $x = u - u_o$, and the actual relation between u and r is

$$u = r^2/40 \qquad (4)$$

we can deduce the range of linearity in terms of radial distance. This and the overall measurement range are given in Table 1 for each of the three objects shown in Fig. 3.

It is seen that the range of linearity and the total range of measurement are not large, being of the order of $\pm 3\%$ and $\pm 5\%$ of the overall radius respectively. The total range is ultimately the more relevant figure, since up to this value there is a one-to-one relation between response and radius which can be interpreted accurately from a calibration curve. In addition, it will be noted that the 10% non-linearity in x referred to earlier corresponds to only ~0.3% when computed as a non-linearity in the overall radial measurement of the object.

Though relatively small, the values for range of measurement should be adequate for measuring the small variations in sizes of products as they come off a production line, typical variations in foodproducts (such as biscuits) usually being in the region of $\pm 5\%$, and those for

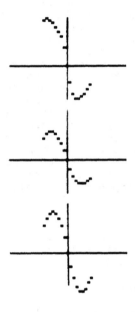

Fig. 6. Response curves for radial measurement of the
three objects

<u>Table 1. Range of linearity and total range of measurement</u>

	(a)	(b)	(c)
range of linearity	±1.0	±1.4	±1.3
total range	±1.5	±2.0	±2.0
central radius	40.0	40.0	40.0

This table lists the range of linearity and the total
range of measurement for the objects shown in Fig. 3.
Results are in pixels rather than absolute distance units.

precision metal parts (spindles, washers, coins, etc) being less than ±1%. Thus the new approach described here seems a reasonable one for the purpose of product radius measurement.

The other important factor to be investigated is the ultimate accuracy of the approach. To determine this, the technique was set up to measure the radius of accurately machined aluminium discs of known diameter. The algorithm was trained with the aid of a disc 54 mm in diameter presented in various positions and orientations, and a zero-mean template similar to that in Fig. 5(a) obtained. The system was then calibrated using two discs of size 53 and 55 mm respectively. Finally all three discs were measured by the algorithm in a number of positions and orientations. The results gave standard deviations of 0.13, 0.04, 0.09 mm respectively for the 53, 54 and 55 mm discs, indicating a limiting accuracy better than 1 part in 400. Since the radius of the 54 mm disc corresponds to 40 pixels, this represents a measurement accuracy of 0.1 pixels. (It is relevant that this accuracy was only obtained after allowance was made for non-linearity of the vidicon camera used to observe the discs, by restricting observation to the central part of the camera tube where the nominal linearity is 1%.)

Unfortunately, it is not easy to make absolute checks on the accuracy with which non-ideal products such as biscuits are measured, since they are prone to be elliptical in shape, rough at the edges, and somewhat non-uniform in luminance. However, present indications are that this approach gives reliable results in such cases, and the method is currently being optimised for a foodproduct application.

5.   CONCLUDING REMARKS

The criteria that should be fulfilled by an ideal inspection technique for measuring product radius include ease of application, good sensitivity and resistance to noise, high accuracy, and adaptability to different types of product. These criteria are not really met by the 'obvious' approach of differentiating the radial intensity distribution and locating its peak, since this procedure is rather susceptible to noise. However, they are achieved by a new technique based on matched filtering.

The fact that the new technique incorporates matched filtering ensures optimal sensitivity and adaptation to the shape of the relevant radial profile. Ultimately this is because it has proved possible to find an automatic means of forming a suitably weighted average of all parts of the radial histogram that can contribute to the radial measurement being performed. This means that it can be used with unusual radial profiles while retaining high accuracy and incurring minimal computational effort.

When products are sufficiently accurately made for such a measurement to be meaningful, radius can be determined to within 1 part in 400 by the new approach. Against this advantage is the fact that the technique gives a limited range of radial measurement. In typical cases this is in the region of 3-5%. Non-linearity can be overcome by using calibration curves - the really important factor being reproducibility. Tests with accurately machined metal discs confirm that reproducibility can be maintained to 1 part in 400, as indicated above.

Application of the method involves computation of a special zero-mean template; then calibrating the measurement system; and finally performing radial measurements. The purpose of adjusting the applied template to zero mean is to eliminate the effects of changes in background illumination: further compensation may be obtained by periodic checks of system calibration.

Overall, the approach seems highly competitive where a large range of measurement is not required. This occurs in particular where a great many nominally identical round products are to be inspected and their diameters are to be checked rapidly and accurately.

Acknowledgement

The author is grateful to the SERC and to United Biscuits and Unilever for financial support during the course of this research.

REFERENCES

1. Davies, E.R., 1983, "Image processing - its milieu, its nature and constraints on the design of special architectures for its implementation", pp. 57-76 in Duff, M.J.B. (editor) "Computing Structures for Image Processing" (Academic Press: London)

2. Davies, E.R., 1985, "Radial histograms as an aid in the inspection of circular objects", IEE Proceedings D (in press)

3. Ballard, D.H., 1981, "Generalising the Hough transform to detect arbitrary shapes", Pattern Recognition, 13, no.2, pp. 111-122

4. Davies, E.R., 1984, "Design of cost-effective systems for the inspection of certain foodproducts during manufacture", Proc 4th Conference on Robot Vision and Sensory Controls, London (9-11 Oct 1984), pp. 437-446

5. Pavlidis, T., 1982, "Algorithms for Graphics and Image Processing" (Springer-Verlag: Berlin-Heidelberg)

6. Davies, E.R., 1982, "Image Processing", pp. 223-244 in Sumner, F.H. (editor) "State of the Art Report: Supercomputer Systems Technology" (Pergamon Infotech: Maidenhead)

7. Davies, E.R., 1984, "Circularity - a new principle

underlying the design of accurate edge orientation operators", <u>Image and Vision Computing</u>, <u>2</u>, no. 3, pp. 134-142

8. Hough, P.V.C., 1962, "Method and means for recognising complex patterns", US Patent 3069654

9. Davies, E.R., 1984 (December), "A glance at image analysis - how the robot sees", <u>Chartered Mechanical Engineer</u>, pp. 32-35

10. Rosie, A.M., 1966, "Information and Communication Theory". Blackie, London

11. Turin, G.L., 1960, "An introduction to matched filters", <u>IRE trans Information Theory</u>, <u>IT-6</u>, pp. 311-329

Visual feedback control for orientating parts in an assembly robot cycle

R.E. Jones and P.M. Hage

ABSTRACT

In order for an industrial robot to handle components correctly during an assembly process, the components are usually presented to the robot in a known orientation using some form of jigging. If the component cannot be orientated using jigging or simple sensory devices then the assembly process is usually performed by a human operator.

In order to overcome the problem of component orientation a vision system can be interfaced to an industrial robot which can identify and calculate the position and orientation of the component. This paper discusses the application and flexibility of such a robot vision system.

1.   INTRODUCTION

In the past a major drawback in using industrial robots in a manufacturing process was due to the problem that the robot could not detect any positional changes in its working environment. In order to overcome this various sensory devices have been developed which enable the robot to detect the presence of components and hence calculate the position of the component relative to the robots coordinate system. However during many assembly processes more than positional data of the component is required, indeed, the system may have to check the overall shape of the component and also check its position and orientation at a certain time within the work cycle.

Many components which are assembled to produce finished product have identification characters stamped or moulded onto them by the manufacturing process. Generally these identification characters will always be in a known position on the component and it is possible using a vision system to be able to recognise the component and the components orientation from these characters.

This paper illustrates how a 30mm diameter disc which may be part of an assembly can be orientated correctly. The disc has alphanumeric characters printed onto it and a vision system is used to identify and analyse the disc. The data extracted by the vision system is used to pass control information to an industrial robot for the purpose of positioning the disc with the required orientation. A diagram of the layout of the test system is given in FIG.1.

2.   MAGISCAN VISION SYSTEM

The basic layout of the vision system used is shown in FIG.2.
The system has an image storage of 1024 x 1024 pixels by 8 bits per
pixel, but for the purpose of industrial applications generally it
is normal to use grey level images of 512 x 512 by 6 bits per pixel
resolutions.

The image processing and image analysis programmes are written
in the PASCAL language which has the advantage over other high level
languages such as FORTRAN and BASIC in that it has the facility for
representing complex data structures in a simple form.

The principal stages of image analysis are listed in FIG.3
and will now be briefly described:-

An image of the component to be analysed is captured using a
suitable TV camera and the image is stored as a grey image of the
required resolution. Having stored the grey image Segmentation or
Thresholding may be performed on the grey image to produce a binary
image of the component. From the binary image the object detection
is then possible and also measuring and analysis classification of
the object can be performed. Alternatively the initial grey image
may be processed to produce a histogram of the image which also may
lead to a description of the object and yield all the required data.

Having identified and calculated the necessary information
about the object the vision system can then pass control data to an
industrial robot to perform the specific task required within the
assembly or manufacturing cycle.

3.   HOST PROCESSOR/ROBOT COMMUNICATIONS

A standard 6 degree of freedom Unimation Puma robot system
was used for the handling of the test discs. The robot system uses
a terminal for input of commands to the robot controller and a
floppy disc unit for storage of robot programmes. In order for the
vision system to control the robot, the host processor of the
vision system has been linked to the robot controller via the
terminal port. This link enables serial data to be transmitted
between the host processor and the robot controller using the
standard RS232 interface operating at a data transmission speed of
9600 BAUD. The host processor is hence used as the system
supervisor controlling the vision analysis and processing and also
the robot movements required after the vision analysis. A diagram
showing the communications link is given in FIG.4.

The robot may be positioned in space by two ways, these are
as follows:-

1.   Joint Coordinates

The robot is instructed of the required individual axes
rotations necessary to position the robot at a point in space with
the required orientation of the robots wrist.

2.   World Coordinates

The robot is instructed of the X,Y and Z coordinate positions
relative to the robot's datum and also the wrist orientation which
is given as three angles (EULERS ANGLES).

For a detailed discussion of coordinate systems and their derivation refer to REF.1. The disc orientation problem discussed later uses the method of joint coordinates for the positioning of the robot.

The major functions of the host processor when linked to the robot can be summarised as follows:-

(A) Instruct the robot control to execute the necessary robot programmes.

(B) Determine the position of the robot in space.

(C) Instruct a robot position to be changed to a new position.

(D) Control the vision processing and analysis.

(E) Instruct the required robot programmes to be loaded from the disc system.

(F) Monitor system errors and interrupts to the system:

e.g. Emergency stop activated at robot control.

4. APPLICATION TO DISC ORIENTATION

4.1 OVERALL SYSTEM

The system receives discs with varying amounts of alphanumeric data printed onto them which are picked up by the robot and presented to the inspection camera. Typical details of the discs are given in FIG.5. The robot picks up the disc using a standard suction cup operated by a pneumatic air supply. At the inspection station vision analysis of the disc is carried out and the robot is then instructed to place the discs with the correct orientation to the receiving station. The following operations are carried out during the work cycle.

(A) Host processor instructs the robot to pick up a disc from the input feeder and place it in front of the inspection camera.

(B) The host processor records the inspection position.

(C) Image analysis is performed at the inspection station.

(D) From the image analysis data the amount of rotation the disc needs to achieve correct orientation is calculated.

(E) The host processor instructs the robot to move the disc to the receiving station and place it with the correct orientation.

(F) The cycle is repeated until the number of discs required for the receiving station has been achieved.

(G) Vision analysis is carried out using a camera mounted above the receiving station in order to check that the discs have been processed correctly.

The layout of the system is given in FIG.1 and the typically required orientation of the discs is given in FIG.6.

4.2 VISION ANALYSIS OF DISC

The vision system takes the image of the disc via the system inspection camera. Using suitable thresholding levels the vision system detects four edges of the disc which enables it to calculate the centre of the disc FIG.7. From the centre of the disc the lettering around the surface of the disc follows a fixed radius and reading grey levels through the centre of the lettering FIG.8 yields a distinct pattern or grey level distribution FIG.9.

From the distribution of the grey levels it can be easily seen where the lettering around the disc begins and ends and using this information it is possible to calculate by how much the disc needs to be rotated to achieve the required orientation.

4.3 TEST SYSTEM RESULTS

The vision system can calculate the disc orientation to an accuracy of one degree and carries out this analysis in a time of less than one second. With the robot operating at high speed the total cycle time for picking up and placing the correctly orientated disc is approximately three seconds.

A problem that occurred with the test system was that the robot may pick up the disc slightly off centre which reduces the accuracy to which the robot will orientate and place the disc at the receiving station. It was found that due to this problem that the robot will place the discs at the receiving station to within 0.5 milli-metres which for industrial applications such as packaging of components is adequate.

The system has been tested over typical running periods i.e. eight hours per day over a five day week and was found to continually perform within the above mentioned tolerances.

5. CONCLUSIONS

By interfacing a vision system to an industrial robot the problem of orientation of components in an assembly process may be solved, thus eliminating the need for any component jigging or releasing a human operator from the ardous task of orientating individual components.

The application of the vision analysis system previously discussed operates very quickly and reliably and the system can be easily integrated into a Computer Aided Manufacturing (CAM) system. The use of the vision system can also increase the flexibility of the manufacturing system for the expansion to assemble or process other components of similar dimensions.

6.   REFERENCES

REF.1.   RICHARD P PAUL
           ROBOT MANIPULATORS: MATHEMATICS, PROGRAMMING
                                AND CONTROL.
           (THE MIT PRESS 1981)

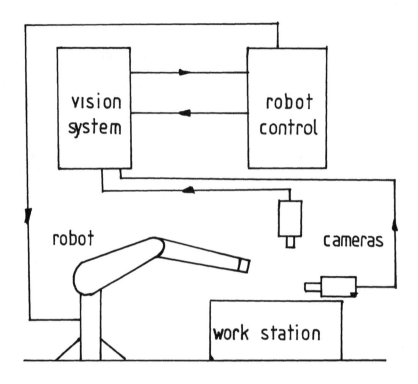

Plan view of work station.

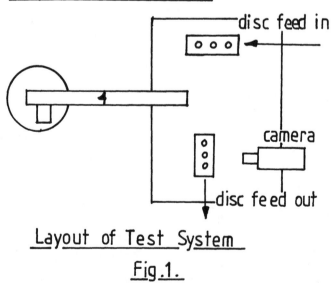

Layout of Test System

Fig.1.

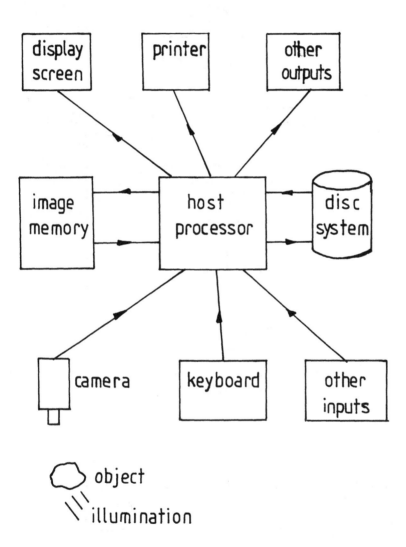

Layout of the Magiscan Vision System.

Fig. 2.

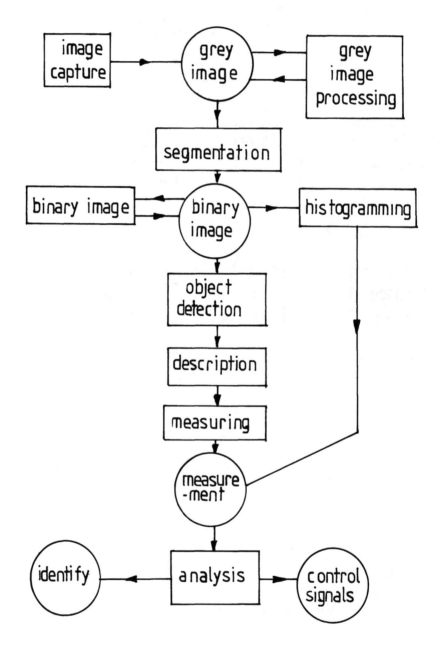

Principal Stages of Image Analysis.

Fig.3.

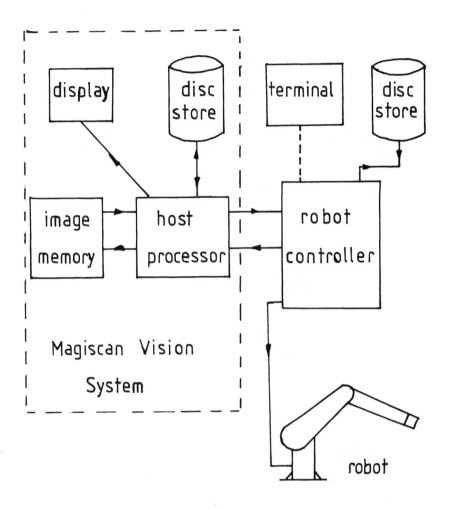

Host Processor - Robot Communication.

Fig.4.

Detail of Discs.

Fig.5.

Typical Required Disc Orientation.

Fig.6.

Scan to Find Edges

Fig.7.

Scan Thro Lettering

Fig.8.

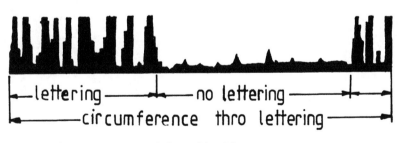

Grey Level Distribution

Fig.9.

Automation and robotisation in welding—
the UK scene

J. Weston

1.1 INTRODUCTION

The development of MIG/MAG processes made possible machine welding. Robots have allowed these machines to become more flexible, firstly because of their freedom to move the welding tool, and secondly because of the robot control system which has permitted interfacing with joint following and peripheral equipment.

Although robots were first used with the resistance spot welding process, particularly in the automotive industry, this paper will concentrate on arc welding developments. Comment will also be made on the application of robots to other welding processes and of related research and development activity.

While the rate of application of welding robots within the United Kingdom has been slower than in much of the developed world, their range of application has been wide. There have also been a number of significant and leading innovations.

1.1.1 Mechanised, Automatic or Robotic Welding?

Some definitions are required to give some meaning to these terms. There have been some attempts made within both the UK (BS499: 1983) (1) and USA (AWS 1985) (2) to provide standard definitions. The march of technology has placed some limitations on these and led to some new proposals by Salter (3). Comparison of these references suggests the following definitions:

Machine welding: The generic term used to describe 'welding in which the welding appliance is carried by a machine'.

Mechanised welding: 'Machine welding' using a process in which adjustment of both joint tracking and welding parameters is under a continuous operator control. The important feature is the direct influential involvement of the operator.

Automatic welding: 'Machine welding' in which joint tracking and welding parameters are predetermined and there is no opportunity for manual adjustment during welding. Adjustments may be made between welding operations to accommodate components of similar geometry but different dimensions.

Robotic welding: 'Machine welding', where the machine carrying out the process is a robot.

These last two definitions essentially differ by the mechanical dexterity of the 'welding machine'. A robot is defined as: '....a reprogrammable device designed to both manipulate and transport parts,

tools or specialised manufacturing implements through variable programmed motion for the performance of specific manufacturing tasks': British
Robot Association, (4).

Supportive terms which are added to the description of the welding
machine are those which define the additional or additive functions such
as

a   Adaptive control: a generic term defining the addition of a system
    to a welding machine which monitors changes in a particular welding
    variable (e.g. arc voltage, current, joint location etc.) and
    arranges the appropriate corrective action.

b   Joint tracking: a form of adaptive control which monitors changes in
    the location of the joint to be welded and instructs the welding
    machine to take the appropriate corrective action.

c   Joint recognition: a form of adaptive control which recognises the
    joint to be welded, detects changes in the joint geometry and
    instructs the welding machine to take the appropriate corrective
    action.

d   Weld recognition: a form of adaptive control which recognises varia-
    tions in the geometry (including penetration depth) of the weld or
    weldpool being made and instructs the welding machine to take the
    appropriate corrective action.

e   Joint searching: a form of adaptive control which seeks the joint to
    be welded (particularly the start of the weld) and instructs the
    welding machine to take the appropriate corrective action.

An important items of these adaptive control systems is the 'sensor'
which gathers or receives the data on the environment it is designed to
study. For example the television camera may be the sensor for a vision
based joint tracking system.

These definitions are linked and related as shown in Fig.1.

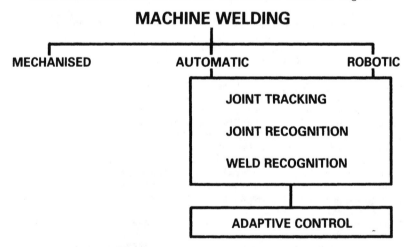

Fig. 1 Associated definitions

1.1.2 The Robots

Industrial robots still bring to mind the picture of an arm on a post, which is the form of the first machines, e.g. Fig.2. As such they were the first tool handling machine which could to some extent emulate the behaviour of the human arm. The welding robots seen within this country have almost all come from outside the UK. Developments ranged from the hydraulic Unimate to the electric-drive ASEA, a trend which has been so successful that virtually all arc welding robots are now electrically driven.

Fig. 2 Unimate Robot (Photograph courtesy NEL)

From the arm on the post machine we saw the introduction of a gantry system, a concept which is now developing in two directions. The first incorporates the arm into the gantry so changing the 'shape' of the robot; this trend has been most used for resistance spot welding machines. The second is to mount the arm-on-a-post robot on to a floor, wall, or gantry supported slide (or slides) mechanism(s).

The early robot arc welding installations were supplied with simple fixed point rotational/tilting workpiece manipulators. Increasingly these workpiece manipulators are controlled and synchronised with the robot arm motions.

The first robots were unable to adapt to the changing environment. Today most arc welding robots have the potential to joint search and joint track: their ability to do so is only limited by the availability of suitable sensor and adaptive systems. The most common method for tracking at present uses the 'Thru-the-arc' sensing technique. There have been significant vision developments in the United Kingdom, (5,6). Vision systems have the ability to provide joint recognition information and so lead to the possibility of adaptive control of the welding process. However as yet their use in a production environment is relatively limited

These robot developments and trends are typical of those seen in the rest of the world.

2.1 APPLICATIONS

It was not until the arrival in the UK in 1975 of the all-electric drive ASEA robot, fitted with welding equipment by ESAB, that the expansion and practical utilisation of robots for arc welding really commenced. Quickly the range of applications studied within the UK broadened. In 1978 robots were being used to weld mild steel in the manufacture of automotive components, items of mining equipment, supermarket trolleys, domestic stoves and agricultural equipment. At that time arc welding robots comprised some 10.4% of the total UK robot population. (125 robots).

About this period, and later in conjunction with BOC and then GEC, Hall Automation carried out considerable robotic development to enable their spraying arm to be used for welding applications. While having some useful control features such as 'permanent' programme retention on power failure and programme building facilities it did suffer in relation to the competitors by being less flexible in weld path editing capability and by having a hydraulic drive. In spite of these limitations they were to be found arc welding both in UK and abroad. The problems associated with hydraulic drives, essentiallly low repeatability, also hampered Unimate, Cincinnati Milacron and Trallfa at this time.

The spacer welding study commenced in 1978 by NEI Nuclear Systems led to the construction of a large system (Schoeters 7) comprising three separate gantries each supporting two robots and moving on a 86m track, Fig.3. This installation required considerable development both by the equipment suppliers and by NEI themselves. The NEI studies of the equipment, the jigs and the process (Henderson 8) led to the successful welding of some eight million production spacer welds in materials ranging from $9\%Cr\frac{1}{2}\%Mo$ - $9\%Cr1\%Mo$ to 316 stainless steels. This installation pre-empted the present trend to mount the robot on some form of track to increase working range.

Fig. 3 Robot gantry system (NEI)

In 1980 British Leyland commenced a development programme (Industrial Robot 9) aimed at arc welding the chassis of Land Rover vehicles utilising 16 robots. Within two years the line was in production. This would be one of the early applications of arc welding directly to the production line and is significant also because it involves chassis construction. Chassis construction in car production has been an almost exclusively resistance spot welding application area, also increasingly undertaken by robots.

Also at this time APV Co.Limited was studying the robotic welding of stainless steel components using the TIG process. They were successful in getting the system to a production working stage.

There have been a number of multi-robot arc welding systems installed. One of the more notable being at Tallent Engineering where in 1983 11 Yaskawa robots were used to produce suspension units for the Ford Sierra (Industrial Robot 10).

Elsewhere in industry robots have been installed in ones or twos, often with repeat orders following. By December 1984 there were 341 arc welding robots installed. The application range had increased and now covered, in addition to the list given earlier, the manufacturing of items such as:

automotive components (sub-assemblies, sub-frames, seats etc.).
fork hoist components
pallet trucks
hospital beds and fitments
taillift hoists for trucks
electrical switchgear enclosures
chairs seats and stools
motor cycle exhausts
cycle frames
pedestrian barriers
trailer manufacture
earthmoving equipment
fuel cans and containers

The majority of these tasks involving the use of the MIG process for welding mild steel. There are one or two instances where selfshielded cored wire is being used for welding mild steel and the MIG process for the welding of aluminium. There have been one or two reports of the use of TIG welding for the manufacture of weapons components.

Resistance spot welding applications largely relate to the automotive industry and to car production lines in particular. One notable exception is the introduction by the Ford Motor Company of a system comprising four Unimates to weld truck cabs (Roswell 11). The major challenge met and overcome related to the physical size of the truck cab. More recently robots have been employed for resistance spot welding in other areas such as the manufacture of supermarket trolleys.

Robots are also being used to manipulate lasers, plasma torches or water jet tools for cutting, welding and weld dressing operations. We can expect these application areas to increase and expand in the future.

3.1 STATISTICS

Since its formation in 1977 the British Robot Association has provided as one of its services a collation of robot statistics. In 1981 they commenced holding an annual conference at which the British data was presented and it is this information that has been used in this section.

Following the introduction of robots at the Olympia Exhibition in 1975 the rate at which robots were installed was at first slow, Fig.4, as researchers, users and potential users went through the learning process. By 1980 with 48 arc and 59 resistance spot welding robots installed confidence was growing and since then the rate of installation has steadily increased.

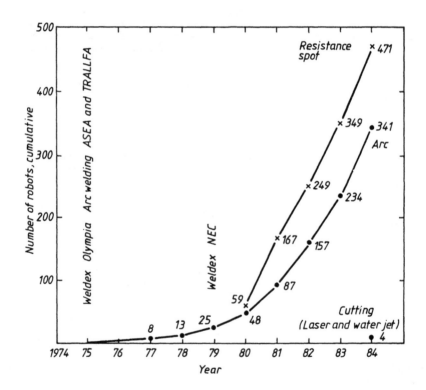

Fig. 4. Growth of Welding Robots in the UK

The number of welding robot suppliers have also increased from three (ESAB/Trallfa/ Unimate) in 1975 to some 18 suppliers of resistance spot welding robots (approximately 3 UK manufacturers) and some 20 arc welding robot suppliers (approximately 5 UK manufacturers). In a number of instances the same organisation supplies both resistance spot and arc welding robots with most of the manufacturing activities associated with licensing agreements (the designing company originating outside the United Kingdom.)

In themselves these statistics might look impressive, with the number of arc and resistance spot welding robots installed during 1984 increasing by 45.7 and 35.0% respectively. However in absolute terms the numbers are low, especially if comparisons are made, say, with West Germany, Fig.5.

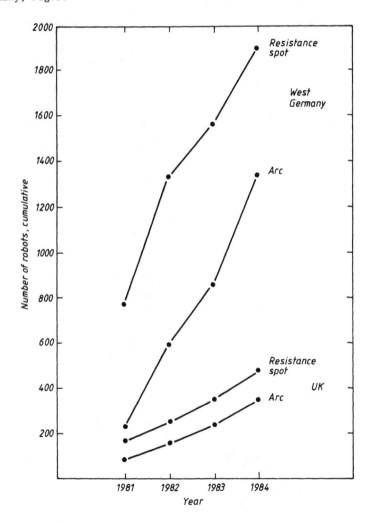

Fig. 5 Rate of installing robots in UK and West Germany

The rate of installing arc welding robots has been steady and is now overtaking resistance spot welding robots. The reason proposed is that the resistance spot welding market is of finite size and that it is approaching saturation. On the other hand the potential for arc welding robots is much greater and growing as robot and sensor systems improve.

4.1 GOVERNMENT SUPPORT

The British Government early recognised the potential of robots for increasing productivity and they instigated several schemes which aimed to foster implementation and research. The Department of Trade and Industry ran a programme giving support in three areas.

Firstly, support towards feasibility studies in which independent consultants would study a company's products and working practice to establish whether robots could make an economic contribution to their production success. This was funded at a 50% rate for up to 15 man days of effort.

A second, and perhaps the most important area, followed the study and supported the installation of robots. Grant support (upto 33% in May 1982) could be claimed to cover capital costs, development costs and the time needed to get the installation working to a reasonable level of production efficiency. These grants aimed at encouraging initial robot application, particularly those applications which were ambitious in scope and innovation.

The third area of support was available to companies wishing to develop or manufacture robots.

These three areas of support undoubtedly contributed to the introduction of arc welding robot installations over the past years. While present government policy is to reduce the level of support under these schemes, encouragement and assistance remains available where leading edge technology is concerned such as in the development of flexible manufacturing systems.

The Science Engineering Research Council industrial robot initiative was launched in June 1980 to research and develop the technology needed to produce second generation robotic devices. The main thrust of this activity has been to forge links between academic researchers and industry.

Also administered by SERC is a teaching company scheme which was initiated in 1974/5. The scheme aims to develop active partnerships between academic bodies and industry in order to raise industrial performance by effective use of academic knowledge and capacity, improve the implementation of advanced technology, train graduate engineers for careers in industry, and develop and re-train existing company and academic staff.

In operation young graduates are guided and paid by the academic body while they work closely with the particular industrial partner to solve a specific problem or to implement a new technology. There are some 160 schemes in operation with more than ten having direct relevance to welding.

These academic related programmes have also had a considerable influence on the introduction of robots to welding.

5.1 RELATED STUDIES AND DEVELOPMENTS

5.1.1 Standardisation and Safety

The government, industry and research organisations have all shown interest in the safety of robot installations and in the standardisation of robots and working systems.

The Health and Safety Executive (HSE) formed a Robot Safety Liaison Group in 1981, composed of interested parties representing manufacturers, users, employees, academics and government. The major objectives of this body were to:

a   provide a focal point for contact on technical, practical and legal issues involving robot safety

b   formulate advice which would ensure that safety is given proper consideration in the development of robotics and

c   consider the various national and international safety standards through contact with the appropriate national bodies.

This body has remained active since its inception and is able to present on an international basis the opinions of the United Kingdom industry.

A related activity was the compilation and publication by the Machine Tool Industry Research Association of a guide entitled Safeguarding Industrial Robots, Part I, Basic Principles (12). It was recognised that this most useful guide required support from application related supplements and welding was considered high priority. 'Part II, Welding and Allied Processes' (13) is scheduled for publication during 1985.

5.1.2 Sensor Development

There has been considerable development, particularly in vision based sensors by CEGB (Smith 14) BOC (Wright 15) Liverpool (Lucas 16), NEI (Hage 17), Welding Institute (6) and Oxford (5). Studies at the latter University led to the development of the Metatorch which contains a vision system with the potential for both joint tracking and recognition during an actual welding cycle. This unit is now a commercial product.

CEGB have also made extensive studies of process automation. The underbead observation technique being used in practical application by them and, in the commercial field in a related development, by BOC (15) who used the technique in the production of road tankers.

5.1.3 CAD/CAM/simulation

Today the designer can generate data directly within a computer. The aim is then to be able to manipulate this data by various computing processes to give the desired robot motion, select the correct welding process parameters, and to initiate, under adaptive control, the welding or cutting task.

An important aid to ensuring that this activity can and does occur safely and effectively is the computer simulation of the robot cell. Work in this area commenced at Nottingham (Bonney 18) with the development of a'System for Aiding Man-Machine Integration Evaluation' (SAMMIE). This programme was used to evaluate the Ford truck cab welding study (11) and developments of this system have continued at Nottingham, PERA and Loughborough. The GRASP version studied at Loughborough has been used in a number of welding studies (17, Monckton 19).

Joint activity by ICL, Rediffusion (Bolton 20) and Werner and Sarll (Anon 21) is leading to a fully integrated 3D CAD system for sheet metalworking. The system is being installed to promote a flexible manufacturing system which will allow the customer's needs to be met with reduced lead times. Robot welding and cutting stations form important parts of this system.

Design. Mechanised welding systems are not as free nor as universal as the human welder. The constraints imposed by the mechanisms, for example the physical size of the robot wrist or the lack of vision interpretation, demand a fresh look at the component design. Useful work in this area has of necessity been carried out by all of the robot users and at research centres such as Loughborough (Wadsworth 22) and The Welding Institute (ESPRIT Programme 22).

The Welding Institute. The Welding Institute has maintained an active interest in all facets of mechanised and robotic welding. Robotic studies commenced in 1974 when a joint project was undertaken with NEL. The study introduced both organisations to robots and produced useful basic work on the relationship between weldbead shapes and welding parameters which led to work on adaptive control (Hunter 24) of the arc welding process. The robot used at NEL was the Unimate 2000. In 1978 TWI obtained its first robot installation, an ESAB system based on the ASEA IRb6 robot. Then followed robots from Hall Automation, Yaskawa (courtesy GKN Lincoln Electric) and more recently OTC LK (courtesy Butters) and Komatsu Tomkat (courtesy NEI Thompson.) This equipment has given the Institute and its Members experience of electrical and hydraulic powered arc welding robots and of robot control systems which represent the range of technology encountered in the past decade. These robots were used for application studies (Weston 25) which allow the formulation of the basic guidelines which should be followed if successful implementation is to be achieved.

Current studies cover the following topic areas:

the combination of robots with various welding process options, demonstration of multipass weld quality obtainable with robots investigations of robot use with high deposition rate welding techniques
development of sensing technique including vision
study of techniques for adaptively controlling the arc welding process
examining the communication problems encountered when linking CAD systems and process data bases to robotic welding

This latter phase is being undertaken as a trans-European activity within the ESPRIT programme.

6.1 COMMENT

Since robots first became available UK industry has explored the use over a wide range of applications. The learning pains have been considerable and not every installation has been a success, but there have been notable and leading achievements particularly in the welding field. The rate of installation has continued to rise in spite of the continuing financial stagnation. However, when compared with other countries, and West Germany in particular, the innovation in application range has not been matched by the quantity installed. If as the writer believes, robot utilisation is one of the measures of industrial activity then the prognosis for United Kingdom industry is not good.

However there is no doubt that more and more welding tasks will be performed by robots. Already the advent of joint searching and joint tracking, used in conjunction with track or gantry mounted robots, has brought a big increase in the weight and size of assemblies which can be welded. Acceptance and further evolution of vision base sensors will see this trend continue.

The decreasing cost of computing power will increase the availability of computer simulation of robot systems and will in turn lead to CAD and offline programming techniques to control the welding cells.

One of the most important aspects of the introduction of robots to UK industry has been the knock-on effect that they have caused. The introduction of high capital cost items into areas which traditionally use little capital has forced a new look at production techniques. To obtain return on investment the robot installation must work fast and efficiently and this has demanded component design studies which recognise the constraints imposed by the welding machine. Work flow and work handling studies demand new approaches to component preparation and scheduling. Robots are providing the impetus for dramatic changes within the UK welding industry.

7.1 REFERENCES

1   AWS (1985) Standard Welding Terms and Definitions AWS A3.0-85 American Welding Society.

2   BS499:1983 Welding Terms and Symbols, Part I, Glossary for Welding, Brazing and Thermal Cutting, British Standards Institute.

3   Salter, G.R. (1984) Automation - What do we mean? Metal Construction Vol.16.

4   British Robot Association (1984) - Annual Statistics

5   Clocksin, W.F. Davey, P.G. Morgan, C.G. Vidlera (1983) Progress in visual feedback for robot arc welding. Robot Vision Ed: A Pugh Publ. Kempston MK42 7BT, UK.

6   Doherty, J. Holder, S.J. Baker, R. (1983) Computerised guidance and process control. Robot Vision and Sensory Controls, 3rd Int. Conf Cambridge Mass.

7   Schoeters, T. (1982) Svetsaren 2.

8   Henderson, L. Smith, P.H. (1984) Exploiting robots in arc welding fabrication. Seminar Handbook The Welding Institute.

9   Industrial Robot (1980) Vol. 17. No. 3. p197 BL Landrover chassis line.

10 Industrial robot (1982) Vol.9 No. 4. pp.218-220 Investment in automation pays

11 Roswell, S.L. (1980) Automated welding of truck cabs using robots. Developments in Mechanised Automated and Robotic Welding WI Int. Conf. London.

12 MTTA (1982) Safeguarding industrial robots Part I Basic principles

13 MTTA (1985) Safeguarding industrial robots Part II Welding and allied processes (To be published).

14 Smith, C.J. (1974) Self-adaptive control of penetration in TIG welding. Advances in Welding Processes 3rd Int. Conf. Harrogate, TWI.

15 Wright, R.R. (1980) Welding Automation in Action Developments in Mechanised Automated & Robotic Welding TWI Conf. London.

16 Lucas, J. (1984) Low cost seam tracking for TIG welding. Seminar, Exploiting Robots in arc welded fabrication. TWI, Coventry.

17 Hage, P.M. Hewit, J.R. (1983) The robot welding of boiler tube platen assisted by vision sensing. Robot Vision and Sensory Controls, 3rd Int. Conf. Cambridge, Mass.

18 Bonney, M.C. (1974) Using SAMMIE for computer aided work place and work tool design. SAE Automotive Congress Detroit.

19 Monckton, P.S. Hawthorn, R.W. Jones, R. (1984) Simulation of a robotic welding cell to be used in small batch production. Proceedings 7th BRA Annual Conf. Cambridge UK.

20 Boulton, J B. Yeo, R.G.B. (1984) Welding in flexible manufacturing systems. Seminar, Exploiting robots in arc welded fabrication. The Welding Institute, Coventry.

21 Anon (1983) How 3D CAD can reap benefits in sheet metal work Numerical Engineering Vol. 4, No.5. Oct/Nov 1983 pp25-27.

22 Wadsworth, P.K. Middle, J.E. (1984) Case studies in robotic arc welding. Metal Construction Vol.16. No.4. pp213-218.

23 ESPRIT Program - Reports to the European Commission

24 Hunter, J.J. Bryce, G.W. Doherty, J. (1980) On-line control of the arc welding process. Developments in Mechanised, Automated and Robotic Welding TWI Conf. London.

25 Weston J (1980) Arc Welding Robots - A Welding Engineers Viewpoint. Developments in Mechanised Automated & Robotic Welding TWI Conf. London.

Practical industrial low cost vision for welding robots and automated welding machines

M.A. Browne, E.R.D. Duckworth, R.A. Wainwright and J. Ashby

1.   SUMMARY

The potential benefits of equipping arc welding robots with vision to provide adaptive control for seam tracking have been widely accepted.   A limited survey of industrial users' requirements suggested a more robust and cost effective solution was needed.   This paper describes continuing work at UMIST to build a practical industrial device to meet this requirement, based on earlier work using a single linescan, and avoiding the use of laser illumination.

2.   INTRODUCTION

This paper deals with the continuation of work at U.M.I.S.T. into the application of CCD linescan sensors to seam tracking in automated and robot MIG welding.

Having earlier established that the weld line may be tracked by a CCD linescan device, using ambient light and illumination from the weld arc alone, the work has been continued with the help of a Wolfson Industrial Research Fellowship and S.E.R.C. support with the aim of producing a practical industrial device using this principle.

This paper describes the design principles and implementation of such a design directed initially at automated welding applications, and the experimental results achieved with it.

3.   SURVEY OF POTENTIAL INDUSTRIAL USERS

To evaluate the potential demand for yet another device for providing adaptive control for MIG welding robots, a small cross-section of actual and potential users were approached, together with some robot manufacturers, to discuss their requirements and observations.   These covered automotive components and assembly, the nuclear industry, heavy batch fabrication to defence standards, and consumer durables.

Feedback appeared to confirm the priority need for a robust cost effective solution for mainstream work rather than the further sophistication and refinement of some of the more complex and high cost solutions already available or under development.

4.   DESIGN OBJECTIVES

4.1   Approach

In the face of industrial feedback that a robust and
cost effective device was most needed, and the delays costs
and complexities associated with an integrated approach, we
concluded that a self contained cascade controller would
provide not only what industry was demanding, but also
provide some interesting additional features.   In addition
the Group's success in tracking weld joint lines under
illumination from the weld arc alone added to the prospect
of an effective low cost solution (ref.3).
Such an approach not only appeared to meet the needs of
current and prospective robot users, but also meet the needs
of automated welding applications reported to be an order of
magnitude greater in number than suitable robot
applications.   Further, such a device, being independent of
the internal control system of the host could be both retro-
fittable to non-vision robots of diverse manufacture, and
also mobile, being available to be installed temporarily on
any of a number of hosts to handle batches of out of
tolerance components and other problem tasks as the user's
need arises.
Within this broad objective, the aim was to take
advantage of the higher update frequency made possible by
the smaller processing overhead of a 1x256 pixel CCD array
over that of a 256x256 or larger array, and of the
associated reduction in hardware costs.

4.2   Design And Performance Compromises

Cascade adaptive control as proposed using a single
scan line does however carry some penalties, some of them
shared with the integrated approach.
Adaptive control will not be possible during small
radius curves due to the need to focus the scan line a
discrete distance ahead of the weld pool.   This is done to
obtain an image unaffected by the weld pool or by glare from
the arc itself.   The result is due to a single scan line
providing insufficient data for interpolation.
In addition the single line scan provides no data for
stand-off distance, so that reliance must be placed on the
self adaptive effect of weld wire feed controlled by a
constant current weld set. Failing this, a second scan line
would enable stand-off to be measured directly (ref.5).
Finally, the consequence of keeping communications
between the Device and its host simple, ideally limited to
existing interlock facilities, the role of the
setter/operator becomes critical in setting up the means of
synchronising the Device with a programmable host.   The
design of the operator interface is seen as key to the
success of the Device as a practical and useful piece of
robot industrial equipment.
For automated hosts or where the Device is to be

dedicated to a single application, dependence on the operator can be reduced to a simple minimum.

4.3 Performance Objectives

The types of work for which the Device has been designed covers the bulk of automated and robot MIG welding, and in particular those applications where poor fit-up, loose tolerances and distortion are significant factors. The typical work envisaged was volume production using flame cut blanks and pressings.   This has meant measuring weld joint width and providing an appropriate degree of weave. Provision has also been made for offset from the gap centre-line to balance the heat capacities of the welded components.

The equipment was designed to handle butt, outside corner and edge joints in addition to fillet, inside corner and lap joints, in straight lines and curves of 50mm radius or more. Techniques have been developed to handle small rectangular lap and fillet welds.

No operation was permitted that would involve extending the welding cycle time, for instance through previewing each work piece before beginning to weld.

The equipment has been been designed to withstand the normal abuse likely to be encountered in manufacturing environments, as well as the ambient temperature, dust and R.M.I. associated with welding shops.

5.   SYSTEM OVERVIEW

5.1 Principal Components

The system consists of a Control Computer, a Wrist Unit containing control servos and CCD camera, and holding the weld torch, and a Driver Module located close to the Wrist Unit (Fig.5.1).   Operator interface is provided through an LCD display and keyboard.   Host interface and synchronisation relies on the host's existing interlock channels.

The Control Computer contains two Motorola 6809 8 bit processors.   The Driver Module contains both drive circuitry for the CCD sensor and also RF filtering, fast A/D conversion, and line driver circuits.

The Wrist Unit allows standard robot and automated welding torches to be used, of both air and water cooled type giving current capacities up to 600 amps.

5.2 Modes of Operation and Operator Interface

5.2.1.   Relationship to Host.   The Device is operated in three modes corresponding to the teach/edit/playback cycle used to set up the host.   For automated welding machines, the teach phase consists of set-up and hardware programming in addition to more limited programmable control.

For both robot and automated hosts, the programming phase involves inserting the interlocks by which the Device is synchronised with the host.

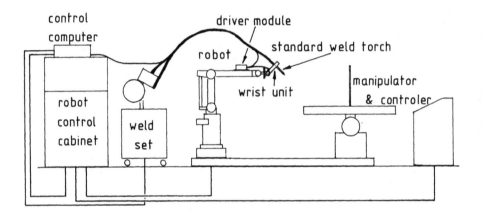

Fig.5.1   The principal system components

5.2.2   Teach/Edit Mode. During teach/edit mode, the
Device is programmed in parallel with the host, driving the
Device scan line to approximately the correct orientation to
the weld path.   This is at right angles to the weld path
and ahead of the weld point.

In addition, weave may be enabled or disabled, and
lateral offset selected to cope with differing heat
capacities of the welded materials or with multiple bead
welds if desired.

When editing an existing program, it may be played back
on both the Device and the host until the section to be
amended is reached.   Driving the scan line to another
position overwrites the previously recorded position.

5.2.3   Rehearse Mode. An additional phase is needed
beyond that theoretically required to program the host.
During this, the approximate orientations recorded by the
Device during the teach/edit phase are refined using
feedback from the Device's optical system.

In practice this phase is unlikely to be a serious
overhead as, typically, programs are already rehearsed as a
check for correct programming.

5.2.4   Playback Mode (Normal Operation). Finally the
Device will under normal production conditions play back the

recorded axial orientations in synchronisation with host movements, adjusting the lateral position of the weld torch to eliminate observed error between taught position and actual, and providing both weave according to programmed criteria and the offset established at the Teach or Rehearse phases.

5.2.5   Operator Interface. Particularly in the Teach/Edit and Rehearse phases, the actions of the Setter/Operator are critical to the correct operation and synchronisation of the Device and of the robot.   To ensure, as far as possible, fool proof operation, the operator is prompted and reminded of a limited set of options open to him through the control screen and keyboard.

5.3 Image Formation and Analysis

The imaging system described already is based on the earlier UMIST work (refs.2 & 3).   Its physical form has been modified to miniaturise the sensor and to remove drive electronics to a less vulnerable position where its bulk has little impact on access to the workpiece.

A Thompson-CSF 256 pixel CCD line scan device has been adopted in preference to the earlier Fairchild unit, with a resulting simplification of drive electronics.   After A/D conversion each pixel's digitised grey level is loaded into RAM for later analysis.

The image is focused using a simple convex lens protected by a replaceable broad band filter designed to favour reflected light from the workpiece and limit radiant light from the arc and from incandescent particles.   The filter is replacable and so sited to be shielded from the main rain of particles from the weld site.

Examples of typical CCD scan grey level plots for differing types of weld are shown in Fig.5.2.

Three levels of software filtering have been found to be more than sufficient to obtain reliable tracking.   The basic techniques of time and spatial averaging together with continuity testing were fully reported by Browne et al (ref.3).

5.4 Wrist Unit Electro-Mechanical Design

5.4.1.   Design Criteria. The aim of the mechanical design of the Wrist Unit was to minimise both avoidable bulk near the torch head, and the inevitable weight penalty, as well as to achieve a robustness suitable for the industrial environment and low manufacturing costs.

In addition the objective of a 0.2mm resolution at the weld tip required the careful control of aggregate backlash in the Wrist Unit.

It was also thought that the eventual industrial users would probably prefer to use a torch of their own choosing based on their own experience, and to suit their particular work.

Thus the design objective was to replace the standard torch holder with a compact, robust and backlash free unit

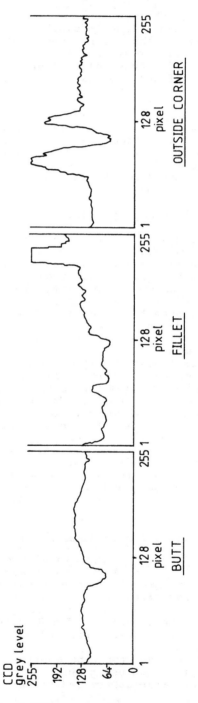

Fig.5.2   Typical CCD grey level plots for butt,
           fillet and outside corner welds

providing control over both scan orientation and lateral error correction. In addition, both the weight and cost objectives applied.

The overall system was expected to be heavily over damped. This was due mainly to the heavy filtering applied to the CCD image. More significant than conventional frequency response were the problems created by measuring error 15mm ahead of the weld point. This required relating the observed lateral error measured some 25 to 120 update cycles ahead of the weld point, and also allowing for the effects of changes in orientation between the moment of measurement and the arrival of the weld arc at the point of measurement.

5.4.2 Achievement of Design Criteria. The resulting unit had a mass, in addition to the standard torch and allowing for the standard cable support provided by the weld set, was just acceptable within the limitations of the 6kg nominal carrying capacity of the IRB6 robot.

Backlash in the development prototype has been measured at less than 0.2mm at the torch head.

The dynamic performance of the Device is discussed in 8.0 below.

5.5 Control Computer and Software.

5.5.1 Hardware. The application involves two substantially independent processes running concurrently.

Two 8 bit processors were chosen because using two processors permitted parallel processing and the decoupling of the regular cycle of optics refresh and analysis from the less predictable operator and host interface operations. Separating the processes also made implementation easier. Most of the data was to be to eight bits resolution, and where necessary the MC6809 had well developed 16 bit facilities.

Parallel processing made it possible to keep the position update time sufficiently short to achieve the design resolution. The application lent itself to a modular approach with additional axes and facilities being running on separate processors, with limited communications between them. The chip system was cheap, familiar, and likely to produce a low system cost.

Upgrading to say the MC68000 would entail developing a multi-tasking system of avoidable complexity in order to offset the very much higher chip cost.

Given weld speeds of 5-25mm/sec, the design resolution of 0.2mm would require an update frequency of 25-125Hz. In practice an update cycle time of 25ms (40Hz) has been achieved giving resolutions of 0.1mm to 0.6mm. Welding tests have so far indicated an acceptable performance.

5.5.2 Image Processing and Seam Finding. The main development here was to improve upon the techniques that had been successful with straight seams (refs.1 & 3), to enable the Device to follow curves and more substantial deviations.

Learning and fading techniques were developed to

overcome possible confusion with parallel line features, by
providing a form of continuity testing (ref.3).

In designing the algorithms, rapid processing has been
found to give more satisfactory and smoother line following
than more complex solutions that reduce the update
frequency.

The other major area of development has been in the
reliable measurement of seam width, needed for the selection
and control of weave, including the associated control of
wire feed, weld current and voltage. At present width can
be measured given standard illumination and surface
finishes. Work continues on non-standard conditions.

5.5.3   Lateral Error Control. The output from the image
analysis and seam finding modules is a smoothed and filtered
error measurement and direction. The lateral servo control
will move at a traverse rate of 8mm/second to reduce the
error.

5.5.4   Orientation Control. Control of the orientation
axis depends on the phase within the Teach/Edit, Rehearse,
Playback cycle which in turn is tied in with the host
programming cycle. Control input may be derived from
respectively the operator, the vision system, or the stored
weld program.

Motor control simulates DC servo control. The
stepping motor is ramped up to a speed proportional to the
filtered error with negative feedback proportional to rate
of change of error. This approach is necessary to achieve
fast repositioning between weld legs.

5.5.5   Host Interface. Typically industrial robots
suitable for arc welding are equipped with a number of 24v
current-sinking interlocking inputs and relay driver
circuits. These are in addition to program and emergency
start/stop rings.

In the case of the IRB6/1, these number 16 and 14
respectively. In addition the IRB6/1 has connections for a
15v supply and provision for 24v current sinking search/stop
function. Of these, most are required for control of the
weld set, the manipulator, and guard and jig interlocks.

The aim has been to operate using the minimum of
interlocks and to monitor the weld set control lines as a
further source of data.

6.   EXPERIMENTAL RESULTS ACHIEVED

6.1 Host System Used

The Device was tested on a robot host simulating both
automated welding rigs, and conventional robot operations.

The robot was an ESAB A30A arc welding station
consisting of an ASEA IRB6/1 with an ESAB Lincoln LAH500
weldset and MHS 150 manipulator. A standard Binzel RoBo
450 water cooled torch was used, with a rated capacity of
450 A continuous duty.

The whole unit had been acquired through ESAB Lincoln

from an industrial company who had been using it as a complete manufacturing unit typical of that likely to be used by target users for the Device.

6.2 Welding Tests

6.2.1 Lateral Dynamic Response.
To measure frequency response a sinusoidal line of 10mm amplitude was marked on 16swg sheet wrapped round a 300mm diameter mandrel and rotated at varying speeds. A pen attached to the weld torch recorded the actual path traced by the torch. These showed a peak response around 1 Hz. However more significant was the phase effect of the scan being 15mm ahead of the weld tip.

As a result the control algorithm was modified to provide a delay corresponding to a normal welding speed. Presented with a 5mm sideways step in observed seam position, the Device produced an over damped response stabilising to within 0.5mm in 0.6s. At normal welding speeds this corresponds to a distance travelled of 3 to 15mm.

6.2.2 Stand-off Dynamic Response.
The equivilent test for stand-off error response and the associated weld parameter effects were undertaken for varying wire feed rates and weld currents and voltages.

These confirmed the expectation that high feed rates would give rise to good error correction and vice versa. Impact on weld quality confirmed that weld parameters needed to be under at least partial control from the Device. Work was continuing in this area.

6.2.3 Straight Welds - Mild Steel/CO_2.
Straight welds were undertaken both radially and tangentially to the base axis of the host robot, using materials of varying reflectivity. These covered fillet, lap, outside corner and butt joints with and without groove preparation. Workpieces were displaced in three axes from the taught position, and with varying fit-up. Tracking errors of between up to 1mm were observed, with the best performance achieved on Vee preparation butt joints on dull finish materials displaced laterally or vertically from the taught seam position. Closely fitting components of bright finish displaced longditudinally were least reliably followed.

6.2.4 Curved Welds - Mild Steel/CO_2.
Two types of single plane curves were tested, cylinders and circles on flat plates, using butt and lap welds, these being found to be the most demanding during straight weld tests. The results confirmed that using the forehand welding position a minimum radius of 30mm on flat plate and on cylinders could be followed satisfactorily, while backhand welding reduced this minimum to 20mm radius.

6.2.5 Right-angles and Small Radius Bends.
The test pieces consisted of small and large rectangular plates lap welded onto larger plate sections using varying side

dimensions and radii.   The aim was to evaluate seam
following under these conditions and to develop effective
techniques.

The most effective technique was found to be to weld
each straight section separately.   If a continuous weld was
essential, then a backhand technique gave the best seam
tracking result on outside lap and fillet welds, and
forehand technique on inside welds (see Fig.6.1).

A key factor in the success of these tests was was the
angle from which the seam track was viewed.   By viewing a
right angle bend from the offset position required for lap
or fillet welding, and by using the appropriate forehand or
backhand technique, the observed track remained in view and
departed progressively in the direction of the turn, rather
than suddenly disappearing from view.   In the Forehand (or

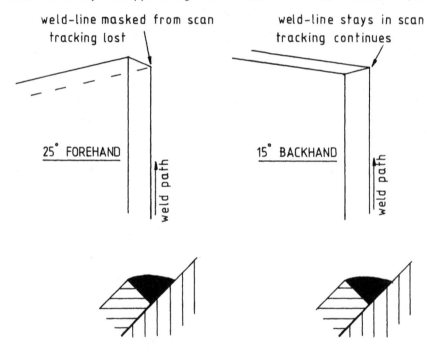

Fig.6.1   Techniques for continuous tracking around
right-angle weld paths

pushing) technique the weld gun is angled from the vertical
to point forward towards the direction of travel, whereas in
the backhand (or pulling) technique the gun is angled to
point back towards the newly made bead.

The results were that provided radii of more than 3mm
were used on squares of greater than 15mm sides, acceptable
tracking was achieved on lap and fillet welds.   Butt welds
of similar configuration required longer minimum radii for
continuous welding.   Unradiused square butt joints tend not
in any case to be welded continuously, so that there would

be time for the Device to reorientate and settle before the
start of a new leg.

7.   ASSESSMENT OF RESULTS AND CONCLUSIONS

Industrial feedback confirms the priority need for a
sound cost effective and robust solution to robot vision,
rather than on the heights of sophistication and
performance.   The cascade approach does appear to offer a
way of avoiding some of the limitations of cost and
inflexibility suffered by the integrated approach.
From the laboratory tests undertaken so far, it appears
that the Device offers useful performance, in most respects
comparable with those achieved with the aid of laser
illumination and without the penalty of the processing
overhead required to handle a 64k pixel array.
Doubts about the feasibility of relying on the self-
adaptive characteristics of wire feed rate combined with
currently available constant current weld sets have been
dispelled.   Acceptable results have been obtained in all
but the most extreme conditions, provided the finest
suitable weld wire is used to permit a high wire feed rate.
There are however evident limitations to the Device
which are the price both of simplicity and of portability.
There are reasonable grounds for believing that these
limitations will be less severe in industrial application
than laboratory testing might suggest.   This is the key
area for further development, with several industrial
undertakings showing interest in evaluating and reporting on
the unit.   One of the key objectives of these trials will
be to gain the industrial experience needed to develop
techniques of host and Device programming and of jig and
component design that will minimise the impact of such
limitations.
There are also plans to explore the potential benefits
of increasing to a limited extent the number of scan lines
to see whether the additional information can be used
significantly to increase performance or reduce dependence
on the setter/operator.

8.   ACKNOWLEDGEMENTS

The authors wish to acknowledge the financial support
received from the Wolfson Foundation in the form of an
Industrial Fellowship, and from SERC.   The advice and
assistance received from members of the Departments of
Instrumentation and Analytical Science and of Mechanical
Engineering and Machine Tools are also acknowledged with
appreciation.   Special appreciation is offered to
Dr. J.L. Falkowski of the Institute for Automatic Control,
Warsaw Technical University for his help and guidance.

9.   REFERENCES

1.   Wainwright, R.A., 1983, 'A Study of a Passive Visual
Sensor for Single Pass Operation of an Arc Welding

Robot', M.Sc. Thesis, University of Manchester.

2. Ashby J., 1984, 'A Design of Sensor for Seam Tracking in Arc Welding Control', M.Sc. Dissertation, University of Manchester.

3. Browne, M.A., Falkowski J.L., and Wainwright, R.A., 1984, 'Visual Seam Tracking with High Noise Content - Learning and Fading'. I.Mech.E C472/84, Proc. Conf. U.K. Robotics Research.

4. Falkowski, J.L., and Browne, M.A., 1983, 'Vision Sensing for Arc Welding Robots - a New Approach', J. Phys., E. Sci. Instrum., Vol.16.

5. Browne, M.A., and Falkowski, J.L., 1983, 'Passive Visual Tracking for Robotic Arc Welding', Proc. Intl. Conf. on Optical Techiques in Process Control, Hague.

A high speed pattern recognition system for robotic applications

A.M. Dean

1   INTRODUCTION.

With the large increase in small batch production and
the need for greater productivity and reduced costs,
automation is taking over. Dedicated automation , such as
is used on large production lines, is unable to cope with
quick product changes hence the use of robots is increasing.
At the moment these reprogrammable assembly devices have
little intelligence and little or no sensing capabilities.
If robots are to expand into other production areas they
will require complex and intelligent sensors.

1.1   Vision And Sensory Feedback.

The mastery of vision and language is very high on the
list of challenges which robotics must overcome if they are
to become truly adaptive and versatile. Industrial robots
will have to see what they are doing if they are ever to
break out of the sightless world of pre-programmed
repetition in which they currently exist. Slightly less
obvious, perhaps, is the importance of sight and speech to
the more general problem of creating a significant level of
machine intelligence.

Despite all the research, papers and conferences on
robot vision, the world still waits for a commercially
acceptable unit to be produced. This lack of credibility
means that only about one hundred of the world population of
over forty thousand robots has a vision system, the majority
of these are being used in experimental robot installations.
It can be seen that there is a very large potential market
for vision within current robotic applications, vision would
also open new areas for robotics. It is this potential that
fuelled the research for the development of the cheap,
reliable, fast and simple image processing unit developed in
this paper.

2 SYSTEMS SPECIFICATION.

2.1 Main Design Criteria.

A study of possible applications for robotic visual feedback systems and visual inspection systems suggested a number of basic design criteria. These basic criteria include a) the ability to be capable of identifying components, that have previously been taught to the system, when they are introduced at random into the field of view. b) the ability to be able to determine the position and orientation of these pre-taught components when they are introduced. c) high speed, the time for a) and b) above to take place, should be comparable with human visual interpretation. This time must not be greater than the time from initial robot limb movement to the robot's requirement for its final coordinates. Both these time criteria require recognition and orientation computations to take place in less than one second and preferably far less. d) commercially competitive system. A total components cost in the order of £2000 was a reasonable amount at the start of the project, other vision systems retail in excess of £13000.

These criteria were considered to be of prime importance and are necessarily general, as is the nature of the problem.

2.2 Secondary Design Criteria.

Other design criteria relating to the achievement of the above four, making the system user friendly and easy to use, were considered as desirable but not as important as the main criteria listed above. These included e) high percentage recognition success to avoid the necessity of re-processing the same scene, which would result in delays. f) an accuracy at least comparable with that of the robot or inspection application in use. g) easily interfaced to any one of a number of industrial robots or inspection systems. h) easy to use by an operator not trained in computing, this being a function of user friendly software. The software should also have built in fault recovery routines allowing fault free operation even if an unexpected event occurs. i) be able to store a reasonable number of components within its "learnt list" and must also be able to handle a reasonable number of components within its "field of vision". A "reasonable number" in an industrial application could vary from one in a simple application to tens in a complex recognition problem.

3 SYSTEMS OVERVIEW.

A considerable amount of research has been reported into methods of pattern recognition and scene analysis, for general texts see Ullman (1), Rosenfeld (2) and Devijver (3). Most of this work being for slower applications such as character recognition or slow speed scene analysis for mineral location purposes. Whilst most of the techniques are useful for robotic vision system applications, the time to carry out the complex algorithms is often not available.

Having considered the main systems specification, the hardware and software design were studied and some decisions made as to the design of the first image processing system. The hardware design is dependant upon the pattern recognition approach adopted, the number of pixels and grey levels and the type of inputting device.

3.1 Pattern Recognition Methods.

The many varied methods of pattern recognition and image processing can be grouped into three main categories which are detailed in many texts, Devijver (3), Batchelor (4), Watanabe (5), Fu (6) and Aleksander (7). The texts which develop and discuss these recognition methods often refer to them using differing names. The most commonly used methods are:-
a/ the template matching approach.
b/ the syntactic or structural approach.
c/ the decision theoretic or discriminant approach.
In the template matching approach sets of images are compared pixel by pixel until a match is found, this method proved to slow and required large amount of memory to store the images. The structural approach requires a string, tree or graph of pattern primitives and their relations to be extracted and compared, the rules for obtaining these individual generic properties for lists of twelve or more taught components became almost impossible and recognition time becomes very long. The decision theoretic approach represents each component as a set of features, decision is based upon comparison of these features. The decision theoretic approach was adopted as features can be extracted and compared within the recognition time available.

3.2 Pixel And Grey Level Density.

Consideration was next given to the number of pixels in both the horizontal and vertical planes. Various factors must be considered when choosing the resolution for the pixels, the main factor is the accuracy required. This accuracy is dependant upon the resolution of the inputting device and the accuracy of the image processing algorithms. The two main inputting devices under consideration were a standard CCTV camera and a CCD camera. The CCTV camera having a horizontal resolution of 1 in 500 at the tube centre and a vertical resolution of 625 lines i.e. 1 in 625. The CCD camera having resolutions of up to 1024 in either line or array form but having the disadvantage of being very expensive. As an accuracy of approximately +/- 1mm was considered a reasonable value for the 250 mm work area decided upon, it was decided to have a work area of 256 by 256 mm at an accuracy of 1 mm i.e. to use 256 by 256 pixels giving an accuracy of 0.39%. This meant having a transfer rate of 7.5 Megabits per second, if using one grey level, and a RAM picture store array of 8K Bytes.

3.3   Central Processing Hardware and Software Considerations.

Having made some preliminary decisions about the image processing required and the type of input device to be used, the hardware and software most capable of achieving these objectives must be identified. It should be noted at this stage that the possibility of using a dedicated parallel processor, as detailed by Duff et al (8), was ruled out due to cost. The Motorola MC68000 working at 8MHz was chosen as the market leading 16 bit microprocessor, it was also over five time faster than the LSI-11 used in other commercial image processing systems.

Having decided upon the MC68000 as the main CPU, consideration was given to the software available for this machine. As speed was of prime importance the structured assembler was the natural choice for the recognition section as it is faster than any high level language. Pascal was used as the high level language as it was ideally suited for user communication, control and assembly routine linking.

3.4   Choice Of Data Acquisition Method.

The main design criteria have been specified as cost,speed, and versatility. The optimum combination between them is attained using a closed circuit television camera (CCTV) as the data acquisition device, with the option of using a more costly charge coupled device camera. As a commercial CCTV camera must be used this gives a transfer rate of 7.5MBits or 0.469MHz at 16 bits per transfer. The CPU can not move data at this speed so dedicated hardware had to be used, it was decided to use a direct memory access (DMA) device the Motorola MC6844.

4   SYSTEMS HARDWARE.

The hardware sections that, when combined, make up the complete system can be functionally split into convenient sections. These sections do not however relate to the physical location of the hardware but to its operation. The system splits into the five sections which are detailed below, a block diagram of the complete system is shown in figure 1.

1/ Camera and Video monitor.
2/ Camera Interface and System Timing.
3/ Direct Memory Controller and Interface.
4/ Graphics module and monitor.
5/ Main Central Processing Card and Terminal.

The combination of camera and video monitor make up the remote sensing station . This station can physically be a large distance from the remaining sections but must be situated viewing the robots work area. The monitor is optional but useful to enable the operator, at the remote robot work area, to see the section of the work area being viewed by the robot.

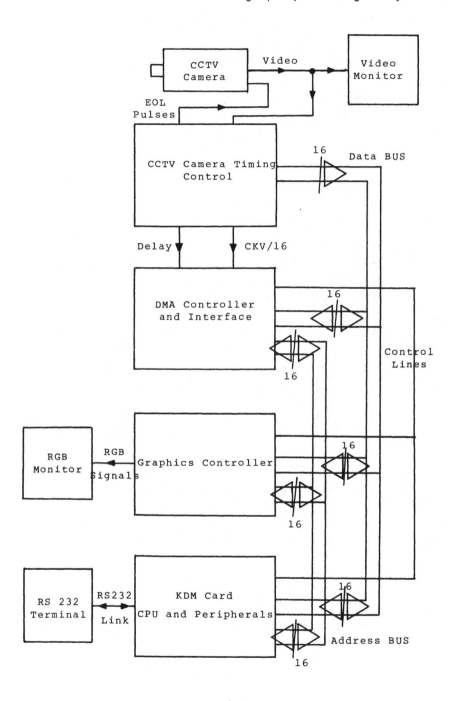

FIGURE   1   SYSTEMS BLOCK DIAGRAM

The camera interface card is physically located on the same printed circuit boards as the DMA controller, and handles all the circuit timing, the sampling and quantization of the video signal, and provides the binary video signal to the data bus latches ready for the DMA transfer process when requested. To achieve this the video signal synchronisation pulses must be separated, and the analogue video signal sampled and quantized into black or white levels. All the system timing is derived from a central clock within this section, with the exception of the CPU and graphics card clocks.

The DMA controller interfaces to the CPU via the data and address buses and controls the transfer of the digital binary video data to the picture memory on the KDM card. There are two modes of operation for this section, the asynchronous mode in which the DMA registers are set up ready to transfer the data, and the synchronous mode in which the CPU is closed down and the DMA transfers data from the latches to the picture store RAM.

The graphics card and associated monitor interface to the CPU via the data and address buses, enabling all the image processing stages to be displayed visually. The graphics card is also optional, its normal mode of operation is in complete isolation to the main vision system, thus it does not slow down the image processing algorithms. The visual display is updated by the vision system by simply writing to the graphics RAM, the graphics card then automatically updates the display.

The final block contains the CPU and all its associated peripherals including the systems memory, input/output, etc. The KDM card which holds the CPU and its peripherals communicates with the user via a standard RS-232 terminal. This terminal also has some additional ,all be it slow, graphics capabilities for displaying the image processing results.

5 SYSTEMS PERFORMANCE.

The systems objectives were specified at the commencement of the project and are detailed above. When the vision system was completed and the final software programmed into the EPROM's a series of detailed tests were performed on the system. These tests were structured to measure the level of achievement of the basic and secondary design criteria and a summary of the results are detailed below. Tests 1 to 3 measure the time of the individual recognition routines and also the total recognition time and percentage recognition success when all models are selected and all components are recognised. Test 4 and 5 give numerical data and errors around the robot frame and as a component is rotated.

Test 1 Test Component = Small Circle, Limits Set = 13%.

TEST	AVERAGE TIME MSECS	MAXIMUM DEVIATION MSECS	% RECOGNITION %
Input	24	12	–
Trace	73	0	–
Chain	12	2	–
Basic	1.2	0.8	–
Hole	0	0	–
Radius	2	0	–
Total	119	13	100

Test 2 Test Components = Small Circle, Medium Square and Large Square with Hole, Limits Set = 13%.

TEST	AVERAGE TIME MSECS	MAXIMUM DEVIATION MSECS	% RECOGNITION %
Input	27	5	–
Trace	74	1	–
Chain	100	8	–
Basic	31	1	–
Hole	39	0	–
Radius	28	1	–
Total	340	20	100

Test 3 Test Component = Sixteen Test Components Limits Set = 13%.

TEST	AVERAGE TIME MSECS	MAXIMUM DEVIATION MSECS	% RECOGNITION %
Input	24	7	–
Trace	75	0	–
Chain	425	14	–
Basic	94	1	–
Hole	200	1	–
Radius	89	0	–
Total	902	47	99

Test 4 Test Component = Medium L, Limits Set = 13%, All models used.

POSITION	PERIMETER (mm)		AREA (sq mm)		Rmin (mm)		Rmax (mm)		OFFSET (deg)	
	Meas	Dev	Meas	Dev	Meas	Dev	Meas	Dev	Meas	Dev
Central	153	1	856	12	7	0	30	0	0	0
Top Left	153	0	847	19	7	0	29	1	0	1
Centre Left	152	2	855	17	7	1	30	1	0	1
1/4 Down - 1/4 Left	151	1	838	4	7	0	30	1	0	0

The true data for the Medium L is given below, along with a maximum percentage error for each position.

POSITION True Value	PERIMETER 153 mm	AREA 850 mm	Rmin 7 mm	Rmax 30 mm	OFFSET 0 degree
Central	0.6 %	1.4 %	0 %	0 %	0 %
Top Left	0 %	2.3 %	14 %	3.3 %	0.3 %
Centre Left	1.3 %	2.1 %	14 %	3.3 %	0.3 %
1/4 Down - 1/4 Left	0.6 %	1.6 %	0 %	3.3 %	0 %

Test 5 Test Component = Large L, Limits Set = 13%, All models used.

TRUE DATA			WORST MEASURED DATA			MAXIMUM ERRORS		
X mm	Y mm	Angle degrees	X mm	Y mm	Angle degrees	X mm	Y mm	Angle deg
113	130	0	112	131	0	1	1	0
116	109	90	116	110	89	0	1	1
129	132	180	129	132	180	0	0	0
133	140	270	133	141	270	0	1	0
30	56	0	31	56	0	1	0	0
34	37	90	33	36	91	1	1	1
46	60	180	46	61	181	0	1	1
52	62	270	51	62	269	1	0	1
73	96	0	72	95	359	1	1	1
75	75	90	75	75	90	0	0	0
86	94	180	85	94	180	1	0	0
92	103	270	91	102	271	1	1	1
31	92	0	32	93	0	1	1	0
35	74	90	35	75	90	0	1	0
44	91	180	43	92	181	1	1	1
46	99	270	45	98	269	1	1	1

These results lead to the following comments on the design criteria :

a) Identification of pre-taught components when introduced at random into the field of vision. The system achives recognition success, with correct models selected, of 99 to 100%.

b) Calculation of position and orientation of pre-taught components. From test data accuracies of +/- one pixel in position and +/- one degree in angular offset were attained. These errors come from a number of sources which include camera geometry faults such as non-linearities and drift, thresholding and image quantization errors, parallax errors and errors in the vision to robot coordinate transformation. The errors produced by the camera could be reduced by using a better CCTV or a CCD camera, parallax errors could be removed by using planar components or a frunel lens to produce a planar image and robot to vision transformation errors can be reduced by regular calibration. The only method of improving the thresholding and quantization errors, which are 1 pixel, or 1 in 256, is to increase the grey level and/or pixel density.

c) Recognition, position and orientation calculation time to be comparable with a human operator and not to cause long delays in the assembly process i.e. less than one second. This time varies from 115 mSecs for a single component and model to a maximum of 902 mSecs or 0.9 of a second for the recognition and calculation of data for 16 components using all the available models. It is envisaged that the average recognition time for an industrial assembly scene having eight components and models with optimal selection will take about 0.6 of a second.

d) The complete system to be commercially competitive i.e. to cost less than £2000. The system uses a cheap CCTV camera and could be minimised to use the robots CPU thus requiring only the addition of the DMA transfer circuitry. The reduced system contains approximately £200 worth of components and the full system including CPU, graphics card and DMA system contains about £400 worth of components. It is envisaged that the complete system could be produced well within the original design price of £2000.

REFERENCES

1. Ullman,J.R.,1973,Pattern Recognition Techniques,
    Butterworth.

2. Rosenfeld,A.and Kak,C.A.,1976,Digital Picture Processing
    Academic Press.

3. Devijver,P.A.and Kittler,J.,1982,Pattern Recognition
    A Statistical Approach,Prentice Hall International.

4. Batchelor,B.,1978,Pattern Recognition Ideas in Practice,
    Plenum Press, London.

5. Watanabe,S.,1978,Methodologies of Pattern Recognition,
    Academic Press,London.

6. Fu,K.S.,1976,Digital Pattern Recognition,Springer-
    Verlag,New York.

7. Aleksander, I., Stonham, T.J.,and Wilkie,B.V., Computer
    Vision System For Industry, IEE.Colloquium Diegest
    No.1982/36,1982.

8. Duff,M.J.,et al,1981,Languages And Architectures For
    Image Processing,Processing,Academic Press.

A vision system for the identification of motor car headlamp reflectors for robotic assembly

P. D. Chuang and J. P. Cosmas

ABSTRACT

Imperial College and Lucas Research Centre have a collaborative research program whose aim is to investigate the feasibility of automating the assembly of motor car headlamps. It was decided that a Flexible Manufacturing System (FMS) is required which would not only identify and assemble various types of headlamps but also be easily reprogrammed to identify and assemble new or modified headlamps. This is an essential requirement since headlamps are cosmetic items with a limited product life cycle.

The FMS would include independent Robot Cells executing independ nt tasks of the headlamp assembly (the insertion of fixing 'tulips' onto reflectors, the insertion of lamps, casing assembly and so on). A cheap, fast and flexible Vision System was required within six months to identify headlamp reflectors prior to assembly. This paper will describe such a system for the purpose of 'tulip' insertions.

1. INTRODUCTION

The Tulip to Reflector Assembly Cell is essentially a single robot assembly cell where reflectors are fed into the system by a conveyor from a manual inspection station. Tulips and fixing screws (fig.1.1) are fed from bowl feeders into a sub-assembly machine prior to being airveyed[1] to a screwdriver mounted on the robot.

The basic requirement of the cell is to fasten a predetermined number of tulips onto a variety of DMC reflectors. The system has been initially configured to handle the following reflector types:

   i) Montego - Home/European (Right Handed and Left Handed);
  ii) Maestro/Rover - Home/European (RH and LH);
 iii) Metro   - Home/European (unhanded);
  iv) Sierra   - Home/European unhanded).

---

1) airveyed - in this context, a hollow flexible tube is used to transport the tulips to the screwdriver by compressed air.

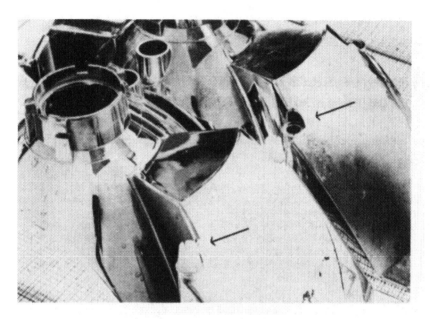

Fig. 1.1 Reflectors with and without tulips inserted

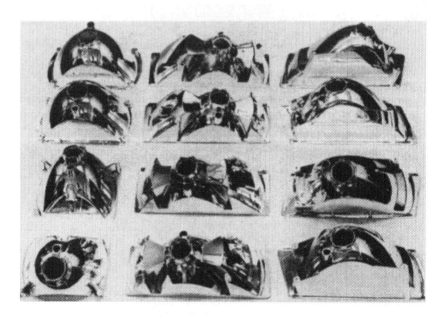

Fig.1.2 Twelve permutation of reflectors

Home and European differences do not effect the robot sequence as the tulip locations are common to both. As such, there are six different reflector types, ie. two for the Montego, two for the Maestro/Rover, one for the Metro, and one for the Sierra. However, the reflectors can be firmly placed on the conveyor in one of two possible longitudinal orientations giving twelve positional permutations (Fig.1.2).

It is proposed that a Vision System is used to verify the type and orientation of reflectors entering the robot cell. The camera will be located above the conveyor to inspect the reflectors before they enter the robot workstation. Structured lighting has been introduced to highlight the features to be checked (pilot and main bulb holes) and to reduce the effect of ambient light variations. In addition, the Vision System must have the attribute of being reteachable for the introduction or modification of new reflector types.

2. PROBLEMS

2.1. Time Factor.

The Vision System is required to reliably recognise any one of six headlamp reflectors in any sequence within the robot cycle time of twelve seconds. Furthermore, the System has to be ready for 'factory' trials within six months.

2.2. High Reflectivity and Feature Selection.

As can be seen from figure 1.2, the highly reflective nature of the reflectors presents considerable problems in recognition. Light variations, reflected images of moving background objects amongst other factors, contribute towards a hazardous environment. To recognise reliably, it is important that the reflectors are shielded during recognition, that a constant light source is applied inside the shield and that the most reliable features are considered for recognition. By experimentation, it was found that the pilot and main bulb hole patterns could uniquely identify each reflector.

2.3. Flexibility.

Due to the product nature of the headlamp reflectors, the Vision System has to be easily reprogrammed for new reflectors. The flexibility, simplicity and ease of reprogramming implies that the production 'down' time is minimal during product changes and that production could meet the widely fluctuating customer demand. Considerable savings in time and money are possible as no recalibration is required due to changes in production schedules.

3. SYSTEM HARDWARE

The target Vision System consists of a Vidicon CCTV camera, a nine inch B/W monitor, a digital frame store (Eltime Image III), a M68000 SAGE microcomputer with a single floppy disc drive and 256K RAM, VDU, keyboard and interfaces. Figure 3.1 shows the block diagram of the system.

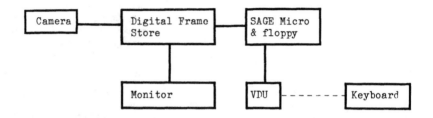

Fig. 3.1 Block Diagram of the Vision System

The Image III digital frame store is configured for four planes of 256 by 256 pixels with 64 grey levels per pixel. Under software control any of the planes can be used to store a TV frame, accessed by the microcomputer or displayed on the monitor.

For security reasons, the keyboard is unconnected and separately kept. The keyboard is required only for the purpose of reprogramming the reflectors. Finally, with the exception of the camera, the whole system is fitted into a closed and locked industrial cabinet (fig.3.2).

4. SYSTEMS TASK

The Vision System is designed as an independent workstation and requires little or no human supervision. On power up, the microcomputer autoboots from the floppy drive, executes the recognition program, sets up the robot protocol through the interface and waits for the robot controller.

During assembly, the reflectors are placed on the ingoing conveyor belt. Due to the design of the belt, the reflectors are forced in either one of two longitudinal orientations and waits in a queue. Using photosensors, the robot controller then feeds the Vision System with individual reflectors, stopping and clamping each reflector before a TV frame is sampled by the digital frame store. The image sampling station is shrouded by a sheet metal tunnel with a camera mounted at at the top and is backlit by fluorescent lights to accentuate the main and pilot bulb holes (figure 4.1 and 4.2). The inside of the metal shroud is lightly painted to reflect light so that a grey level image is obtained rather than a binary image.

Once a decision is made by the microcomputer, the robot controller is signalled and the reflector is released to continue along the conveyor. It is then either clamped again for tulip insertions by the robot or rejected for being faulty. After the insertion of tulips, the reflector is released and removed from the outgoing conveyor.

Fig. 3.2 Target Vision System

Fig. 4.1 Tulip to Reflector
Assembly Cell

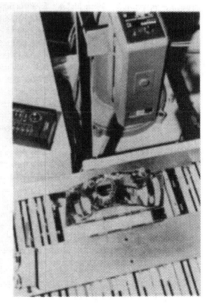

Fig.4.2 Top View of Reflectors
without the shroud

5. SYSTEMS SOFTWARE

The vision program is written in 'C' under the CP/M68k Operating System. The software is developed on a larger SAGE IV microcomputer with 1Mbyte RAM, single floppy disc drive and a 12Mbyte Winchester with comprehensive software tools. As both the development and target system have basically the same processor board and Operating System, no software alteration or system generation is required.

5.1 Recognition Methodology.

As it was found that the main and pilot bulb holes could be used to uniquely identify each of the twelve permutations, existing vision algorithms were experimented with to determine which algorithm would be most suitable. The two most promising algorithms seem to be feature extraction and template matching.

5.1.1 Feature extraction.

Using existing feature extraction techniques, the main and pilot holes could be isolated and various features, for instance, location, area, perimeter and so on, could be extracted to correctly classify the reflectors. However, a recognition time of about 7 seconds is required to isolate the two bulb holes, extract the features and classify the reflector. Although the time required for recognition could be improved, an alternative method using template matching was investigated concurrently and found to be more satisfactory.

5.1.2 Template matching.

Although template matching has been heavily criticised for the necessity of accurately positioning and orientating the object for recognition, in this application this is not a real disadvantage as the position and orientation of the reflectors are fixed. Furthermore, the reflector clamp in the vision station has been designed to imitate the clamp used by the robot during tulip insertions. If the reflector is 'faulty' due to burrs or other constructional faults, the vision system will reject the reflector as the same fault will also affect the tulip insertions.

Initial trials of cross-correlating reflectors with stored templates took less than two seconds to recognise one out of six reflectors. This benchmark was subsequently improved to give a 'best' time of less than half a second and a 'worst' time of four seconds using various time saving methods (see Section 7.).

Besides the simplicity, speed and accuracy of template matching, extensive 'factory' trials have shown that cross correlation is extremely robust to noise and positional deviations of the reflector during clamping. Signal noise includes variations in lighting, video noise, vibrations and noise due to the accepted constructional variances of reflectors.

5.2 Setting Up Templates.

The software allocates one of the four image planes as the template plane. Using rectangular windowing commands, the operator will define a feature which is unique to the reflector and store this 'typical' feature in the template plane. As there are twelve reflector permutations in this application, twelve templates are defined. Selection of these features are based on heuristics and the typical window area is about 400 pixels. The window dimensions vary according to the feature selected. For simplicity and flexibility, the setting up procedure is menu driven and the template plane is easily adapted to include new reflectors or modified to recognise new versions of the same reflector.

5.3 Cross Correlation.

The Normalised Cross Correlation coefficient is used for decision making, and is calculated from:-

$$C_{gt}(m,n) = \frac{\sum_i \sum_j g(i,j)t(i-m,j-n)}{[\sum_i \sum_j g(i,j)^2]^{1/2}} \qquad \ldots\ldots(5.1)$$

where $g(i,j)$ is the digital image,
$t(i,j)$ is the template and
$(i-m,j-n)$ is within the domain of definition
of the template.

The decision criteria is based on the Correlation Percentage Error or acceptance level which is calculated from:-

$$\%Error = (C_{act} - C_{cal}) / 100 \qquad \ldots\ldots(5.2)$$

where C_{act} is the normalised cross correlation coefficient of the template correlated with itself; and
C_{cal} is the normalised cross correlation coefficient of the template correlated with the sampled image.

6. OPERATION

During assembly, photosensors will detect the presence of a reflector entering the vision station. This will activate the clamp and the sampling of a TV frame. The Vision System then sequentially compares each of the windowed templates at the same positional window in the sampled frame.

If the first template and the windowed feature of the sampled image is exactly the same, the correlation error should be zero. However, due to ambient noise and other factors, the correlation error is Gaussian and distributes around this mean. Time consuming statistical teaching by repetitive sampling to determine the maximum error and confidence levels have been considered. Fortunately it was found that this is not required. The decision criteria has proved to be extremely robust by simply setting the maximum error or acceptance level to an arbitrary level, for instance, 10%.

If the correlation error of the first comparison is above 10%, the comparison fails and the next stored template is compared with the corresponding window of the sampled image. The templates are sequentially compared until either the error falls below the stipulated level (match) or until the templates are exhausted (reject).

When a 'match' occurs, the Vision System sends the template number to the robot controller which will execute the appropriate sequence of actions. If none of the templates match, another TV frame is sampled and the comparisons repeated. If this fails again, a third TV frame is sampled and if yet another 'reject' occurs, the reflector is labelled 'faulty' and will not be assembled. Multiple frame sampling is theoretically not neccessary as a single frame is found to be extremely reliable, but due to the surplus time available it is incorporated as a 'failsafe' sequence.

Using multiple samples, the worst case of 'faulty' would require three frame samples and twelve correlation for each sample, or thirty six comparisons in total. The 'best' time would then be the time it takes to match the first template.

Trials have shown that the percentage error is typically less than 5% for a 'match' and more than 20% for a 'reject' when wrongfully correlated with the other eleven templates. As such, the static discrimination criteria has proven to be quite effective.

7. TIME SAVING METHODS

To reduce the recognition time, the templates are correlated at half resolution using only alternate pixels and rows, thus reducing the recognition time by a quarter. Although a decrease in resolution increases the percentage error, trials have proved that the marginal increase is not significant and is still within the 10% decision criteria. If for any reason mismatching occurs, the template window could be increased to correlate more pixels for a more accurate decision.

Further improvement in time is possible by storing the template sequence in an ordered list. The latest template that 'matches' is stored at the top of this list and is the first to be compared when the next reflector enters the vision station. As reflectors are normally assembled in production batches and are more likely to be in one of two orientations, with the ordered list the most probable template will propagate to the top of the list. Henceforth, the recognition time is typically 0.3 seconds at half resolution for the 'best' case; 3.6 seconds for a 'reject', and about 11 seconds for 'faulty'.

However, the most effective time saving method is in the extensive use of rectangular windows. It is easy to assume that identifying window templates are taken of the complete feature, whether it is a main or pilot bulb hole. Unfortunately, the main bulb hole has a typical diameter of 55mm and the pilot hole has a diameter of about 22mm. This is roughly equivalent to a complete template area of about 2600 pixels and 600 pixels respectively. Even at half resolution, 2600 pixels per main bulb hole template requires an expensive amount of processing time and memory. However, a complete feature is not necessary. It is possible to visualise sub-features within this complete feature. For instance, a unique indentation of an edge within the main bulb hole and so on. Windowing upon these unique sub-features will identify the reflectors with the same reliability as before.

8. CONCLUSION

The Tulip to Reflector Assembly Cell has been operational since June 1985 in Cannock, West Midlands. The Cell is seen as the first stage towards totally automating the assembly of headlamps for Lucas. Research into vision as an effective sensor for robotic assembly is presently a continuing programme within Lucas Research Centre and Imperial College.

REFERENCES

1. Duda and Hart, 1973, 'Pattern Classification and Scene Analysis'.

2. Batchelor, B.G., 1978, 'Pattern Recognition'.

ACKNOWLEDGEMENTS

We would like to thank Dr. C Besant and Dr. R Hibbard from Imperial College for their assistance, D Cawthray and J Tilley from Lucas Research Centre for their support, and especially J Lawrence for his expertise.

Index

Emotion Dysregulation and Outbursts in Children and Adolescents: Part I

Editors

GABRIELLE A. CARLSON
MANPREET K. SINGH

CHILD AND ADOLESCENT PSYCHIATRIC CLINICS OF NORTH AMERICA

www.childpsych.theclinics.com

Consulting Editor
TODD E. PETERS

April 2021 • Volume 30 • Number 2

ELSEVIER

1600 John F. Kennedy Boulevard • Suite 1800 • Philadelphia, Pennsylvania, 19103-2899

http://www.theclinics.com

CHILD AND ADOLESCENT PSYCHIATRIC CLINICS OF NORTH AMERICA Volume 30, Number 2
April 2021 ISSN 1056–4993, ISBN-13: 978-0-323-76251-9

Editor: Lauren Boyle
Developmental Editor: Arlene Campos

Child and Adolescent Psychiatric Clinics of North America (ISSN 1056-4993) is published quarterly by Elsevier Inc., 360 Park Avenue South, New York, NY 10010-1710. Months of issue are January, April, July, and October. Business and Editorial Offices: 1600 John F. Kennedy Boulevard, Suite 1800, Philadelphia, PA 19103-2899. Periodicals postage paid at New York, NY and additional mailing offices. Subscription prices are $348.00 per year (US individuals), $844.00 per year (US institutions), $100.00 per year (US & Canadian students), $388.00 per year (Canadian individuals), $899.00 per year (Canadian institutions), $446.00 per year (international individuals), $899.00 per year (international institutions), and $200.00 per year (international students). International air speed delivery is included in all *Clinics* subscription prices. All prices are subject to change without notice. **POSTMASTER:** Send address changes to *Child and Adolescent Psychiatric Clinics of North America,* Elsevier Health Sciences Division, Subscription Customer Service, 3251 Riverport Lane, Maryland Heights, MO 63043. **Customer Service: 1-800-654-2452 (U.S. and Canada); 314-447-8871 (outside U.S. and Canada). Fax: 314-447-8029. E-mail:** JournalsCustomer Service-usa@elsevier.com **(for print support) or** journalsonlinesupport-usa@elsevier.com **(for online support).**

Reprints. For copies of 100 or more of articles in this publication, please contact the Commercial Reprints Department, Elsevier Inc., 360 Park Avenue South, New York, New York 10010-1710 Tel.: 212-633-3874; Fax: 212-633-3820, E-mail: reprints@elsevier.com.

Child and Adolescent Psychiatric Clinics of North America is covered in *MEDLINE/PubMed (Index Medicus), ISI, SSCI, Research Alert, Social Search, Current Contents,* and *EMBASE/Excerpta Medica.*

Contributors

CONSULTING EDITOR

TODD E. PETERS, MD, FAPA
Vice President/Chief Medical Officer (CMO), Chief Medical Information Officer (CMIO), Sheppard Pratt Health System, Consulting Editor, *Child and Adolescent Psychiatric Clinics of North America*, Baltimore, Maryland

EDITORS

GABRIELLE A. CARLSON, MD
Professor of Psychiatry and Pediatrics, Renaissance School of Medicine, Stony Brook University, Putnam Hall-South Campus, Stony Brook, New York

MANPREET K. SINGH, MD, MS
Associate Professor of Psychiatry and Behavioral Sciences, Director, Pediatric Mood Disorders Program, Stanford University School of Medicine, Stanford, California

AUTHORS

ROBERT R. ALTHOFF, MD, PhD
Associate Professor, Department of Psychiatry, University of Vermont, Vermont Center for Children, Youth and Families, Burlington, Vermont

MERELISE AMETTI, MA, MPH
Graduate Student, Department of Psychiatry, University of Vermont, Vermont Center for Children, Youth and Families, Burlington, Vermont

MARIANNA ASHUROVA, MD
Consult Liaison Child and Adolescent Psychiatrist, Cohen Children's Medical Center, Queens, New York

TAMI D. BENTON, MD
Psychiatrist-in-Chief, Department of Child and Adolescent Psychiatry and Behavioral Sciences, Children's Hospital of Philadelphia, Associate Professor of Psychiatry, University of Pennsylvania Perelman School of Medicine, Philadelphia, Pennsylvania

JOSEPH C. BLADER, PhD
Meadows Foundation and Semp Russ Professor of Child Psychiatry Research, Department of Psychiatry and Behavioral Sciences, Joe R. and Teresa Lozano Long School of Medicine, The University of Texas Health Science Center at San Antonio, San Antonio, Texas

BEVERLY J. BRYANT, MD
Associate Professor, Department of Psychiatry, The University of Texas Health Science Center at Tyler

CATHY BUDMAN, MD
Clinical Professor of Psychiatry, Donald and Barbara Zucker School of Medicine at Hofstra/Northwell, Hempstead, New York; Director, Long Island Center for Tourette, Manhasset, New York

GABRIELLE A. CARLSON, MD
Professor of Psychiatry and Pediatrics, Renaissance School of Medicine, Stony Brook University, Putnam Hall-South Campus, Stony Brook, New York

BARBARA J. COFFEY, MD, MS
Division Chief, Child and Adolescent Psychiatry, Professor, Department of Psychiatry and Behavioral Sciences, University of Miami Miller School of Medicine, Miami, Florida; Incoming Chair, Department of Psychiatry; Director, Tourette Association Center of Excellence

DANIEL F. CONNOR, MD
Emeritus Professor, Department of Psychiatry, Division of Child and Adolescent Psychiatry, University of Connecticut School of Medicine, Farmington, Connecticut

JUDITH A. CROWELL, MD
Professor and Division Director, Child and Adolescent Psychiatry, Stony Brook Outpatient, Department of Psychiatry, Stony Brook Hospital, Stony Brook, New York

LEONARD A. DOERFLER, PhD
Professor and Director, Clinical Counseling Psychology Program, Assumption College, Adjunct Professor, Department of Psychiatry, University of Massachusetts School of Medicine, Worcester, Massachusetts

SUSAN J. FRIEDLAND, MD
Assistant Professor, Ann and Robert H. Lurie Children's Hospital of Chicago, Northwestern University Feinberg School of Medicine, Town and Country Pediatrics, Chicago, Illinois

ELIZABETH R. GARGARO, MD
Instructor, Departments of Psychiatry and Pediatrics, The University of Utah, Salt Lake City, Utah

AMANDA GORECKI, DO
Clinical Assistant Professor, Stony Brook Outpatient, Department of Psychiatry, Stony Brook Hospital, Stony Brook, New York

JASON D. JONES, PhD
Research Assistant Professor of Psychiatry, University of Pennsylvania, Perelman School of Medicine, Research Scientist, Department of Child and Adolescent Psychiatry and Behavioral Sciences, Children's Hospital of Philadelphia, Roberts Center for Pediatric Research, Philadelphia, Pennsylvania

BROOKS R. KEESHIN, MD
Associate Professor, Departments of Pediatrics and Psychiatry, The University of Utah, Salt Lake City, Utah

JENNIFER KELUSKAR, PhD
Clinical Assistant Professor, Stony Brook Outpatient, Department of Psychiatry, Stony Brook Hospital, Stony Brook, New York

JASON LEWIS, PhD
Department of Child and Adolescent Psychiatry and Behavioral Sciences, Children's Hospital of Philadelphia, Assistant Professor of Clinical Psychiatry, University of Pennsylvania Perelman School of Medicine, Philadelphia, Pennsylvania

CARLA MAZEFSKY, PhD
Associate Professor, Department of Psychiatry, University of Pittsburgh, University of Pittsburgh School of Medicine, Pittsburgh, Pennsylvania

JON McCLELLAN, MD
Professor, Department of Psychiatry and Behavioral Sciences, University of Washington, Lakewood, Washington

ELI MUHRER, MD
Child and Adolescent Psychiatry Fellow, Department of Child and Adolescent Psychiatry and Behavioral Sciences, Children's Hospital of Philadelphia, Philadelphia, Pennsylvania

JERRY PB, EdM
Senior Administrator, New York City High School, Founder and Director, New York City Parent Support Network, Trustee, National Mental Health Nonprofit, Private Practice, Career Coach for Adults w/Mental Health Concerns, DSMB, A Longitudinal Study, University of Pittsburgh Medical Center

TARA S. PERIS, PhD
Associate Professor in Residence, Division of Child and Adolescent Psychiatry, University of California, Los Angeles, Los Angeles, California

DEBRA REICHER, PhD
Clinical Assistant Professor, Stony Brook Outpatient, Department of Psychiatry, Stony Brook Hospital, Stony Brook, New York

ADITI SHARMA, MD
Assistant Professor, Department of Psychiatry and Behavioral Sciences, University of Washington, Seattle, Washington

LAUREN SPRING, MD
Program Director, Psychiatry Residency Training, Clinical Assistant Professor, Division of Child and Adolescent Psychiatry, Department of Psychiatry and Behavioral Health, Stony Brook, New York

JEFFREY R. STRAWN, MD
Associate Professor of Psychiatry, Pediatrics and Clinical Pharmacology, Associate Vice Chair of Research, Department of Psychiatry and Behavioral Neuroscience, University of Cincinnati College of Medicine, Cincinnati, Ohio

ARGYRIS STRINGARIS, MD, PhD, FRCPsych
Chief, Section of Clinical and Computational Psychiatry, National Institute of Mental Health, National Institutes of Health, Bethesda, Maryland

PABLO VIDAL-RIBAS, PhD
Postdoctoral Researcher, Social and Behavioral Science Branch, Eunice Kennedy Shriver National Institute of Child Health and Human Development, National Institutes of Health, Bethesda, Maryland

JOHN T. WALKUP, MD
Margaret C. Osterman Professor of Psychiatry, Chair, Pritzker Department of Psychiatry and Behavioral Health, Ann and Robert H. Lurie Children's Hospital of Chicago, Northwestern University Feinberg School of Medicine, Chicago, Illinois

DEBORAH M. WEISBROT, MD, DFAACAP
Clinical Professor of Psychiatry, Distinguished Fellow, American Academy of Child and Adolescent Psychiatry, Renaissance School of Medicine, Stony Brook University, Stony Brook, New York

Contents

This article examines two themes in my 25-year journey raising a son with severe mood dysregulation, attention deficit with hyperactivity, and learning disabilities (MAL). Raising children with MAL significantly alters parents' own development, having to manage their children's chronic rages and aggression from toddlerhood through young adulthood. I propose a framework of periods parents go through, and describe a years-long progression of internal and behavioral adaptations necessary to parent these children effectively. The article suggests that more study is needed to understand how parents fare and what happens to them over time, and considers support that would improve their trajectory.

Explosive outbursts in children and adolescents have been long identified by clinicians and have been described using many different conceptualizations and terms. The topography of explosive outbursts is complex, heterogeneous, and includes the interactions of different emotional and behavioral constructs. Included here are pre-existing central nervous system vulnerabilities including psychiatric and neurologic diagnoses, various contributing emotions that generally carry a negative valence, and aggressive behaviors that are usually overt and reactive. Emotional impulsivity and deficient emotional self-regulatory mechanisms may contribute to episode severity and duration.

Outbursts (severe temper loss) in children are a common reason for treatment referral. However, the diagnostic system has not classified them in a way that expands knowledge. Outbursts are nested in the concept of irritability, which consists of a feeling and a behavioral dimension. Both need to be identified but kept separate. This review summarizes the phenomenology of outbursts normatively and clinically. Severe temper loss needs a consistent label, an operationalized way of classification and measurement, and an assessment approach independent of diagnosis until other data are gathered to more accurately determine what condition provides the most accurate diagnostic home.

This article highlights the breadth of measures available for measuring emotion dysregulation, or facets thereof, in children and adolescents, and reviews in detail a subset of these measures. We describe broadband measures and measures that are specific to emotion dysregulation, including observational tools, clinical interviews, and rating scales. Furthermore, we discuss the strengths, weaknesses, and psychometric properties of each approach and specific contexts or populations in which certain methods may be particularly useful. Finally, recommendations for thorough assessment of emotion dysregulation in future studies are provided.

Dysregulation by Disorder/condition

This article provides a comprehensive review of emotion dysregulation (ED) in individuals with autism spectrum disorder (ASD). The authors describe ED from a developmental perspective, and highlight how aberrations in social development and a restricted-repetitive repertoire of behaviors, render individuals with ASD more vulnerable to ED. The article also summarizes how ED in children and adolescents with ASD has been measured and conceptualized in research and clinic settings. Evidence-based pharmacologic and cognitive behavioral interventions targeting ED in ASD are summarized, with a focus on how such approaches are tailored to the developmental needs of individuals with ASD.

Individuals with attention-deficit/hyperactivity disorder (ADHD) frequently experience strong reactions to emotionally evocative situations. Difficulties modulating anger and other upsets have clinically significant behavioral consequences. Those with ADHD may have anomalies in emotion generation, emotion expression, or both that predispose to these problems. The association between ADHD and emotion dysregulation raises Important clinical and research issues, including possible heterogeneity in the mechanisms by which they are related. Although first-line treatments for ADHD often help to resolve emotional dysregulation symptoms as well, the evidence base for widespread practice of combination pharmacotherapy remains sparse. Psychosocial treatments that engage processes underlying emotional dysregulation are in development.

Tourette disorder is a complex neuropsychiatric syndrome of childhood onset characterized by multiple motor and phonic tics and is associated with high rates of psychiatric comorbidity. Symptoms of impulsive aggression (explosive outbursts or "rage") are commonly encountered in the clinical setting, cause significant morbidity, and pose diagnostic and treatment challenges. These symptoms may be multifactorial in etiology and result from a complex interplay of illness severity and psychosocial factors. Treatment strategies require careful differential diagnostic evaluation and include both behavioral and pharmacologic interventions.

Traumatic experiences, subsequent traumatic stress, and other trauma reactions are common among youth who experience emotional dysregulation. This article highlights key considerations for the delivery of care to emotionally dysregulated youth with histories of trauma. An initial, trauma informed assessment is critical to identify those youth with emotional dysregulation best served by evidence-based, trauma-focused treatments trauma-informed approaches to severely emotionally dysregulated youth, including youth in in-patient and residential settings, can improve emotional and behavioral outbursts while maintaining the safety the milieu. Finally, incorporating awareness of trauma is considered when prescribing psychopharmacologic interventions in severely emotionally dysregulated youth.

Suicide rates continue to rise among children and adolescents; suicide is the second leading cause of death in the United States. Although research studies have identified factors associated with suicide risks for youths, none distinguishes those who have suicidal ideation from those who most likely will make an attempt or die by suicide. Most studies focus on psychiatric diagnoses associated with suicide risks. Recent studies suggest that cross-cutting symptom profiles may be a stronger predictor of risks for suicide than diagnosis. This article provides an overview of emotional dysregulation as it relates to suicidal ideation, intent, and behaviors for youth.

Based on its course over time, irritability is linked to depression cross-sectionally and longitudinally. Cross-sectionally, irritability takes an episodic form as a symptom in pediatric depression; yet, irritability in the absence of depressed mood or anhedonia is rare. Longitudinally, chronic irritability has been shown to predict depression rather than bipolar disorder or externalizing disorders. Evidence suggests that the link between

CHILD AND ADOLESCENT PSYCHIATRIC CLINICS

SERIES OF RELATED INTEREST

Psychiatric Clinics of North America
https://www.psych.theclinics.com/
Pediatric Clinics of North America
https://www.pediatric.theclinics.com/
Neurologic Clinics
https://www.neurologic.theclinics.com/

THE CLINICS ARE AVAILABLE ONLINE!
Access your subscription at:
www.theclinics.com

Preface

Emotion Dysregulation and Outbursts in Children and Adolescents: Part I

Gabrielle A. Carlson, MD   Manpreet K. Singh, MD, MS
Editors

We would like to introduce you to some children that are like ones you have seen or likely will see at some point in time. More detail about these cases are in the appendix of this issue.

Ten-year-old Tyreke was referred for banging his head on the desk and walls, stabbing himself with a pencil, and saying when he gets agitated that he wants to die. He was upset by noise other children make, the things they say, and any kind of deviation from his routine, and these major outbursts occur at least once a week, sometimes more often.

Nikki had been anxious for a whole year before she presented for psychiatric consultation at age 9 for frequent irritability, particularly in anxiety-provoking situations. In some of these situations, Nikki would become so "enraged" she would at times be "violent" according to her mother.

A foster mother brought 8-year-old Mark to clinic because he grabbed a knife and threatened another foster child in the home. Mark has a history of "abuse," but no one has details because he is in his fifth foster home with his third Child Protective Services worker.

Also age 8, Kyle has motoric hyperactivity, impulsivity, and inattention, but also poor frustration tolerance, anxiety, sensory sensitivities, and tics. Though he frequently rolls his eyes, has facial twitching, snorting, throat clearing, head turning, and repeated touching, it is his increased episodes of explosive anger from his poor frustration tolerance at home with his mom, that has drawn clinical attention.

Seven-year-old Jared's second-grade teacher and parents need help with his terrible "meltdowns." He landed in the emergency room the day he refused to do math, threw his school work onto the floor, and was sent to the principal's office, where

Child Adolesc Psychiatric Clin N Am 30 (2021) xiii–xvi
https://doi.org/10.1016/j.chc.2020.10.014
1056-4993/21/© 2020 Published by Elsevier Inc.

he tried to flip a table and punched and kicked some adults. Jared's frequent outbursts last for several hours.

Lest you think these problems are confined to young children, Jenny is a 17-year-old adolescent rehospitalized with a history of suicidal behavior, self-harm, aggression, and reports of hallucinations. She has had difficulties with poor impulse control, oppositional behaviors, and anger outbursts since elementary school. While hospitalized, she has engaged in frequent aggressive, assaultive, and self-harming behaviors, at times requiring mechanical restraints multiple times a day.

Corrections officers at the state adolescent prison sought evaluation for 16-year-old Stephon, where he had been incarcerated for 8 weeks. Cell block officers were concerned about Stephon's unpredictable explosive temper, aggression, and dangerousness and noted that he was placed in "lockdown" for "beating down" his cell mate.

Maddy, who is 15, initially seemed to have symptoms of attention-deficit/hyperactivity disorder and depression. However, after a hospitalization for suicidal behavior, she began to do uncharacteristic things like stay up all night trying to make furniture so she could become a star of a TV show on home improvements. She became irritable when her mother asked her to do anything and started losing friends because she was so nasty to them.

Cody is yet another teen who became physically and verbally aggressive toward his mom to the point that she had to call the police to calm him down. Frequently disrespectful and sometimes violent with others, he made no eye contact with others, refused to go out, stayed in bed all day long, and hadn't showered for over a month. He looked very sad. His mood switched quickly, and he became easily upset about random things.

What these children and adolescents have in common is a problem regulating their emotions—most prominently, their anger, though not always. They become too angry, too fast, too often, and for too long. They have a range of psychopathologies, including anxiety, autism, Tourette syndrome, posttraumatic stress disorder, disruptive mood dysregulation disorder, depression, and mania. Their psychiatric diagnosis does not really explain their outbursts. The appendix provides more information about them.

Not only does the outburst behavior cut across a variety of disorders but it also doesn't have a consistent name, an accepted way to measure it, a way to code for it, or a specific treatment.

There are 2 issues of the *Child and Adolescent Psychiatric Clinics of North America* dedicated to the challenges that outburst behavior in children presents. The first issue deals with outburst behaviors and how to treat them in the context of specific psychopathologies. We wanted to put a face on the complex issues with which these youngsters and their environment present. Thus, we begin Part I with an article by a father who is a special education teacher who has helped raise an explosive child who has now grown up to become a young adult with outbursts. His narrative provides an almost moment-to-moment experience of how children with severe outbursts complicate family, educational, and community life.

The next 3 articles in Part I deal with the vicissitudes of naming and describing the phenomenon in question. Are outbursts conceptualized as simply aggression? Irritability? Dysregulated emotion? The behavior is not new, but the many somewhat overlapping ways it has been labeled are defined by Connor and Doerfler. Spring and Carlson then review the literature on the phenomenology of outbursts as well as presenting data on the aspects of outbursts that appear to be most relevant (e.g. the severity of the behaviors engaged in, their frequency, and duration). Finally, Althoff and Ametti summarize some of the literature on measures that touch on outburst behavior.

The rest of Part I is devoted to many of the conditions in which one frequently sees outburst behavior. The first part may be said to be developmental and behavioral conditions. Keluskar and colleagues describe dysregulation in autism from a developmental perspective. Next, attention-deficit/hyperactivity disorder is probably the disorder with which outburst behavior is most often associated (though usually comorbid with oppositional defiant disorder). Blader addresses mood dysregulation in that context. When tic disorder is added to hyperactivity, impulsivity, and compulsive behaviors, as addressed by Ashurova, Budman, and Coffey, children have a double handicap. That is, their tics appear to get worse when their behavior gets worse (or vice versa).

The second part of the phenomenology/diagnosis section contains conditions that are more clearly in the mood and behavior domain. Benton and colleagues present the dual complexities of outbursts and suicidal behavior. Children who threaten to kill themselves when enraged are a significant health concern. Keeshin, Bryant, and Gargaro write about issues related to trauma and aggression as well as provide a cogent summary of trauma-informed therapeutic interventions.

Other mood disorders in which severe outbursts can occur are depression and anxiety. Vidal-Ribas and Stringaris discuss irritability and its relation to depression and why they are related. The question about whether irritable/explosive behavior is a prodrome of bipolar disorder is convincingly dispelled by the data they present. Counterintuitively, perhaps, anxiety and dysregulated behavior have a meaningful association. Anxious children who have outbursts act as if they are fighting for their lives (hence the term catastrophic reactions) as opposed to merely being angry about being thwarted. Walkup, Friedland, Peris, and Strawn explain this association and provide a framework for treatment. Finally, Sharma and McClellan take on the complexities of dysregulation in mania, schizophrenia, and borderline personality disorder. While outburst behaviors help define borderline personality disorder, they are neither necessary nor sufficient for bipolar disorder and schizophrenia. The behavior in those contexts, however, presents significant diagnostic and treatment challenges.

The last article in Part I is dedicated to children that Weisbrot and Carlson call "diagnostically homeless." The child/young adult described by Jerry P-B is one of those children. They have a little bit of a lot of things, including severe outbursts. Some symptoms sound autism spectrum–like, but children fail to meet criteria for autism. We can say they have disruptive mood dysregulation disorder, or intermittent explosive disorder, but those conditions don't tell us much about the child. Children who fall between our diagnostic cracks are especially frustrating to parents, teachers, and clinicians, who can't read up on a condition that doesn't have a name. Thus, this issue provides key considerations while in pursuit of finding a common label and understanding of explosive outbursts transdiagnostically.

The second volume begins with developmental considerations in the treatment of emotion dysregulation. Through a developmental lens, the needs of a child can be carefully considered to formulate a treatment plan. Articles in the second volume then critically evaluate the benefits of existing psychosocial and pharmacological interventions. There are recommendations for how these interventions can evolve with an improved understanding of the youth with emotion dysregulation, the etiological factors that contribute to perpetuating the dysregulation, and the various settings in which their dysregulation occurs. Besides striving to better approximate which treatments fit well for which youth, the second volume also explores the timing of the interventions along primary, secondary, and tertiary prevention strategies.

Part II addresses developmental considerations and treatment of emotion dysregulation more broadly.

Gabrielle A. Carlson, MD
Renaissance School of Medicine at
Stony Brook University
Putnam Hall-South Campus
Stony Brook, NY 11794-8790, USA

Manpreet K. Singh, MD, MS
Stanford University School of Medicine
Stanford, CA 94305-5719, USA

E-mail addresses:
Gabrielle.Carlson@StonyBrook.edu (G.A. Carlson)
mksingh@stanford.edu (M.K. Singh)

Walking on Eggshells

A Parent's Journey Raising an Emotionally Dysregulated Son

Jerry PB, EdM[a,b,c,d,e],*

KEYWORDS

- Raising child with severe mood • Parenting troubled teen
- Child adolescent mood disorders • Severe mood dysregulation
- Child and adolescent emotion dysregulation • Adolescent bipolar disorder
- Explosive child/adolescent • Parent's journey • Diagnostically homeless

KEY POINTS

- Parents experience severe, chronic stress and isolation from the mainstream managing children's unpredictable aggression. Factors may perpetuate early reactive behaviors, putting healthy adult development at risk.
- Preserving parents' options for their children's education and later employment becomes severely limited for children with MAL by constant disruption and discontinuity.
- Parenting these children requires special skills training, therapies, and psychoeducation to serve the interests of their children, plus safeguarding their own interests.
- A framework shows how parents progress psychologically and behaviorally as they adapt to raising children with MAL, who do not fit neatly within a diagnosis and treatment.
- Even parents who master being strategic, intentional, and skillful (what the author calls professionalized parenting) have to walk on eggshells; it is how they walk that matters.

INTRODUCTION
Why This Article?

Being neither an MD nor a PhD, my qualifications are as an honorary principal investigator, of my own son. Call me a parent investigator, I am invited here to present my data as one who survived 25 years raising my severely mood dysregulated child. Charts and figures are hand drawn, urgently made in wee hours. Data are taken directly from my field notes, school reports, and clinical records over 25 years.

[a] A New York City High School; [b] New York City Parent Support Network; [c] A Preeminent National Mental Health Nonprofit; [d] Private Practice; [e] A Longitudinal Study, University of Pittsburgh Medical Center
* Corresponding author. A New York City High School.
E-mail address: jerrypb.pub@gmail.com

Child Adolesc Psychiatric Clin N Am 30 (2021) 283–298
https://doi.org/10.1016/j.chc.2020.10.012
1056-4993/21/© 2021 Elsevier Inc. All rights reserved.

childpsych.theclinics.com

Interspersed, 15 boxed vignettes place you at the scene of events with my son. I use first person singular and plural pronouns interchangeably to include my spouse.

VIGNETTE 1:

Age 15 years. School calls. "After suicidal comments, he climbed out a second-story window, refused help and jumped. An ambulance took him to the ER [emergency room]."

Severe Mood and Impulsivity: Will They Kill Our Son?

Our young adult son made his first suicidal threat at age 5 years. Out for dinner with his grandparents and me, he wanted his dessert first. Grandpa's blunt "First eat your chicken" resulted in our son's swift, full-throttle tantrum. My parents had heard about, but had never seen, this. Our little boy grabbed his butter knife, held it to his own neck, and shouted, "I'm gonna kill myself! I'll slit my throat!" The restaurant fell silent. I grabbed the knife back. Where the hell had that come from? He had never heard it from any source I knew. Disturbing as it was, he eventually calmed down, contrite, weepy, and sweet. We moved on.

But our son never really has.

His suicidal ideation, with and without a plan, and at least 1 attempt, has been a chronic part of our lives. At 18 years old, following social media, he tattooed his arm with a semicolon, a symbol for suicidal thinking among youth that this thought does not end with a period; it goes on. Child psychiatrists say that our son's triple-threat of severe mood dysregulation, attention-deficit/hyperactivity disorder (ADHD), and learning disabilities, which I refer to as MAL, makes treatment especially difficult. That my son's issues have defied diagnostic consensus, and therefore delivery of effective treatment, has made my parenting journey as terrible and as painful as any symptoms his clinicians have tried unsuccessfully to treat. I am still searching unstoppably to find what none of us has yet uncovered, for the sake of my health, and on behalf of my son—my intermittently loving and beautiful boy. I want, like you, to discover the key that unlocks effective treatment of this ugly triumvirate of conditions.

VIGNETTE 2:

Age 23 years. Son calls. "I'm kind of in trouble. Can you get me a lawyer?"

The medical community never successfully treated our son. Not with polypharmacy in combination with talk therapies, not in therapeutic settings with bells and whistles. Any progress that looked like it had legs was cut off at the knees. A drug tranquilized intensity of moods, but not mixed states. Therapeutic campuses helped some, but not for long. MAL and its treatments brutalized his endocrine and autonomic systems, object permanence and identity formation, academic progress, family attachment and social networks, ambition, and self-esteem. My spouse and I are a quarter century into parenting and still in space in between worlds of his wellness and illness. With failed treatment plans one after then next, the medical community recognizes us and says "We tried." So, we hobble forward indistinctly with our only child, adopted by us on the day he was born, in between that joyful, precious day and what has happened since.

A Difficult Boy Becomes a Difficult Adult

Difficult boy to difficult adult puts him, us, and unsuspecting communities at significant risk. Because he is an adult, laws prevent me from fully caring for him when he struggles with MAL symptoms. Disgruntled, he is off meds again, has fired his therapist, quit work in his field, trashed his apartment, and was fired from his fifth part-time job in eight months. Courting police in his small town, I fear they will become his next parent figures. Parents like me have little choice but to walk on eggshells (WOE) raising their mood-labile children. How we learn to be strategic and intentional, even while walking on eggshells, is the subject of this article.

VIGNETTE 3:

Age 10 years. Grocery. He paces. "These people are weird," (neutral shoppers).
Covers ears (store is quiet). Ties hoodie over eyes. "Get me out of here!" Smacks items from my hands. Outside: "I'm going to push you into a car, Daddy." Taxi, wailing, "Can I smash the TV?" Punches window repeatedly. "You took my pets away, Daddy, you took my LIFE away! I hate you!"

Will our son survive the intensity of emotional dysregulation? His erratic behaviors have been startlingly consistent since age 2 years. I fear less his suicidal intentionality than I do his impulsivity and poor judgment when he escalates. How many youths are suspected of killing themselves accidently, desperate for relief from emotional pain? I knew his birthmother's hospital record indicated prenatal exposure to substances so, I have constructed epigenetic ideals since then to ensure that he stays an outlier to his own statistical likelihoods, an exception to the rule by research data. I have lived for those small percentages that escape depressing trajectories, want him in those, managing his mental health, problem solving, and getting on relatively well. I may not get what I want; frankly, at this writing, I am not.

VIGNETTE 4:

Age 9 years. Airport gate. Lands toy plane on podium, falls through hole. He
freezes, heaves to floor swearing, stomping, kicking over travelers' bags. TSA (Transit Security Administration): "If you can't control your child, you won't be allowed on-board." Angry traveler, bag kicked over: "Get control of your kid!" We are a public nuisance. Parents everywhere, but no help.

BABIES, LAW AND MEDICINE
Off the Rails

I began walking on eggshells (**Table 1**, misperception) when my sweet-hearted, irritable 8-month-old woke and pulled himself onto his feet using the safety rail from inside his sturdy crib. After 15 minutes of screaming, he pulled himself over the tall rail, toppling onto the spongy lettered floor. He crawled screaming to the closed bedroom door and squeezed his tiny hand through the space underneath. Attentive on the other side, we were not supposed to respond to his screaming, but to see his little hand suddenly shove through the crevice of the under-door space, like some crazy horror movie, was shocking. There was no way I could wait out 30 minutes according to the book of the reigning doctor then of baby sleep routines, Richard Ferber, MD.

"Open the door!" I hissed. My spouse snapped, "Don't!" But I did, and scooped our son off the floor, returning him. "What were you waiting for?" I inveighed later. "Did I miss the chapter, 'When Your Infant Jumps the Rails?'" Spousal conflicts continued and, looking back, it was about more than just parenting style. The hurt on both sides felt increasingly sharp and isolating. The real fight was within, as each struggled to keep deep-seated systems of belief in place (see **Table 1**, denial).

A fluke, my spouse thought, until our baby jumped out again, the infant soldier crawling to the battle front, then the mouse dislocating bones to fit under anything. There came his hand under the door and that was the end of "Ferberizing" our son. We purchased a zippered crib top made of mesh netting to keep him safe, and agreed to a 20-minute wait time before entering. But we did not understand his behavior. Well, we decided, he was in the 99th percentile in height and weight; maybe he was just unusually strong and active? (see **Table 1**, misperception). A nagging feeling suggested that explanation was insufficient.

VIGNETTE 5:

Age 5 years. Breakfast, a book. "Daddy, Darth Vader's my real father. I'm from the Dark Side."

Judge Confers Adoption, and Psychiatry Follows

Jumping 20 years past the crib story, past initial periods of our parental misperception and denial, we eventually entered into periods of acceptance and what I term professionalized parenting (see **Table 1**). Through this development, I professionalized further to become a school leader for adolescents with mental health challenges, and run a support network for parents raising mood dysregulated children. These activities are the rewards of parenting into the periods of acceptance and professionalized parenting, which have kept me healthy. Fellow parents, who initially arrived to parent meetings on their knees, distraught, angry, exhausted, and overwhelmed, also stabilized by being with kindred others, learning through peer networking, and relieved not to be on this grueling parent journey alone.

However, all of this is a byproduct of my job at home, appointed by destiny, and perhaps by the New York City Family Court judge, who granted my spouse and me the right of petition to adopt jointly in the late 1990s. No one understood that our son's skittering up onto the judge's robed lap and kicking his legs as if she were a swing was a harbinger of more activation to come. This dazzling, 14-month-old towhead charismatically chirped as the judge conferred rights of parenthood; Little did we know those right would be used to direct intervention by an expanding army of professionals responding to his hot, unpredictable reactivity.

VIGNETTE 6:

Age 8 years. Table. "I'm like The Hulk, Daddy. When people make me mad, I turn into a monster."

First, we needed language. Early onset? Neuropsychopharmacology? What were they? Before I learned perceptual impairment, intergenerational sequelae of mood dysregulation; before active ignoring, motivational interviewing, or avoiding symptom accommodation, we spoke mainstream, neurotypical English.

Assumed language, assumed worldview. We were just having birthday cake; just serving dinner; just playing softball. Unbeknownst to us (see **Table 1**, denial), our son perceived disjointed unfairness not perceived by anybody else that triggered his anxiety, lit a short fuse, and ended in outrage in every situation. His receptive and expressive delays frustrated him. With high impulsivity, so lacking little-boy wait time, he exploded. Smashing the slice of birthday cake; clearing the dinner plate with his angry arm; stomping off with the ball, midgame, parents glaring at me as they followed their children following our son in puzzled pursuit.

With Nothing to Home In on, He Is "Diagnostically Homeless"

At 5 years old, our son was diagnosed with ADHD, combined type, inattentive and hyperactive, comorbid with language-based learning disabilities. Scores with wide variability among subtest percentiles ranged from very superior to at risk. Executive functions lagged, and processing speed was in the single digits. At seven years old, mood disorder not otherwise specified (NOS) was added. That changed at eight years old to severe mood dysregulation.

VIGNETTE 7:

Age 9 years. Babysitter will not pet rabbit. Gets butcher knives, says he will kill her, kill himself, chasing her. We walk in. He crams himself in rabbit cage. Sitter, crying, exits, "I don't want your money, that's blood money!"

At 9 years old, a third psychiatrist diagnosed bipolar disorder I, severe. At 10 years old, a fourth second opinion assessed that he had been misdiagnosed: "Not card-carrying anything." She said he had "an ugly triumvirate" of disorders difficult to treat: mood disorder NOS with chronic irritability ADHD, combined type; learning disorders (MAL). "Features of this and features of that," but nothing to home in on.

Then came the pièce de resistance from the diagnostic hinterlands, the left field of all findings: "He's diagnostically homeless," she concluded. Ten years old, and homeless. The primary assurance as adoptive parents we promised to provide: a sense of home, family, emotional security. If he had nothing else, he would have himself. Hearing her diagnosis stung, but it is the truest by far. "We're using the same medications we used 50 years ago, tranquilizing our kids, and we've got to do better," she said. Without fitting any diagnosis, and therefore any clear treatment, where the hell were we? (**Fig. 1**)

Lions and Tigers and TD!

People have nightmares about being attacked by lions and tigers. I had nightmares of my son being attacked by tardive dyskinesia, an incurable movement disorder brought on as a side effect of some psychotropic prescriptions. Local parents had approved meds for their son following doctor's orders. His neck, tongue, and trunk were in constant motion. I was consequently fearful, resistant to giving my kid meds (see **Table 1**, denial), which created tension at home. Meds target a symptom, but, with so many symptoms, what was our target? Rage? Excessive silliness? Inattention? Irritability? Anxiety? Which meds? How many, in what combination? In psychotherapeutic treatment, I weighed the issues. My spouse was separately in treatment. We added couples therapy, aware of what happened to other couples torn asunder by chronic difficulties of parenting their mood-labile children. I came to accept that meds could help our son. My spouse and I forgave each other and underscored that we wanted first and foremost to love and support each other, then our child. Sound obvious? Prioritizing the strength of the couple vanishes under strain in our particular parent population.

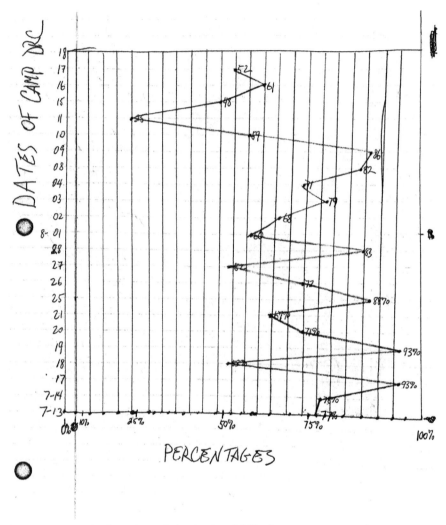

Fig. 1. Parent-made chart shows behavioral variability in summer camp tasks.

We carried heavy burdens of worry taking risks on drug reactions. As our child got worse, we studied classifications and medications suggested by our psychopharmacologist. By age eight years, low doses of several medications were considered better than monotherapy. We provided informed consent for 5 medications to be used at once: a neuroleptic, mood stabilizer, antidepressant, hypertensive and stimulant (**Fig. 2**). Noticeable side effects upset everyone. Mouth and shoulder tics started, clicking noises, and leg tics affected his gait. "Biologically med resistant and med sensitive," our doc said. We couldn't tell what was working, and what was not. This situation went on for years. How far I had traveled from denial and resistance, to what? To diagnostic homelessness, to training in physical restraint, to dispensing benzodiazepine as needed when rages endangered him, or us; this was what good parenting had become. This was love.

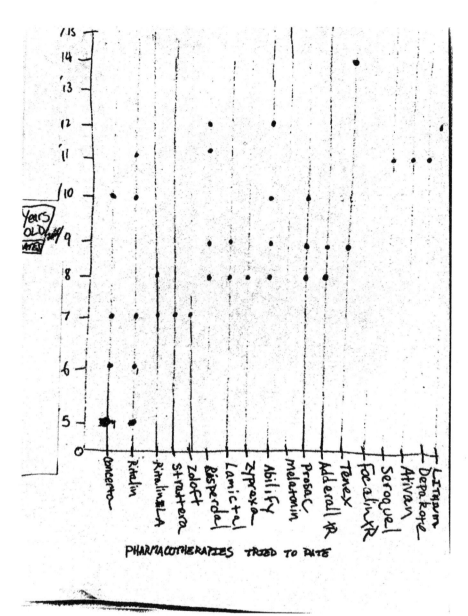

Fig. 2. Parent-made chart shows 18 different psychotropic medications from age 5 to 15 years. Note, ages 8 and 9 years, polypharmacy trials with titration up/down, at times cross-tapering, at times on as many as 5 medications at once.

VIGNETTE 8:

Age 10 years. Guatemala, night, jungle lodge, screen windows. Paces, says he cannot breathe in a little room (spacious). Begs, "Let me out!" Hits, throws. Screams, blood cur-dlingly, "Help! Police! They're trying to kill me! They're hurting me, Somebody, please!" We're terrified authorities might react, parentally vulnerable outside our country.

Life Savings Spent, Debt for Many

We had grown tired, anxious, and brittle. I upgraded psychotherapy to analysis, itself taxing, certainly financially. Among parents raising children with MAL, college funds are often used for high school; costs are staggering. Our son's twice-weekly therapy (and our own); his evaluators, tutors, speech pathologist, and school tuition added to typical enrichment: lessons for swimming, guitar, and gymnastics. Costs skyrocketed once education attorneys, consultants, and a bevy of new evaluators, tutors, and tuition for therapeutic residential schooling began. My spouse sued the Department of Education every year for 5 years to recover some expenses. We fretted for many parents in our network, who lost their jobs to be home with school-refusing children, and others who refinanced home mortgages and cleaned out their 401Ks to pay for specialized services.

Preparation? What Preparation?

In what parenting book, or family conversations, did anyone learn about pediatric and adolescent mental health, and to prepare financially? We were fortunate to have savings and jobs, but what about parents without them? Agencies such as the Department of Education deny appropriate placements and services so they do not have to pay. It is dangerously short-sighted and promises more costly long-range problems for society. Parents largely seek legal recourse annually, or go without proper providers.

VIGNETTE 9:

Seven years old. Principal's letter. "...his disruptive behavior ... severely subverted instruction ... his inability to control his impulses ... will result in dismissal from our school program."

K-12 YEARS
Navigating the Turbulence of School

Our son attended four schools in four states before graduating high school. Between nursery and seventh grade, schools did what they could, until they couldn't. We joined an N-8 private school, relieved not to think about a transition until ninth grade. I got a job teaching there. When our son had tantrums no one could soothe, teachers would bring him to my classroom, or pull me out. Teaching was one job; helping teachers with him, another. Still, I loved dropping him off mornings and heading to my classroom. He loved popping by to connect. By pre-K, administrators persuaded us to have him evaluated and, in kindergarten, to meet with a child and adolescent psychiatrist. He started first grade on methylphenidate and the school asked us to hire an outside speech and language pathologist, and, separately, a learning specialist 3 times a week.

Nobody Wants Me, Everybody Hates Me, Guess I'll Go Home School

It is hard for a child to be disliked by a peer in class. Imagine an entire school rejecting him. Our son tried, but could not control his impulses to avoid suspensions. By second grade, the school required an outside behaviorist, half days, every day. By third grade, they would only allow him to return if we hired a full-time interventionist. I hired an Applied Behavioral Analysis (ABA)-certified teacher because they are trained to take copious notes. I wanted data for myself, his psychiatrist, and the twice-weekly psychologist he saw from age 5 years until he left for residential at 11 years old.

By first semester of third grade, after 2 suspensions for pushing and kicking, he was expelled. We had been there for 7 years. Administrators had tried valiantly. I had become a valued teacher, while taking time to shepherd my son: time away from my students. My family had pushed the system to failure. I felt grateful, but rejected, like him. I had failed, too, to answer the problems caused by our son's intense mood swings (**Box 1**).

Before the ax fell, school parents grew cranky. Complaints were filed. "He sucks up all the air in the classroom!" They pressured administrators. We felt it too. That year in second grade, not one parent invited him to a birthday party. Concerned that no one would come if we invited children for our son's birthday, we gave a lunch with family friends. At school, angry parents sought to isolate us from the community. If our son had behaved meanly, these adults were no better. Privileged mainstream parents treated us as badly as their children complained that our son treated their children. Only 2 families supported us anyway. They are still friends today.

School Two. Going, Going, Gone

Fourth through sixth grades were unmitigated disasters. We switched to a K to 12 private learning disabilities school. No therapeutic services (1 psychologist for the whole population), we focused attention on learning strategies and smaller classroom sizes, meaning we were still in a period of denial (see **Table 1**). Soon came suspensions. The principal called weekly, "Pick him up." By March, "Keep him home until he's stable," she said. "What does 'stable' look like?" "When I don't fear his being near my office windows." "What are you doing to accommodate his needs," I asked? He's your kid, too. "We're not equipped. It's not what we do." "Why not?" I asked, demoralized.

I hired instructors through Craigslist for 3 months, prepared them to match the curriculum at school so our son could return. Me, still in denial, resisting. In September, with ongoing in-school anger, crying, a threat to self and others, he was home, for seven months. I managed a team of six teachers, scheduling daily arrivals between nine and three, Monday to Friday. The master teacher supervised a learning disabilities (LD) instructor, athletics; a yoga teacher and a Reiki master soothed his body, racked with rage. In May, the principal called, "We've made a decision...."

I applied to seven LD schools and a therapeutic day school; all rejected him. There was no improvement in his MAL symptoms during home schooling. Without a school acceptance, and resolved to support social development, we swallowed hard, and my spouse spoke the words: "Residential therapeutic school."

Box 1
Excerpts from fourth neuropsychological evaluation, age 17 years, showing both exceptional ability and significant disability

Snapshot of cognitive functioning

- Verbal Comprehension Index in the superior range.
- Notable contrast: Perceptual Organization Index lower segment of the average range. Discrepancy of this size between the 2 scores is statistically highly significant, seen in only 4.5% of the norming population of the test.
- Significant deficit: Processing Speed Index in single digits, below the average range. Discrepancy between verbal comprehension and processing speed of this size seen in only 4.0% of the norming population.

> **VIGNETTE 10:**
>
> *Five years old. Teacher's note: this was his journal entry. He would not change it and said he was serious.*
>
> *"When I grow up I want to be in the Army because I want to get shot in the face."*

Residential Therapeutic Versus Loss of Primary Parenting

"Over my dead body." No one would parent as carefully, understand him as deeply. I could not bear it, such a tender age. Children left for college, not middle school. We planned for years to adopt a child. "No. No, no; no way. Not on your life." "That's about you, Sweetheart," my spouse braved, softly. "We have nowhere to educate him. Home doesn't work for him or for us." I hated hearing that, hated him, hated myself. I felt tears, choked. "We can't give him up, I won't!" "We're not giving him up, Honey, we're giving up what doesn't work, and we need a school that can handle him." "What would he hear as an adopted child, exquisitely sensitive, who feels his birth mom abandoned him? Please, I'll change teachers. I'll do more. He can't go! Why aren't we enough? We're his parents, for God's sake!" I heard my admission. We were not enough. He needed more hands, hearts, services. I sank, still, blank, stopped, and wept. And wept. Crying a river over my beautiful boy, angry as hell with myself.

I made 3 decisions: ensure his education, follow his needs, and match those needs with responsive services (**Fig. 3**; see **Table 1**, acceptance).

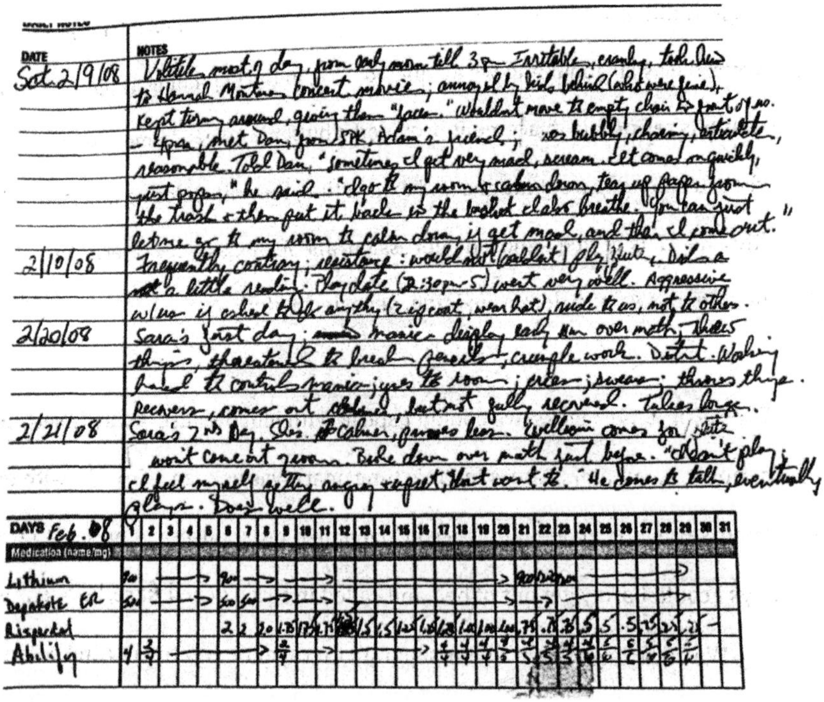

Fig. 3. Parent-made chart correlating medication titration with daily behavioral field notes. For planning professional consults.

Long Journey to Acceptance, and Residential Therapeutic School

After careful research, we chose a therapeutic residence two hours away. A 12-month calendar, weekend visits, and what they called transfer of treatment; several 10-day vacations allowed international family travel to develop interests and talents. Still, year 1, although residential placement provided space to recover from parent fatigue and exhaustion, the pain of losing him from our home life gave us a diagnosis of depression. Feelings of bewilderment, disorientation, and demoralization cycled.

The five-week settling-in period required no communication. In the silence, my guilt and remorse swelled; some days I barely got out of bed, calling in sick. We continued individual therapy and couples treatments. Following months, our long period of denial began to change. Weekly family therapy began by phone. We attended monthly parent trainings on campus, and got used to 4-hour roundtrip travel. I started attending national conferences on child and adolescent mental health.

On campus, in the company of dozens parenting like us, we connected resonantly. Not only was our son healing, we were too. We entered the period of acceptance (see **Table 1**). Our family strengthened as months passed. The courage to shift to this school model, which irrationally I feared would ruin our family, improved it.

I missed parenting at home. Wake-ups, read alouds, task charts, cooking, kisses, cuddles, activities. Six years followed, transforming primary parenting routines to campuses, but not everything transferred. Normal family life, assumed by mainstream culture, was lost forever. He never did return to live at home and we grew expert at working with program leaders on his presenting needs. Year 4, a therapist told us, "Everything your son arrived with is still present, but more contained." A crystalline moment, it remains true today.

VIGNETTE 11:

Age 12 years. Mixed states. Silly, anxious, raging. Hits head, punches wall, slams doors, swears. Cries sadly, talks about old apartment. "Go back and throw them out! Kick their asses! F**kers!"

THE BIG PICTURE
A Theoretical Framework for Clinicians

Table 1 is a sequenced matrix validating how parents raising children with MAL progress over time. Four periods (1) misperception, (2) denial, (3) acceptance, and (4) professionalized parenting present parent development as progressive, dynamic, and each period as supportable and necessary. Denial is an essential experience for parents like me. No fellow parent skipped denial, although some remained stuck longer by refusing needed services. If moving into acceptance is the result of disrupting and resolving inner resistances, then dignifying denial as constructive work provides the full scope of this parenting journey.

Four stages pair with these periods, depicting the long WOE around these volatile children: reactive WOE, based on previously held parenting beliefs that do not match the needs of children with MAL; defensive WOE, based on unconscious resistance; intentional WOE, based on strategic intention, acquired skills, and readiness to execute plans; and deliberate WOE, mastery of skills making strategic choices that hold boundaries and assign roles.

Table 1

Four sequential periods and four stages in parent development, raising children with mood dysregulation, attention-deficit/hyperactivity disorder, and learning disabilities

Four Parental Periods	Societal Responses to Child Behaviors	Four Stages of Walking on Eggshells	Parent Feelings
(1) Misperception: Reactive WOE			
Parent perception, knowledge, skills do not match child's needs. Responses to child's maladaptive behavior biased by assumptions. Misunderstand child's internal system. Mislabeling behavior. Rewards/consequences too big, small	Playdates end badly. Adults reduce/stop contact between children. Teachers complain. School psychologist enters. School admin meet. "We love him, but…" Adults gossip, complain	Parent unsuccessful redirecting maladaptive choices. Child escalates, leading to parental escalation. Recurrent behaviors seen in isolation. Each misunderstands other. Rule-bound punishment/reward	Anger, surprise, indignation, confusion, resentment, worry, insecurity, sorrow, embarrassment. "Don't need help"
(2) Denial: Defensive WOE			
Emotionally unprepared to perceive child's needs. Denial is used to protect parent being overwhelmed, struggling to relinquish long-range parent hopes/plans	Peer parents stop calling. Not invited to parties/events. Family members attack parenting style as permissive. Schools place out child/family	Unconscious resistance. Parents excuse behavior. Argumentative. Doctors/evaluations "wrong" "not my kid." Accommodate to avoid tantrums. Misread maladaptive behaviors; eg, "boys will be boys"	Intolerant. Irritable. Insecure. Holding off nagging questions, feelings. Couples take opposing sides. Fight. Parents accuse each other, may separate or divorce. Isolation, loneliness, profound exhaustion
(3) Acceptance: Intentional WOE			
Parents are ready to handle the emotional weight of following the child's needs, leaving behind old parental plans, observing anew and cooperating with professional team	Mainstream/LD schools place child out. Applications rejected (re. school history). Referrals made to therapeutic schools. Referrals made to education, legal consultants	Research therapeutic services. Add to professional support. Attend child mental health lectures. Meet kindred parents. Set new goals, adjust priorities to family systems	Inner realizations, remorse, guilt, relief. Inspired by dawnings. Renewed energy. Emotionally reconnecting. Hope restored somewhat
(4) Professionalized Parenting: Deliberate WOE			
Parent acquires new skills/methods to network, prioritize communicate, organize, plan, and execute strategically	Parents open to and seek out therapeutic schools and programs	Habits of mind and routines mastered in problem solving, decision making, and prioritizing. Boundaries are clear	Pride in ongoing practice. Feelings of relatedness, interconnection, and feeling able

Abbreviations: LD, learning disability; WOE, walk on eggshells.

WOE is Me

We have all walked on eggshells around someone with a short fuse. It's exhausting, a relief when it is over, but what if, as with my son, it is never over?

What if chronic irritability and severe mood dysregulation are always with him? Then where is the wisdom in not treading cautiously; for example, walking on eggshells? The acronym WOE, furthers our conversation. The feeling of woe has certainly been part of our parenting, and of every parent in our support network, but woe has enough of a whiff of melodrama to represent a warning. If it shows up too often, a sense of "poor me" pervades. "What can I do?"; "I can't force Jonny to take his meds"; "I can't make Lilli go to school"; "Billy won't listen to me; he hates me!" Statements like these emanate from demoralized parents during denial. Until they receive therapeutic interventions, there is insufficient tension to push adult development forward. Those energies are consumed by the demands of parenting. The overused oxygen mask metaphor is apt: parents are no help to their children if they are knocked out.

Doctor-Coaching Moves Parents to Acceptance and Professionalized Parenting

When child psychiatrists strengthen parents, they strengthen their adolescent patients. Without family-system therapies to support parents' self-care, home-care, employment, and so on, problems compound. Parents become unavailable to redirect, structure, and help emotionally regulate their children. Duress forces parents into separate corners, eroding the warmth and comfort that are so desperately needed, given the intensity of raising these children. Sadly, divorce often follows. Cutting-edge psychiatrists now call themselves doctor-coaches, using prescription pads to assign children weight lifting, treadmill, and meditation. Imagine also writing this script: "Couples Therapy, 2X weekly, 1 year." Child psychiatrists could provide this kind of doctor coaching to support parents' journey to professionalized parenting, making the whole family stronger.

Woe *was* us, early on, until we strengthened skills and learned how to follow our child. Is it wrong to walk on eggshells over time? Yes and no. It is how a parent prepares to walk, which took me years to practice and understand (**Table 2**).

Some believe walking on eggshells is wise until you can get the hell out of Dodge. Others say it is wrong to do that for too long: it suggests accommodating bad behavior. I say, if it springs from intimidation, then it is nonstrategic; if it effects change, manages outcomes, it is skill.

Table 2
Timetable of son's age range, progression of skill building to mastery in ways parents walk on eggshells around children with mood dysregulation, attention-deficit/hyperactivity disorder, and learning difficulties

Son's Age Range (y)	Period of Parenting	How We Walk On Eggshells
Birth −4	Misperceiving child's behaviors	Reactive WOE
5–9	Denial peels slowly away in layers	Defensive WOE
10–16	Acceptance allows us to follow our son's needs, new actions taken	Intentional WOE
17–25	Professionalized parenting offers well-informed, skillful planning, organizing, networking	Deliberate WOE

Parents raising a child with MAL should tread carefully, with deliberate intention and purpose. Being clear about my reasons for walking on eggshells around my son means I engage with a plan, a strategic mindset, accepting that he may become irrational, verbally aggressive, and/or explosive. I may choose to wait, ignore unwanted behavior, or prepare other options.

VIGNETTE 12:

Age 14 years. School calls. "Eloped." Staff followed. Stones thrown at the staff. "Attempted to break in" to private home. "Please come up."

Not Just a River in Egypt

A closer look at denial is warranted in this parent population. At risk and often lost are parents' hopes for their children. Dashed, too, are hopes for themselves, planning for future social events personal to their enjoyment and development. I missed many holiday events, religious ceremonies, and family gatherings. Similarly, parents' give up personal aspirations, such as educational and professional advancement, and with them plans for greater material comforts. Perhaps most difficult is the loss of being able to pass on personal values. I had hoped to pass on my love of learning, and could not wait to introduce him to university options. I looked forward to sharing my values for saving, but he could not use them.

Of critical importance, denial kept me functioning: going to work, running the household, and so on, while a growing professional team helped me toward awareness, direction, and action. It was devastating to get off the track we had so carefully lain. However, without an engine to run on them, what good were they?

Although denial is a period commonly met with intolerance, frustration, and complaint, as if parents' resistance has no function, clinicians who anticipate and understand it can treat it as a protective period of coping. They can validate parents, explaining why it is in place.

VIGNETTE 13:

11 years old. Taxi, loses glove. Tantrum. Inconsolable. Sprints over park wall, runs away. I follow. He yells, "I'll cut your head off with a knife!" Biting, kicking. Upstairs, screams out window, "Help, police! They're murdering me!" Then, tender, quietly weeps, remorseful: loving.

Enter Professionalized Parenting: Walking on Eggshells, Strategically

With our son getting such strong support in residences from eighth to twelfth grades, we hoped the need to walk on eggshells would lessen. But no. We learned to walk strategically. Even with growing engagement in goal setting, cognitive behavior therapy incentives, self-monitoring charts, equine and other therapies, talent development, routine schedules, social learning, specialized academics, and enrichment such as art and theater, his biosocial schema was what it was. His adherence to rules wavered; his controls in family sessions were unpredictable, as was his recovery time. For our part, we had a solid presence on campuses, regained balance in personal and professional matters, and strengthened parenting skills. For our son's part, he was safe on campus through high school and spared involvement in drugs, alcohol, and juvenile detention. We credit having the services and consistency that 24/7 therapeutic care offered. Thanks to them, we left the period of acceptance and entered professionalized parenting.

ARRIVAL
Young Adulthood: Professionalized Parenting Is Essential

Following two years of independent living with support, he went off meds, broke down, chose one of two transitional residential programs we offered, and turned the place upside down. He refused their recommendation for a higher level of care, preferring a cockamamie scheme. Mindful that resistance from us would trigger a fight, we strategically walked on eggshells, accepting his decision. Ignoring his accusations, we asked the staff to drive him to a shelter, and wished him luck. This moment defined professionalized parenting: skillful, strategic, intentional decision making to set limits and expectations, calmly. They drove him to the shelter, whereupon he asked to call us, having reconsidered intensive treatment.

The psychiatric director took his case and struggled for weeks, eventually ruling out bipolar disorder (again), borderline personality disorder (again), autism spectrum disorder (again), and settled on diagnosing unspecified episodic mood disorder and generalized anxiety disorder. In immediate aftercare step-down, his diagnoses changed to severe recurrent major depressive disorder without psychotic features and generalized anxiety disorder. In private treatment afterward, before he quit, diagnoses assigned were major depressive disorder, recurrent episode moderate; ADHD/combined presentation; high expressed emotion level within family; adjustment disorder with mixed disturbance of emotions and conduct; unspecified personality disorder (mixed traits of borderline personality disorder and dependent behaviors, attachment disturbance). Diagnostically homeless is he, or does he have homes everywhere?

> ### VIGNETTE 14:
> *Age 21 years. Life coach calls. Son was held up, now wants to buy a BB gun. Life coach expressly says "No, don't." He does, secretly.*

Credit Where Credit Is Due

Our son is exquisitely sensitive. He can do inspiringly well during euthymic states, sometimes calling on superpowers that mere neurotypicals cannot. He has got what a friend calls microcourage: summoning bits of it during myriad anxious moments each day. That it evaporates with terrible speed humbles me. My job is to wait for what I want, listen closely, and offer sound options. Always two, one preferred, one less preferred, so that, when things implode, there are pathways that develop his needs, interests, and talents.

Most parents raising children with MAL hear, "I just want to be normal!" Our son seems desperate not to be who he is, and self-rejection may partially explain his ongoing low self-esteem. He believes meds are bad, his therapist is no help, job supervisors are [expletive]s, and we robbed him of a normal childhood. We hope he will accept his differences and love himself at some point, sooner rather than later, enough to revive self-care.

Full Circle, Back to the Big Question: Will Our Son Survive His Illness?

Clinicians know children can "rule the roost," running households. Even in professionalized parenting, parents of young adults in our parent support network show what can happen when they cannot bear the thought of their children completing suicide "on their watch," and say they couldn't live with themselves: if.

The trouble is, they cannot parent with deliberate intention and follow through on consequences as long as the threat of completed suicide rules their decisions. Parents need to be able to walk on eggshells around their children's suicidal and other poor thinking with deliberate language providing limits, common sense, and steps to prompt them, one step after another. By not parenting strategically, we could also lose them. Clinicians can doctor-coach parents' wellness through each period of their parenting journey.

> **VIGNETTE 15:**
> *Age 22 years. Therapist calls. "Fly in tonight. Girlfriend breakup, he's suicidal and wants hospitalization."*

FURTHER STUDY

Child and adolescent psychiatrists read an abundance of journal articles focused on understanding and treating children with MAL. Far less available are articles describing ways to understand and treat their patient's parents, particularly ones written by the parent of a patient. I am that father.

Raising a child with MAL is profoundly life altering. The rates of divorce and loss of employment, homes, friends, and family need further study. Informally, these events are highly prevalent. Also, although myriad studies show the likelihood of onset of mental health diagnoses in children whose parents have a lived experience with mental illness, no studies investigate onset of mental health diagnoses in parents with no prior history of mental illness before children onset early with severe mood dysregulation, especially with MAL. In closing, I love my son; we believe in him. If ongoing intermittent despair and worry have been part of the fabric of parenting, there is a better blend now having sewn in pleasure and pride in our professionalized parenting. We have learned how to walk strategically on eggshells and remain effective, and affecting, parents.

DISCLOSURE

Names of some institutional organizations, as well as some other names and dates, have been changed or omitted to protect my child's privacy.

The Many Faces (and Names) of Mood Dysregulation

Daniel F. Connor, MD[a],*, Leonard A. Doerfler, PhD[b]

KEYWORDS

• Child • Adolescent • Irritability • Impulsive aggression • Emotion dysregulation

KEY POINTS

• The explosive, dysregulated child poses an important clinical problem.
• Many different terms have been used to describe the explosive child. This may lead to difficulties in diagnosis and treatment of the individual child with explosive outbursts.
• Explosive outbursts are complex behaviors involving emotional and behavioral elements and are associated with difficulties in emotion and behavioral regulation.
• Symptoms of the explosive child are transdiagnostic.
• A strong and negative internal emotional valence may drive dysregulated behaviors and moods.

Although long recognized as a significant problem by parents and clinicians, explosive outbursts in children have gone by many aliases over the years. Included here are emotions and behaviors related to the Diagnostic and Statistical Manual of Mental Disorders-5th Edition (DSM-5)[1] externalizing behavior diagnoses of attention-deficit-hyperactivity disorder (ADHD), oppositional-defiant disorder (ODD), conduct disorder (CD),[2] intermittent explosive disorder (IED),[3,4] obsessive compulsive disorder,[5] Tourette's disorder,[6] and internalizing disorders such as pediatric bipolar disorder[7] and disruptive mood dysregulation disorder (DMDD).[8] Other constructs for explosive outbursts include rage attacks, impulsive aggression, severe anger attacks, and temper tantrums.[9] More recently, these behaviors have been described as part of emotional impulsivity (EI)[10] and as a component of the behavioral repertoire of some highly irritable children.[11] Lack of clarity concerning the many concepts used to describe explosive outbursts may create diagnostic confusion and influence the clinician's ability to optimize clinical treatment for the individual explosive child. This article discusses several emotional factors and behavioral concepts that are important to the conceptualization of explosive outbursts in children.

[a] Department of Psychiatry, MC 1410, University of Connecticut Medical School, 263 Farmington Avenue, Farmington, CT 06030, USA; [b] Clinical Counseling Psychology Program, Assumption College, Department of Psychiatry, University of Massachusetts School of Medicine, 500 Salisbury Street, Worcester, MA 01609-1296, USA
* Corresponding author.
E-mail address: dconnor@uchc.edu

Child Adolesc Psychiatric Clin N Am 30 (2021) 299–306
https://doi.org/10.1016/j.chc.2020.10.002
1056-4993/21/© 2020 Elsevier Inc. All rights reserved.

childpsych.theclinics.com

DIMENSIONAL OVERVIEW

Research on child and adolescent psychopathology has identified 2 major dimensions that describe mental health problems among youth.[12] One dimension includes externalizing or undercontrolled behaviors, which are directed toward the youth's environment and impair or interfere with functions at home, school, or with peers. Externalizing disorders are characterized by temper tantrums, disruptive, hyperactive, and/or aggressive behaviors. DSM-5[1] diagnoses for externalizing behavior disorders in children and adolescents include ADHD, ODD, and conduct disorder. Collectively, behavior disorders are the second most common reason that children and adolescents are referred to mental health treatment.[13]

The second major dimension of child adolescent mental health problems includes internalizing disorders and problematic emotion and affect.[12] Internalizing disorders are characterized by intense emotions, such as anger, fear, anxiety, or depression, along with mood-related behaviors like social withdrawal or avoidance of anxiety-provoking situations that interfere with functioning at home, school, or with peers. Youths who experience intense anxiety may exhibit temper tantrums or similar acting out behavior when a parent or other adult interferes with anxiety-related avoidance behaviors (eg, compulsive rituals) or brings the child into an anxiety-provoking situation.

The distinction between externalizing and internalizing disorders that has guided much research and treatment of child and adolescent psychopathology suggests that youths' presenting problems are characterized predominantly by either undercontrolled behaviors or intense and problematic emotions. However, it is becoming increasingly clear that some youth present with a confusing and complicated combination of internalizing and externalizing symptoms. These youth experience frequent affective storms that are characterized by frequent, prolonged, and aggressive temper outbursts. These youths are irritable and experience marked mood instability. Along with emotion dysregulation, these youth can present with impulsive, defiant, hyperactive, and highly aggressive behaviors. Children and adolescents with extreme irritability, severe temper tantrums, rage attacks, and impulsive aggression (ie, the explosive child) and deficient mood and behavioral regulation are especially difficult to treat.[14] Moreover, DSM-5 does not provide a diagnosis that accurately describes this challenging set of problems, and clinicians are left to debate the merits of diagnoses like disruptive mood dysregulation disorder, complex ADHD, complex post-traumatic stress disorder (PTSD), borderline personality disorder, pediatric bipolar disorder, or the presence of multiple comorbid disorders to describe these youths.

The CBCL Dysregulation Profile (CDP), which identifies youth with high scores on the Anxious/Depressed, Attention Problems, and Aggressive Behavior scales, is thought to identify a syndrome of poor self-regulation.[15] In clinical samples this profile is not rare. For example, in a sample of 310 youth who were referred to the authors' outpatient Pediatric Psychopharmacology Clinic, 18% of youth fit this profile.[16] Youth who fit the CBCL profile exhibited severe dysregulation across multiple affective and behavioral domains of functioning and had frequent episodes of explosive outbursts that were not easily categorized by DSM-IV diagnoses.

EMOTIONAL FACTORS: IRRITABILITY, THREAT SENSITIVITY, FRUSTRATION, ANGER, AND AGITATION

Explosive children often exhibit dysregulation in the control of emotions such as irritability, frustration, fear, or anger, and may be agitated before they explode. These emotions generally carry a negative valence, as contrasted with more positive emotions such as joy or happiness. For these children, emotions may be intense, severe, of excessive duration, and arise suddenly, impulsively, and disproportionally in response to cues from the environment.

Irritability can be defined as proneness to anger.[12] It is conceptualized as a low threshold for experiencing anger in response to individual frustration in goal-directed activity.[17] Irritability is considered to have 2 components: tonic and phasic. The tonic component refers to a daily occurrence of persistently grouchy or angry mood. The phasic component refers to episodic behavioral outbursts of intense anger and rage.[11,17] This 2-component process is the basis for the new DSM-5 diagnosis of disruptive mood dysregulation disorder.[18,19] It is important to note that irritability (an emotion) is not the same as impulsive aggression (a behavior). Many irritable children and adolescents are not aggressive. For example, data suggest that aggression occurs in only a minority of those who report excessive irritability. For example, only 12% of community subjects who report irritable temper outbursts also report aggression.[19] Impulsive aggression is a dramatic and highly problematic outcome for a minority of those with excessive and severe irritability.

Irritability can also be conceptualized as an abnormal approach to threat. Threats are fundamental motivators of behavior across species.[11] Depending on characteristics of the threat, such as physical distance to the threat, imminence, and temporal parameters of the threat encounter, the individual initiates a threat response. Distal and slowly evolving threats invoke a fear response that varies between behavioral freezing and flight. Sudden and imminent threat evokes an aggressive and explosive rage response in the service of individual survival. Irritable youth tend to orient toward threatening environmental stimuli (instead of distancing themselves from the threat), and are more likely to interpret ambiguous environmental stimuli as threatening.[20] They are then at increased risk for inappropriately engaging the presumed threat with an aggressive rage response.[11,20,21]

An exaggerated or dysregulated threat response system has been found with some children who demonstrate dysregulated emotions and behaviors. One group of highly emotionally reactive youth is comprised of children and adolescents who have experienced traumatic stress during their early development.[22] These children and adolescents may be overly sensitive to environmental cues of threat, perceiving threat where none exists. Children and adolescents who have experienced such traumatic stressors may react with rage or impulsive aggression. Social information processing theory[23] hypothesizes that hypervigilance and selective attention to environmental cues of threat emerge, in part, as an adaptive mental response to past life experiences of actual threat, such as physical and/or sexual abuse and victimization. Youth who experience physical and/or sexual abuse or victimization or harsh physical punishment may develop a hostile attribution bias, which is the tendency of some children to routinely attribute hostile intent to relatively benign or ambiguous social cues. Traumatized children who have developed a hostile attribution bias may suddenly become emotionally and/or behaviorally dysregulated when they react to a perceived threat in their immediate environment.[24]

Frustration is an emotional state that arises when an individual is blocked from achieving meaningful goals or achieving an expected reward (frustrative nonreward); it is a normal component of human emotion.[19] Frustration may lead to the contingent use of explosive outbursts by the child to avoid tasks that are not immediately reinforcing.[25] It is characterized by motor overactivity, irritability, anger, and aggression. These responses resemble characteristics of a child rage outburst.[11] For example, a child with ADHD who is asked to stop watching his favorite television show and take out the garbage may erupt in an explosive outburst. If the parent backs down, the child has achieved the goal of avoiding a noxious task (eg, taking out the garbage) and continuing to engage in an immediately reinforcing task (eg, watching television). The child is thus negatively reinforced to continue the use of explosive behaviors to avoid unpleasant tasks.[25]

Anger is also an emotion associated with frustrative nonreward.[11,19] Anger is conceptualized as a negative affect associated with behavioral approach (ie, unlike fear or anxiety, which are associated with behavioral avoidance). Anger is triggered by blocking one's attempts to attain meaningful goals.[26] It is tied to an appraisal of being wronged and an individual's attempts to address the perceived wrongdoing. Anger can range in intensity from mild annoyance to furious rage. Anger can differ in form from resistance to retaliation.[26] Anger may also serve other purposes for the individual. In individuals who experience intense and persistent negative emotions, efforts to escape or avoid these feelings may be associated with the expression of pathologic anger.[27] For example, anger may arise when a depressed or anxious person attempts to avoid feeling painful emotions (eg, better to be mad than sad). Another example is the patient with borderline personality disorder who expresses intense rage as a strategy to communicate deep and pervasive internal distress.[26]

Agitation is a term often used to describe an emotional state characterized by feelings of excessive inner tension often accompanied by irritability and anger and frequently expressed by excessive motor activity.[9] Although not generally a part of the explosive outburst itself, agitation may be a precursor to such an event.[9]

Irritability, threat sensitivity, frustrative nonreward, anger-rage, and agitation are all distinct yet inter-related elements in the conceptualization of the child with explosive outbursts. They exist on a spectrum of severity and will vary in intensity and severity across individual cases. These factors are transdiagnostic, occurring across many different psychiatric diagnoses.[28] Identification of significant emotional elements associated with explosive outbursts may help the clinician better define appropriate treatment targets for the individual child.

BEHAVIORAL FACTORS: IMPULSIVE AGGRESSION

Aggressive behavior is a common reason that children and adolescents are referred to psychiatrists, and it occurs with multiple psychiatric and some neurologic disorders.[9] Aggression is often a part of the behavioral repertoire associated with explosive outbursts in children. Defined by Ramirez and Andreu[29] as "the delivery of any form of definite and observable harm-giving behavior toward any target,"[29] aggression is a heterogeneous and complex construct.[9] When it occurs in the context of neurologic or psychiatric pathology, some types of aggression may be considered maladaptive, as aggression may be expressed in the context of, and possibly the result of, a central nervous system (CNS) that is not functioning optimally under the burden of and vulnerability conferred by neuropsychiatric disease.[30] This type of aggression has several identifiable characteristics. First, it is often impulsive, intense, explosive, sudden, and disproportionate to the environmental context. This may reflect difficulties with emotion generation causing sudden, abrupt, impulsive, and intense emotions and is referred to as emotional impulsivity (EI). Maladaptive aggression may be prolonged and not terminate in a reasonable time frame as a result of deficient CNS regulatory mechanisms that govern emotion regulation and cognitive-social skills. This is referred to as deficient emotional self-regulation (DESR).[10] Because of weak CNS regulatory processes and deficient developmentally appropriate cognitive-behavioral skills, emotionally reactive behaviors can reach high levels of intensity and be abnormally prolonged in duration.[10] Such intense, impulsive, and reactive behaviors are usually abrupt and situationally provoked, and are not spontaneous.[10] Considering explosive outbursts in children, these concepts may help us understand the explosion in all of its severity, as well as the prolonged duration of an individual episode (**Fig. 1**).

Aggression can be subtyped as overt or covert aggression.[9] Overt aggression is characterized by an open and observable response to an environmental cue and includes behaviors such as physical assault, verbal threats of violence, physical fighting, punching and/or kicking others, and property destruction. In contrast, covert aggression is characterized by hidden and surreptitious activity and includes behaviors such as lying, cheating, stealing, and vandalism.[9]

Aggression can be further subtyped as reactive aggression (RA) or proactive aggression (PA).[31] RA is defined as an angry, hostile, or defensive response to environmental cues eliciting frustration, provocation, and/or perceived threat. It is characterized by high emotional intensity, autonomic nervous system arousal, and activation of fight-flight physiologic mechanisms.[9] In contrast, PA is a deliberate, goal-directed

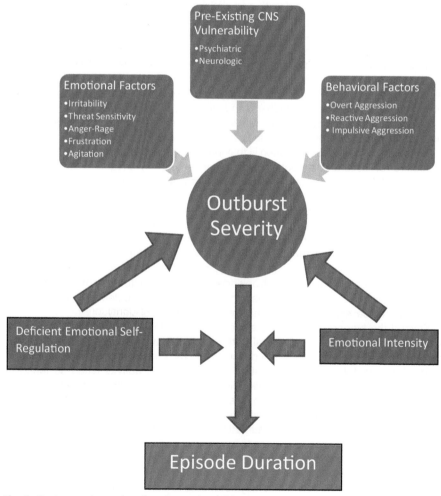

Fig. 1. Factors and severity of outbursts in children.

behavior, characterized by overcontrolled, planned behavior in pursuit of a goal or reward.[30]

Explosive outbursts in children are overwhelmingly characterized by overt aggressive behaviors and have many characteristics similar to RA.[32] Characteristics include a short fuse, sudden escalations of anger and rage, and intense emotion that often lasts for an extended duration. The individual child may have frequent impulsive aggressive episodes and may feel remorse when the episode is over.[9]

DISCUSSION

Explosive outbursts in children and adolescents have been long identified by clinicians and have been described using many different conceptualizations and terms. The authors' brief review of some of these important concepts reveals that the topography of explosive outbursts is complex. Included here are pre-existing CNS vulnerabilities including psychiatric and neurologic diagnoses, various contributing emotions that generally carry a negative valence, and aggressive behaviors that are usually overt and reactive. Emotional impulsivity and deficient emotional self-regulatory mechanisms may contribute to episode severity and duration.

These factors demonstrate a range of severity from normal behaviors to abnormal, depending on episode characteristics such as intensity, frequency, duration, environmental cues and precipitants, and duration. It is important to recognize that explosive outbursts represent the extreme end of a continuum of anger proneness and expression. This intense anger, coupled with highly aggressive behaviors, presents considerable challenges for parents, teachers, and clinicians.

As noted earlier, the current diagnostic system does not provide a diagnosis that accurately describes the constellation of emotional and behavioral features of explosive outbursts. One possible reason is that irritability (which is a proneness to anger) is a feature of several different DSM-5 diagnoses.[11,19] Although a characteristic of multiple DSM-5 diagnoses, irritability is not a required symptom for any specific psychiatric diagnosis.

Because intense temper tantrums or explosive outbursts are not unique to a specific DSM-5[1] diagnosis, clinicians may find it helpful to consider sudden escalations of anger and rage like other problems, such as a child's social isolation or lack of friends, that warrant attention even though they are not captured by a psychiatric diagnosis. Regardless of DSM-5 diagnosis, the presence of strong, intense, and negative emotional valence is a common factor that occurs across episodes of explosive outburst. These sudden and intense escalations of anger and rage, which often last for an extended period of time, have received several labels, which has resulted in a lack of clarity.

Beyond the typography of these explosive outbursts, it is important to recognize that these emotional and behavioral outbursts may have different functions or outcomes. The intense emotion, often described as rage in the explosive child, may be driven by a nontolerance of frustration in goal-directed activity (such as found in some children with disruptive behavior disorders or ADHD), avoidance of uncomfortable or overwhelming internal emotions (such as found in some depressed children or youths with borderline personality disorder), or an exaggerated or misinterpreted fear of threat (such as found in some traumatized and abused children). Regardless of its function, this intense affect threatens to overwhelm cognitive, behavioral, physiologic, and emotional regulatory mechanisms, leading to the risk of disinhibited impulsive aggressive behavior.

By parsing the many factors that may contribute to problematic explosive outbursts in children, it is clear that they contain both internalizing and externalizing elements, and that these behaviors are transdiagnostic in that they occur across many different psychiatric disorders. A clearer understanding of the factors involved in explosive outbursts may help clinicians choose specific targets for treatment therapies and facilitate individual approaches to the explosive child.

CLINICS CARE POINTS

- Identification of any salient emotional elements associated with explosive outbursts may help the clinician identify important treatment targets for the individual child.
- Identification of the contingent use of explosive temper tantrums to avoid noxious tasks may help the clinician identify important behavioral treatment strategies for the individual child.
- The aggressive child with mood and behavioral dysregulation may meet diagnostic criteria for many different DSM diagnoses. The treating clinician must be sensitive to the issue of diagnostic comorbidity for the individual child.

DISCLOSURE

D.F. Connor is consultant for Supernus and Takeda Pharmaceuticals. L.A. Doerfler has not conflicts of interest to report.

REFERENCES

1. American Psychiatric Association. Diagnostic and statistical manual of mental disorders. 5th edition. Washington, DC: 2013.
2. Pliszka SR. Psychiatric comorbidities in children with attention deficit hyperactivity disorder: implications for management. Paediatr Drugs 2003;5(11):741–50.
3. Coccaro EF. DSM-5 intermittent explosive disorder: relationship with disruptive mood dysregulation disorder. Compr Psychiatry 2018;84:118–21.
4. Kulper DA, Kleiman EM, McCloskey MS, et al. The experience of aggressive outbursts in intermittent explosive disorder. Psychiatry Res 2015;225(3):710–5.
5. Villemarette-Pittman NR, Stanford MS, Greve KW, et al. Obsessive-compulsive personality disorder and behavioral disinhibition. J Psychol 2004;138(1):5–22.
6. Budman CL, Bruun RD, Park KS, et al. Explosive outbursts in children with Tourette's disorder. J Am Acad Child Adolesc Psychiatry 2000;39(10):1270–6.
7. Carlson GA, Klein DN. Commentary: frying pan to fire? Commentary on Stringaris et al. (2018). J Child Psychol Psychiatry 2018;59(7):740–3.
8. Carlson GA, Pataki C. Disruptive mood dysregulation disorder among children and adolescents. Focus 2016;14(1):20–5.
9. Connor DF, Newcorn JH, Saylor KE, et al. Maladaptive aggression: with a focus on impulsive aggression in children and adolescents. J Child Adolesc Psychopharmacol 2019;29(8):576–91.
10. Faraone SV, Rostain AL, Blader J, et al. Practitioner review: emotional dysregulation in attention-deficit/hyperactivity disorder - implications for clinical recognition and intervention. J Child Psychol Psychiatry 2019;60(2):133–50.
11. Brotman MA, Kircanski K, Leibenluft E. Irritability in children and adolescents. Annu Rev Clin Psychol 2017;13:317–41.
12. Achenbach TM, Ivanova MY, Rescorla LA. Empirically based assessment and taxonomy of psychopathology for ages 1½–90+ years: developmental, multi-informant, and multicultural findings. Compr Psychiatry 2017;79:4–18.

13. Merikangas KR, He J-P, Burstein M, et al. Lifetime prevalence of mental disorders in US adolescents: results from the National Comorbidity Survey Replication-Adolescent Supplement (NCS-A). J Am Acad Child Adolesc Psychiatry 2010; 49(10):980–9.

14. Connor DF, Steingard RJ, Cunningham JA, et al. Proactive and reactive aggression in referred children and adolescents. Am J Orthopsychiatry 2004;74(2): 129–36.

15. Ayer L, Althoff R, Ivanova M, et al. Child behavior checklist juvenile bipolar disorder (CBCL-JBD) and CBCL posttraumatic stress problems (CBCL-PTSP) scales are measures of a single dysregulatory syndrome. J Child Psychol Psychiatry 2009;50(10):1291–300.

16. Doerfler LA, Connor DF, Toscano PF Jr. The CBCL bipolar profile and attention, mood, and behavior dysregulation. J Child Fam Stud 2011;20(5):545–53.

17. Vidal-Ribas P, Brotman MA, Valdivieso I, et al. The status of irritability in psychiatry: a conceptual and quantitative review. J Am Acad Child Adolesc Psychiatry 2016;55(7):556–70.

18. Avenevoli S, Blader JC, Leibenluft E. Irritability in youth: an update. J Am Acad Child Adolesc Psychiatry 2015;54(11):881–3.

19. Stringaris A, Vidal-Ribas P, Brotman MA, et al. Practitioner review: definition, recognition, and treatment challenges of irritability in young people. J Child Psychol Psychiatry 2018;59(7):721–39.

20. Dodge KA. Social cognition and children's aggressive behavior. Child Dev 1980; 51(1):162–70.

21. Bunford N, Evans SW, Langberg JM. Emotion dysregulation is associated with social impairment among young adolescents with ADHD. J Atten Disord 2018; 22(1):66–82.

22. Dvir Y, Ford JD, Hill M, et al. Childhood maltreatment, emotional dysregulation, and psychiatric comorbidities. Harv Rev Psychiatry 2014;22(3):149–61.

23. Crick NR, Dodge KA. A review and reformulation of social information-processing mechanisms in children's social adjustment. Psychol Bull 1994;115(1):74–101.

24. van Der Kolk B, Ford JD, Spinazzola J. Comorbidity of developmental trauma disorder (DTD) and post-traumatic stress disorder: findings from the DTD field trial. Eur J Psychotraumatol 2019;10(1):1562841.

25. Kingsbury SJ, Lambert MT, Hendrickse W. A two-factor model of aggression. Psychiatry 1997;60(3):224–32.

26. Fernandez KC, Jazaieri H, Gross JJ. Emotion regulation: a transdiagnostic perspective on a new RDoC domain. Cogn Ther Res 2016;40(3):426–40.

27. Cassiello-Robbins C, Barlow DH. Anger: the unrecognized emotion in emotional disorders. Clin Psychol Sci Pract 2016;23(1):66–85.

28. Gross JJ, Jazaieri H. Emotion, emotion regulation, and psychopathology: an affective science perspective. Clin Psychol Sci 2014;2(4):387–401.

29. Ramírez JM, Andreu JM. Aggression, and some related psychological constructs (anger, hostility, and impulsivity); some comments from a research project. Neurosci Biobehav Rev 2006;30(3):276–91.

30. Connor DF. Aggression and antisocial behavior in children and adolescents: research and treatment. New York: Guiford; 2004.

31. Raine A, Dodge K, Loeber R, et al. The reactive-proactive aggression questionnaire: differential correlates of reactive and proactive aggression in adolescent boys. Aggress Behav 2006;32(2):159–71.

32. Bambauer KZ, Connor DF. Characteristics of aggression in clinically referred children. CNS Spectr 2005;10(9):709–18.

The Phenomenology of Outbursts

Lauren Spring, MD[a,b,*], Gabrielle A. Carlson, MD[c]

KEYWORDS

- Outbursts - Irritability - Temper tantrums - Emotion dysregulation - Aggression
- Bipolar disorder - Disruptive mood dysregulation disorder
- Oppositional defiant disorder

KEY POINTS

- Severe outbursts, comprising part of the concept of irritability, have no agreed on label. There are no ways to consistently describe, measure, diagnose, or treat the behavior.
- Outbursts occur as part of the phenomenology of many disorders.
- Disruptive mood dysregulation disorder and intermittent explosive disorder provide a diagnostic home for some children but do not provide classification for all children.
- Although irritable mood is important, frequency of outbursts and behavior during the outburst (eg, verbal, property, or physical aggression) determine the level of impairment.

INTRODUCTION

Emotional outbursts are a common problem in child and adolescent psychiatry[1] and are often the primary reason for referral to child and adolescent psychiatric services with up to 70% of inpatient admissions because of the danger their outbursts pose to themselves and others.[2,3] Nevertheless, there is a concerning lack of consensus on what to call outbursts. We found 10 different terms to describe this phenomenon in the literature over the past 40 years including affective storms,[4] severe irritable mood,[5] rages,[6] rage attacks,[7] anger attacks,[8] and tantrums.[9] Without a consistent label, it is challenging to study this important topic.

DIAGNOSIS

Outbursts present not only labeling but also diagnostic and treatment challenges. Diagnostically, there is no clear home for severe emotional outbursts. In the Diagnostic

[a] Psychiatry Residency Training; [b] Division of Child and Adolescent Psychiatry, Department of Psychiatry & Behavioral Health, HSC T-10, 101 Nicolls Road, Stony Brook, NY 11794, USA; [c] Renaissance School of Medicine, Stony Brook University, Putnam Hall-South Campus, 101 Nicolls Road, Stony Brook, NY 11794-8790, USA
* Corresponding author. Division of Child and Adolescent Psychiatry, Department of Psychiatry & Behavioral Health, HSC T-10, 101 Nicolls Road, Stony Brook, NY 11794.
E-mail address: lauren.spring@stonybrook.edu

Child Adolesc Psychiatric Clin N Am 30 (2021) 307–319
https://doi.org/10.1016/j.chc.2020.10.003
1056-4993/21/© 2020 Elsevier Inc. All rights reserved.

childpsych.theclinics.com

and Statistical Manual of Mental Disorders (DSM)-5,[10] outbursts are part of the diagnostic criteria for disruptive mood dysregulation disorder (DMDD) and intermittent explosive disorder (IED). Irritability, which encompasses outbursts,[11,12] is a symptom of oppositional defiant disorder (ODD), major depressive disorder in children and adolescents, manic episodes in bipolar disorder (BD), generalized anxiety disorder, and the negative affect of post-traumatic stress disorder.[10] It is also an associated symptom in conduct disorder, autism spectrum disorder, and personality disorders.[10]

Unlike DSM-5, the World Health Organization's International Classification of Diseases-11th edition committee[13,14] made chronic irritability a subtype of ODD characterized by a "prevailing, persistent angry or irritable mood, independent of any apparent provocation, often accompanied by regularly occurring severe temper outbursts that are grossly out of proportion in intensity or duration to the provocation... [occur] nearly every day, are observable across multiple settings or domains of functioning (eg, home, school, social relationships), and are not restricted to the individual's relationship with his/her parents or guardians [or] limited to occasional episodes (eg, developmentally typical irritability) or discrete periods (eg, irritable mood in the context of manic or depressive episodes)." However, the criteria do not address outburst severity, account for situations in which outbursts may be the impairing part of another condition, or address children whose mood is fine until they do not get what they want and then they explode.

Regarding treatment of rage outbursts, Carlson and colleagues[6] reported that children who have them repeatedly while hospitalized often remain symptomatic and a danger to themselves, others, and the environment despite treatment with psychopharmacology, behavioral therapy, special education interventions, and family involvement. Clearly, more targeted interventions are needed to best help these children and their families.

Much research has been devoted to figuring out where diagnostically these outbursts belong. In the 1970s a putative relationship to neurologic problems yielded such terms as "episodic dyscontrol," associating the outbursts with minimal brain dysfunction, seizures, or significant brain trauma.[15]

Davis[4] in 1979 proposed that outbursts, referred to as "affective storms," were a "manic-depressive variant syndrome of childhood." These "prolonged and aggressive temper outbursts," which were totally out of proportion to the triggering incident, were believed to be worse than the emotional lability seen in minimal brain dysfunction or hyperkinetic syndrome (the diagnostic labels by which attention-deficit/hyperactivity disorder [ADHD] was known by before DSM-III). Davis also observed that children were often irritable between extreme mood swings. Symptoms sometimes improved after lithium treatment, also suggesting a bipolar spectrum disorder.[4]

OUTBURSTS AND THE BIPOLAR CONTROVERSY

Davis's hypothesis was the prequel to conclusions of the research group at Massachusetts General Hospital, which stated a childhood variant of mania has irritability as the core symptom of the disorder.[4,16,17] Irritability in these children was described as severe, explosive, and aggressive, including such behaviors as throwing things, kicking down doors, hitting, biting, or spitting. Wozniak and coworkers[16] and Mick and colleagues,[17] using the Schedule for Affective Disorders and Schizophrenia-Childhood version (K-SADS[18]) reported that although irritability occurs in other child and adolescent psychiatric disorders, such as ODD, depression, and ADHD, the irritability seen in prepubertal BD is more severe, persistent, highly disabling, and qualitatively distinct based on the frequency and intensity of explosivity. For individuals with ADHD, 76% had ODD-irritability, 38% had depression-related irritability, and had 18% extreme irritability as theorized to occur in BD. Forty-six percent of explosively irritable children with ADHD were thought to meet symptom criteria for mania (usually chronic, not episodic).[17]

There has been significant disagreement about diagnosing BD based on the core symptom of severe outbursts. Because symptoms were double-counted across disorders (ie, hyperactivity for ADHD and mania) there were higher rates of comorbidity such that more than 90% of this explosive form of BD is comorbid with ADHD and ODD. Episodes were poorly defined in DSM-III to DSM-IV-R so "ultra-rapid cycles or ultradian cycling"[19,20] emerged from "a distinct period" of irritability. The K-SADS[18] muddied the waters in that "explosive outbursts" were initially described only in the mania section and only by frequency, not severity (**Table 1**). Even now, if

Table 1
Illustrations of how tantrums/outbursts are elicited and rated in several structured interviews

K-SADS-PL[18]; Updated in 2016	WASH U K-SADS[20]	SMD Module[22]	PAPA/CAPA[36]
Mania	Depression section	SMD	Tantrums or outbursts
Was there ever a time you were so irritable and angry that you exploded? Threshold: occurs 2 consecutive d or 3 separate d	Subjective feeling of irritability, anger, crankiness, externally directed 1 = not at all 2 = slight 3 = mild 4 = moderate 5 = severe; most of the time irritable/ angry, or has thoughts of hurting others, or throws and breaks things 6 = extreme; most of the time extremely angry, or frequent uncontrollable tantrums 7 = rating 6 with homicidal	Is it easy for you to become irritable, angry, or to explode? 1 = not present 2 = more angry than called for, quick to express annoyance, n <3 times per wk in past 4 wk 3 = reaction out of proportion to stressor; verbal rages and/or aggression toward people or property, at least 3 times per wk	Discrete episodes of excessive temper, frustration, or upset, with shouting, crying or stamping, and/or involving violence or attempts at damage directed against oneself, others, or property Intensity 0 = absent 2 = excessive temper, upset, shouting, crying or nondestructive violence directed only against property (eg, stamping, kicking, throwing toys, hitting walls, spitting, holding breath) 3 = with destructive violence (eg, breaking toys) or violence against self or others (eg, hitting, biting, kicking, head banging)

Abbreviations: K-SADS-PL, schedule for affective disorder and schizophrenia for children, present and lifetime; PAPA/CAPA, Preschool Age Psychiatric Assessment/Child and Adolescent Psychiatric Assessment; SMD Module, Severe Mood Dysregulation module from the K-SADS; WASH U K-SADS, Washington University Medical School version of the K-SADS.

the K-SADS is used initially to obtain diagnostic information, the only places in the interview that elicit explosive outbursts are the DMDD or mania sections. Finally, beginning with DSM-III, symptoms of emotion dysregulation for ADHD were relegated to secondary status and kept in the text as "associated symptoms" but as such, did not make it into structured interviews. Thus, the only disorder that could include hyperactivity, impulsivity, and distractibility with explosive outbursts was mania. It is no wonder explosive children with ADHD were called "bipolar."

Two divergent conceptualizations of BD in children evolved.[21] Initially, the narrow and broad phenotypes, articulated first by the intramural mood disorders group at National Institute of Mental Health[22] distinguished episodic from chronic irritability. The former, was defined by discrete mood episodes, clearly different from the individual's baseline function, and observed by multiple informants. The latter was defined by intense and protracted outbursts superimposed on an ongoing irritable mood. Episodic irritability was part of classic mania. Chronic irritability was named severe mood dysregulation (SMD)[11,22] and included irritable mood; frequent, intense outbursts and restlessness; distractibility; intrusiveness; and pressured speech. Again, not surprisingly, SMD frequently co-occurred with symptoms of ADHD and ODD.[23] In DSM-5, SMD became DMDD, which kept symptoms of chronic irritability and frequent outbursts, but eliminated the symptoms commonly seen in ADHD.[10] Nevertheless, DMDD still does not account for all the other conditions and situations in which severe outbursts can occur.

WHAT DO OUTBURSTS LOOK LIKE?

Using a systematic observational scale, Carlson and colleagues[6,24] trained nursing staff to observe the outburst behaviors of psychiatrically hospitalized children. Of 151 admissions, 49 had at least one outburst. Yelling, screaming, and cursing occurred in 92.7% of outbursts; kicking furniture and punching walls in 80.7%; foot stomping and head banging in 77.1%; and tearful, sad, anxious behavior in 45.9%. Manic symptoms were not observed, demonstrating that outbursts were not ultrarapid cycles of mania. In fact, the factor structure and temporal organization of outbursts revealed their similarities to preschool tantrums (**Table 2** for comparison). The

Table 2
Comparison of inpatient outbursts with toddler tantrums

Outburst Behaviors by Two Factors	Inpatient Outbursts	Toddler Tantrums
Agitated, angry		
Scream, shout	X	X
Hit, kick	X	X
Threaten, curse	X	
Bite, scratch	X	
Push, pull, throw	X	X
Pace, run away, stamp, punch wall, bang head	X	X
Distress		
Whine, cry, anxious	X	X
Withdraw, not respond	X	

From Angold A, Prendergast M, Cox A, et al. The Child and Adolescent Psychiatric Assessment (CAPA). Psychol Med 1995;25(4):739-753; with permission.

difference was that the average duration of the child's outburst was 45 minutes versus 5 minutes seen in the average young child's tantrum.[6]

WHAT ARE THE RATES OF OUTBURSTS IN THE GENERAL POPULATION?

Without a consensus on what to call the phenomenon of explosive outbursts, it is difficult to determine their frequency. Studies have been done on the epidemiology of BD, DMDD, and IED, all of which may include explosive children but certainly not all of the youth encompassed by those conditions are explosive. However, data from those studies serve to give some information.

The bipolar controversy has focused on children (vs adolescents), but epidemiologic studies have been done mainly in adolescents. In addition, some studies address the broad bipolar spectrum, others include only bipolar I mania. Overall, studies exploring broad and narrow phenotype childhood BD were conducted so differently methodologically that they are difficult to compare let alone tell anything about outbursts or true mania.[21] A recent, updated meta-analysis by Van Meter and colleagues[25] reported rates of 3.9% for the bipolar spectrum and 0.6% for bipolar I mania. The authors concluded, however, that there was significant heterogeneity across studies, with more recent studies demonstrating lower rates of BD.[25]

Copeland and colleagues[26] used data from the Great Smoky Mountains study to examine rates of DMDD in samples between ages 2 and 17 years. Between 45.7% and 80% of youth (depending on age) displayed severe tantrums at some point, with 6% to 17.7% doing so "frequently." Overall the rate of youth meeting criteria for DMDD in this study was 2.8%. Dougherty and colleagues,[27] in a community sample of 6 year olds, found 47.7% of children had a history of tantrums in the preceding 3 months, 8.2% met full criteria for DMDD, and 3.6% had tantrums with property destruction.[28] No child was physically aggressive.

In clinical samples, Margulies and colleagues[29] reported that 76.2% of psychiatrically hospitalized children had parent-described outbursts and irritability but only 17.4% actually met criteria for DMDD. In an outpatient sample, Axelson and colleagues[30] found that of their 6- to 12-year-old patients, 52% had severe, recurrent tantrums; 35% were chronically irritable; but only 26% met full criteria for DMDD. Lastly, Roy and colleagues[31] examined a sample of 51 children with tantrums, 25.5.% of whom had a predominantly irritable mood, but only 21.6% met criteria for SMD.

In summary, not all children that have significant outbursts are persistently irritable, not all irritable children have significant outbursts, and not all children who have outbursts and irritability meet criteria for DMDD. Furthermore, few of these studies address the level of severity of outbursts. Nevertheless, DMDD at least acknowledges the phenomenon of severe outbursts.

DIAGNOSES IN CHILDREN WITH OUTBURSTS

ADHD is a common co-occurring diagnosis in youth who experience emotional outbursts.[7] Emotional symptoms, including outbursts, have long been recognized as part of the symptomatology of ADHD. In the 1950s and 1960s, children diagnosed with hyperkinetic reaction of childhood were described as irritable and explosive with a low frustration tolerance.[21] In the DSM-III, these symptoms were moved to "associated symptoms" of attention deficit disorder, which included increased mood lability, low frustration tolerance, and temper outbursts. Cross-sectionally, symptoms of ADHD and a manic episode can seem similar, eg, distractibility, increased goal-directed activity, motor hyperactivity, intrusiveness, and interrupting or pressured speech. However, the quality and time-course are different. ADHD is a chronic disorder beginning in childhood. Mania occurs in episodes with functioning returning to baseline behavior between episodes and generally starts in adolescence/young adulthood.[21]

Faraone and colleagues[12] reviewed the current evidence for symptoms of emotional dysregulation in ADHD. They reported that emotional symptoms are commonly appreciated in youth with ADHD, with 40% to 50% experiencing significant anger, low distress tolerance, rages, and irritability. Furthermore, the presence of significant emotional symptoms in ADHD is associated with higher functional impairment compared with ADHD without any emotional symptoms.[12]

Two emotion dysregulation symptoms have emerged as important to the phenomenology of ADHD: emotional impulsivity, which occurs quickly with high reactivity to the provoking stimulus; and deficient emotion regulation, including the inability to prevent emotions from rising to problematic levels and slower than normal return to their emotional baseline.[12] Deficient regulation may occur with negative emotions, such as irritability leading to explosive outbursts, and with positive emotions, such as silly behavior. Faraone and colleagues[12] point out that compared with BD, the emotional symptoms of ADHD are brief and compared with DMDD, emotions return to a euthymic baseline. Authors conclude that ADHD criteria should ultimately include emotional symptoms, that novel measurement tools for emotional symptoms are needed to study and track symptoms over time, and finally, that future research is needed to develop better treatments for these symptom clusters.

Children and adolescents with psychiatric disorders other than ADHD also experience emotional outbursts. Storch and colleagues[32] studied 86 children, 6 to 16 years old, with obsessive-compulsive disorder (OCD) and "rage attacks," that is, recurrent episodes of explosive anger/aggression brought on by minimal provocation. They found that 53% of patients with OCD had experienced a rage attack in the prior month. Rage attacks were characterized by yelling or screaming (46%), threats or hitting (40%), and becoming destructive or violent behavior (9%).[32]

Budman and colleagues[7] reported that 25% to 50% of children with Tourette disorder experience behavioral outbursts with explosive anger or aggression. Their sample of 48 children diagnosed with Tourette disorder (mean age, 11.4 years old) experienced significant rage attacks averaging about 3 per week.[7] (See Marianna Ashurova and colleagues' article, "Ticked Off: Anger Outbursts and Aggressive Symptoms in Tourette's Disorder," in this issue.)

Children with anxiety disorders can also experience recurrent outbursts. In a 107-person sample of 7- to 13-year-old youth with anxiety disorders, Johnco and colleagues[33] found that 55.1% had experienced a past-week rage episode. Rages were associated with higher comorbidity (separation anxiety disorder, ADHD, ODD, and OCD), family accommodation to the child's anxiety symptoms, caregiver strain, poorer interpersonal relationship quality, and more significant functional impairment overall.[33] (See John T. Walkup and colleagues' article, "Dysregulation, Catastrophic Reactions, and the Anxiety Disorders," in this issue.)

Outbursts also occur in adult patients with depression. Winkler and colleagues[8] found 31.5% of 203 adults with depressive disorder experienced "anger attacks."

Clearly, emotional outbursts are not pathognomonic for a specific diagnosis.

TEMPER TANTRUMS IN NORMAL DEVELOPMENT

Temper tantrums are a normal part of development.[14,34] Potegal and colleagues[14] obtained parental narratives of tantrum behavior for a community sample of 335 children aged 18 to 60 months old. A tantrum was defined as an outburst of negative emotion including shouting, screaming, crying, pushing/pulling, stamping, hitting, kicking throwing, or running away. The median duration of tantrums was 3 minutes. More intense tantrums lasted longer. Kicking, throwing, and stamping occurred early in the tantrum; crying behavior progressively increased as tantrums continued. Tantrums had two components: anger and distress. Anger behaviors were, "scream and kick, hit and stiffen, shout and throw, and stamping." The distress behaviors were "whine, affiliate, and cry." The authors concluded that toddler temper tantrums are characterized by anger emotions and behaviors that occur early, peak near the beginning or middle, then decline. Distress emotions and behaviors build over tantrum duration.[14] **Table 2** compares toddler tantrums with hospitalized children's outbursts.

Wakschlag and colleagues,[34] using a thorough rating scale developed in her laboratory, looked at parents' reports of temper loss in a community sample of 1490 preschool children. Although 83.7% of children had some form of a tantrum in the previous month, only 23.5% of children hit, bit, or kicked during a tantrum. Even fewer children (8.6%) demonstrated daily tantrums. Authors concluded that problematic tantrums occurred daily; lasted longer than 5 minutes; consisted of hitting, biting, kicking, or breaking or destroying things; tantrumming to exhaustion; and having a tantrum with nonparental adults. Lastly, temper loss was associated with hyperactivity, anxiety, and depression.[34]

Carlson and colleagues[28] assessed a community sample of 462 children aged 6 years old for co-occurring tantrums and irritability. Symptoms were elicited with the Child Behavior Checklist[35] and the Preschool Age Psychiatric Assessment,[36] which unlike the K-SADS,[18] rates the type of tantrum and frequency (see **Table 1**). Eleven percent had a temper tantrum at least three times a week, and one-third of tantrums lasted more than 15 minutes. However, most of the children expressed their anger verbally, none were physically aggressive toward others during tantrums, and symptoms were only associated with a small increase in impairment.

Copeland and colleagues[37] studied irritability and outbursts in the Great Smoky Mountains community sample of 1420 youth at ages 9, 11, and 13. They conceptualized irritability as having a tonic component that is the persistently angry/grouchy mood and a phasic component defined by emotional/behavioral outbursts. They found that at any given time point, 51.4% of participants endorsed experiencing outbursts, 28.3% endorsed persistent irritable mood, 56.9% endorsed one or the other, and 22.8% endorsed experiencing both at the same time. Both phasic and tonic irritability rates decreased with age and did not differ by sex. Children experiencing only outbursts were more common than those that experienced irritability and outbursts. Least common was irritability alone. Of the participants who experienced outbursts (usually males), 15.9% had them in more than one setting (usually home and school), and 11.3% were violent during outbursts. In terms of severity, the top 95th percentile of the community sample had daily tantrums lasting 45 to 60 minutes. It is not clear whether theses outbursts were verbal, physical, or both. The average tantrum duration, however, lasted 5 minutes and occurred every 2 weeks. Nevertheless, even low levels of phasic or tonic irritability were associated with an increased risk for functional impairment, that is, being suspended from school, use of financial, health and social services, and emotional symptoms.[37]

Similarly, Belden and colleagues[38] also found longer tantrums and increased aggression in mood and behavior disordered versus normal preschool children.

In her study of IED from the National Comorbidity Study of Adolescents, McLaughlin and colleagues[39] used information from the Composite International Diagnostic Interview asking if the teenager had anger attacks during which they (1) "broke or smashed something worth more than a few dollars," (2) "hit or tried to hurt someone," and (3) "threatened to hit or hurt someone." She reported that 63.3% had at least one lifetime anger attack. Common behaviors were threats of violence (57.9%), property destruction (31.6%), and being violent (39.3%). Only 14% had a serious problem, comprised of 7.8% with IED and 6.2% with IED symptoms but also with exclusionary criteria. These latter youth had many anger attacks but no diagnosis fit for them.[38]

IMPAIRMENT

To examine what specifically contributes to impairment in children with outbursts, we re-examined an outpatient sample of 911 children[3] on whom outburst data were available. These data were gathered on a screening questionnaire called the Irritability Inventory completed by parents before their child's psychiatric evaluation (**Box 1**).

With a mean age of 12.6 years, 26.5% female, 83.8% White, and 53% in special education, 125 of 911 children (16.7%) had parents who sought psychiatric help specifically because of their child's severe outbursts and 541 children (59.4%) had anger problems, although parents were not specifically seeking help for them (**Table 3**). Outbursts occurred at home (or home and school) 80% of the time, and 20% were in school only. Physical aggression doubled the odds of help-seeking. Outbursts occurred weekly or daily in half the help-seeking sample. Furthermore, whereas half the help-seeking parents in the outpatient setting said their child hit, kicked, spit, or bit, rates were even higher in inpatient children (77.8%). Whereas about a quarter of outpatient children were said to need to be restrained by a parent sometimes, 47.8% of inpatient children's parents said that. All these children were irritable. What they did when irritable, however, was much more impairing than simply feeling grouchy.

OUTCOME

Adult outcomes of children with irritability or outbursts are complicated by the same definitional problems that confound this topic area in general. In a meta-analysis of longitudinal studies where outcome of irritability was studied, Vidal-Ribas and colleagues[37] found 12 studies where data could be combined; half included young adults. With diagnosis as the outcome, the strongest predictors (odds ratios with 95% confidence intervals) of irritability were depression (1.79 [1.42–2.27]), anxiety (1.72 [2.31–2.26]), and ODD (2.62 [1.42–4.85]). Irritability did not significantly predict BD, conduct disorder, ADHD, or substance abuse. Using continuous measures, irritability predicted future depression or depression and anxiety together with low but significant pooled effect sizes ranging from 0.12 to 0.21. Future impairment was also predicted. (See Pablo Vidal-Ribas and Argyris Stringaris' article, "How and Why are Irritability and Depression Linked?," in this issue.)

In a study tracing developmental trajectories of externalizing behaviors, namely aggression (cruelty, physical fights, threatening) and opposition (arguing, disobedient, irritable, sullen, temper tantrums), Bongers and colleagues[40] used a cohort of 1149 Dutch youth who were 4 to 16 years old at the time of initial evaluation and then interviewed regularly over the next 14 years. Parent report on the Child Behavior Checklist[35] was used to obtain information regarding externalizing behaviors. Using growth curve modeling and semiparametric mixture models, the authors arrived at several trajectories, the two relevant to this review being aggression and oppositional behavior. Findings clearly differed for the whole sample versus the most severely symptomatic youth. Thus, although on average aggressive behavior decreased over time for the whole sample, the top 8% declined from a much higher level. At young adult follow-up, their level of aggression was on a par with that of an average 4 year old and although less aggressive, they remained deviant compared with their same age and sex peers and outcomes were externalizing disorders.

Box 1
Irritability inventory: parents and teachers complete before initial interview

1. WHAT DOES YOUR CHILD DO DURING A SERIOUS TANTRUM/RAGE?

2. *HOW EASY IS IT FOR HIM/HER TO GET ANGRY?* (Please circle the letter on the left for *all that apply*)
   a. S/he rarely gets angry but when s/he does, the explosion is huge compared to the incident that provoked it.
   b. S/he is mostly reasonable but has days at a time where s/he is very touchy and gets very angry very easily.
   c. S/he has always been short-tempered and easily angered.
   d. Other (describe):

3. WHAT CAUSES HIM/HER TO GET VERY ANGRY? (Please circle for *all* that apply)
   a. Feels s/he is being criticized
   b. Misunderstands what others are saying
   c. Demands must be met immediately
   d. Can't handle change in routine
   e. Frustrated because s/he can't do something (task or activity)
   f. Other (describe):

4. WHICH OF THE FOLLOWING DOES YOUR CHILD USUALLY DO? (Please circle for *all* that apply)
   a. Expresses anger in an appropriate way
   b. Argues, whines or sulks
   c. Becomes verbally insulting, swears, shouts
   d. Threatens
   e. Slams doors, punches walls, makes a mess, destroys property
   f. Self-mutilates, bangs head, or otherwise takes it out on self
   g. Throws things
   h. Hits, kicks, bites, spits
   i. Needs physical restraint
   j. Other (describe):

5. HOW OFTEN DOES A SERIOUS TANTRUM/RAGE OCCUR? (Please circle the *one best* response)
   0 Less than 1 per month
   1 Two to three times per month
   2 Weekly
   3 Daily

6. HOW LONG DOES THE ANGER GENERALLY LAST? (Please circle the *one best* response)
   0 Usually back to normal (no longer angry) in less than 15 minutes
   1 Usually back to normal in 15 to 30 minutes
   2 Usually back to normal in 30 to 60 minutes
   3 Usually back to normal in 2 hours
   4 Usually back to normal in half a day (4 hours)
   5 Rages/tantrums last longer than half a day

7. HOW IS S/HE. MOST LIKELY TO FEEL AFTER THE EPISODE IS OVER? (Please circle the *one best* response)
   0 Sorry/remorseful; Understands the problem
   1 Blames others for what happened
   2 Continues resentful and spiteful

8. WHAT HELPS YOUR CHILD CALM DOWN ?

Adapted from Carlson GA, Danzig AP, Dougherty LR, et al. Loss of Temper and Irritability: The Relationship to Tantrums in a Community and Clinical Sample. J Child Adolesc Psychopharmacol 2016; 26(2):114-122; with permission.

Table 3
Behaviors causing parents to seek help in the Stony Brook Children's Outpatient Department
(total sample = 911)

	Any Outburst on the IRRI	Sought Help for Outbursts	Odds Ratio (95% Confidence Interval)
N (%)	541 (59.4)	125 (16.7)	
CBCL Subscale 8 (aggression), % ≥ T 67	43.8%	50.3%	1.953 (1.302– 2.928)
TRF Subscale 8 (aggression), % ≥ T 67	32.8%	52.2%	2.215 (1.474– 3.329)
Argues, whines, sulks	73.3%	74.4%	NS
Swears, shouts, insults	69.6%	80.8%	1.837 (1.132—3.007)
Threatens	37.5%	52.0%	1.806 (1.205– 2.705)
Slams doors, punches walls, destroys property, makes a mess	78.4%	79.2%	NS
Throws things	53.4%	68.8%	1.922 (1.256–2.941)
Self-mutilates, bangs head, or otherwise takes it out on self	22.8%	24.8%	NS
Hits, kicks, bites, spits	30.6%	49.6%	2.228 (1.460– 3.354)
Needs physical restraint	12.7%	28.8%	2.769 (1.706– 4.495)
Weekly or daily	33.5%	51.2%	2.061 (1.362– 3.119)
Duration >30 min	40.4%	40.7%	NS

Abbreviations: CBCL, child behavior checklist; IRRI, Irritability Inventory; TRF, teacher report form.

Oppositional behavior had a different trajectory pattern. The most severe 6% of the sample remained highly oppositional. Whereas most of the sample's diagnostic outcomes predicted internalizing disorders, the outcome for the most severely oppositional behaviors predicted disruptive behaviors (4.6 [1.2–17.7]) and anxiety disorders (3.1 [1.1–9.6]).[41,42,43] Severely oppositional behavior did not disappear.

SUMMARY

Outbursts at all levels of severity may occur in children. When they are severe, they cause significant impairment. They are currently nested in the concept of irritability, which consists of a feeling and a behavioral dimension. Irritability is a prime example of a construct with internalizing and externalizing dimensions.

The diagnostic system has not classified outbursts in a way that can expand knowledge. That is ironic given the frequency of treatment-seeking and impairment they cause. The phenomenon needs a consistent label, an operationalized way of classifying it, and an assessment approach independent of diagnosis until other data are gathered to more accurately determine what condition provides the most accurate diagnostic home.

DISCLOSURE

L. Spring, no disclosures. G.A. Carlson gets funding from NIMH; her spouse is on DSMBs for Pfizer and Lundbeck.

REFERENCES

1. Baker M, Carlson G. What do we really know about PRN use in agitated children with mental health conditions: a clinical review. Evid Based Ment Health 2018;21: 166–70.
2. Connor DF, McLaughlin TJ. A naturalistic study of medication reduction in a residential setting. J Child Adolesc Psychopharmacol 2005;15:302–10.
3. Carlson GA, Dyson M. Diagnostic implications of informant disagreement about rage outbursts: bipolar disorder or another condition? Isr J Psychiatry Relat Sci 2012;49:44–51.
4. Davis RE. Manic-depressive variant syndrome of childhood: a preliminary report. Am J Psychiatry 1979;136:702–6.
5. Biederman J, Mick E, Faraone SV, et al. Current concepts in the validity, diagnosis, and treatment of paediatric bipolar disorder. Int J Neuropsychopharmacol 2003;6:293–300.
6. Carlson GA, Potegal M, Margulies D, et al. Rages: what are they and who has them? J Child Adolesc Psychopharmacol 2009;19:281–8.
7. Budman CL, Rockmore L, Stokes J, et al. Clinical phenomenology of episodic rage in children with Tourette syndrome. J Psychosom Res 2003;55:59–65.
8. Winkler D, Pirek E, Kindle J, et al. Validation of a simplified definition of anger attacks. Psychother Psychosom 2006;75:103–6.
9. Potegal M, Kosorok MR, Davidson RJ. Temper tantrums in young children: 2. Tantrum duration and temporal organization. J Dev Behav Pediatr 2003;24: 148–54.
10. Diagnostic and statistical manual of mental disorders (DSM-5). Washington, DC: American Psychiatric Press; 2013.
11. Leibenluft E. Severe mood dysregulation, irritability, and the diagnostic boundaries of bipolar disorder in youths. Am J Psychiatry 2011;168(2):129–42.
12. Faraone SV, Rostain AL, Blader J, et al. Practitioner review: emotional dysregulation in attention-deficit/hyperactivity disorder. Implications for clinical recognition and intervention. J Child Psychol Psychiatry 2019;60:133–50.
13. Evans SC, Burke JD, Roberts MC, et al. Irritability in child and adolescent psychopathology: an integrative review for ICD-11. Clin Psychol Rev 2017;53:29–45.
14. World Health Organization. International classification of diseases for mortality and morbidity statistics (11th Revision). 2018. Available at: https://icd.who.int/browse11/l-m/en. Accessed July 20, 2020.
15. Mattes JA. Psychopharmacology of temper outbursts. A review. J Nerv Ment Dis 1986;174:464–70.
16. Wozniak J, Biederman J, Kiely K, et al. Mania-like symptoms suggestive of childhood-onset bipolar disorder in clinically referred children. J Am Acad Child Adolesc Psychiatry 1995;34:867–76.
17. Mick E, Spencer T, Wozniak J, et al. Heterogeneity of irritability in attention-deficit/hyperactivity disorder subjects with and without mood disorders. Biol Psychiatry 2005;58:576–82.
18. Kaufman J, Birmaher B, Brent D, et al. Schedule for affective disorders and Schizophrenia for school-age children-present and lifetime version (K-SADS-PL): initial reliability and validity data. J Am Acad Child Adolesc Psychiatry 1997;36(7):980–8.
19. Geller B, Sun K, Zimerman B, et al. Complex and rapid-cycling in bipolar children and adolescents: a preliminary study. J Affect Disord 1995;18:259–68.

20. Geller B, Zimmerman B, Williams M, et al. Reliability of the Washington University in St. Louis Kiddie Schedule for affective disorders and Schizophrenia (WASH-U-KSADS) mania and rapid cycling sections. J Am Acad Child Adolesc Psychiatry 2001;40(4):450–5.

21. Carlson GA, Klein DN. How to understand divergent views on bipolar disorder in youth. Ann Rev Clin Psychol 2014;10:529–51.

22. Leibenluft E, Charney DS, Towbin KE, et al. Defining clinical phenotypes of juvenile mania. Am J Psychiatry 2003;160:430–7.

23. Carlson GA. Who are the children with severe mood dysregulation a.k.a. rages. Am J Psychiatry 2007;164:1–3.

24. Carlson GA, Potegal M, Margulies D, et al. Liquid risperidone in the treatment of rages in psychiatrically hospitalized children with possible bipolar disorder. Bipolar Disord 2010;12:205–12.

25. Van Meter A, Moreira ALR, Youngstrom E. Updated meta-analysis of epidemiologic studies of pediatric bipolar disorder. J Clin Psychiatry 2019;80:e1–11.

26. Copeland WE, Angold A, Castello EJ, et al. Prevalence, comorbidity, and correlates of DSM-5 proposed disruptive mood dysregulation disorder. Am J Psychiatry 2013;170:173–9.

27. Dougherty LR, Smith VC, Bufferd SJ, et al. DSM-5 disruptive mood dysregulation disorder: correlates and predictors in young children. Psychol Med 2014;44:2339–50.

28. Carlson GA, Danzig AP, Dougherty LR, et al. Loss of temper and irritability: the relationship to tantrums in a community and clinical sample. J Child Adolesc Psychopharmacol 2016;26:114–22.

29. Margulies DM, Weintraub S, Basile J, et al. Will disruptive mood dysregulation disorder reduce false diagnosis of bipolar disorder in children? Bipolar Disord 2012;14:488–96.

30. Axelson D, Findling RL, Fristad MA, et al. Examining the proposed disruptive mood dysregulation disorder diagnosis in children in the Longitudinal Assessment of Manic Symptoms study. J Clin Psychiatry 2012;73:1342–50.

31. Roy AK, Klein RG, Angelosante A, et al. Clinical features of young children referred for impairing temper outbursts. J Child Adolesc Psychopharmacol 2013;23:588–96.

32. Storch EA, Jones AM, Lack CW, et al. Rage attacks in pediatric obsessive-compulsive disorder: phenomenology and clinical correlates. J Am Acad Child Adolesc Psychiatry 2012;51:582–92.

33. Johnco C, Salloum A, De Nadai AS, et al. Incidence, clinical correlates and treatment effect of rage in anxious children. Psychiatry Res 2015;30:63–9.

34. Wakschlag LS, Choi SW, Carter AS, et al. Defining the developmental parameters of temper loss in early childhood: implications for developmental psychopathology. J Child Psychol Psychiatry 2012;53:1099–108.

35. Achenbach TM, Rescorla LA. Manual for ASEBA preschool forms and profiles. Burlington (VT: University of Vermont, Research Center for Children, Youth, and Families; 2001.

36. Egger HL, Ascher BH, Angold A. The preschool age psychiatric assessment: version 1.1. Durham (NC): Center for Developmental Epidemiology, Department of Psychiatry and Behavioral Sciences, Duke University Medical Center; 1999.

37. McLaughlin KA, Green JG, Hwang I, et al. Intermittent explosive disorder in the National Comorbidity Survey replication adolescent supplement. Arch Gen Psychiatry 2012;69:1131–9.

38. Copeland WE, Brotman MA, Costello EJ. Normative irritability in youth: developmental findings from the Great Smoky Mountains Study. J Am Acad Child Adolesc Psychiatry 2015;54:635–42.
39. Belden AC, Thomson NR, Luby JL. Temper tantrums in healthy versus depressed and disruptive preschoolers: defining tantrum behaviors associated with clinical problems. J Pediatr 2008;152:117–22.
40. Bongers IL, Koot HM, van der Ende J, et al. Developmental trajectories of externalizing behaviors in childhood and adolescence. Child Dev 2004;75:1523–37.
41. Reef J, Diamantopoulou S, van Meurs I, et al. Developmental trajectories of child to adolescent externalizing behavior and adult DSM-IV disorder: results of a 24-year longitudinal study. Soc Psychiatry Psychiatr Epidemiol 2011;46:1233–41.
42. Reef J, van Meurs I, Verhulst FC, et al. Children's problems predict adults' DSM-IV disorders across 24 years. J Am Acad Child Adolesc Psychiatry 2010;49:1117–24.
43. Potegal M, Carlson GA, Margulies D, et al. The behavioral organization, temporal characteristics, and diagnostic concomitants of rage outbursts in child psychiatric inpatients. Curr Psychiatry Rep 2009;11:127–33.

Measurement of Dysregulation in Children and Adolescents

Robert R. Althoff, MD, PhD[a],*, Merelise Ametti, MA, MPH[b]

KEYWORDS

- Review • Behavior rating scale • Interview, psychological • Aggression • Irritability
- Emotional regulation • Executive function • Child

KEY POINTS

- There are numerous instruments available for assessing dysregulation during childhood and adolescence, including rating scales, clinical interviews, and observational measures; however, no gold standard exists.
- A few measures broadly assess challenges with self-regulation, whereas most focus on specific aspects of emotion dysregulation, such as aggression, outbursts, or irritability.
- To thoroughly assess emotion dysregulation, it is recommended that future studies include measures of broad psychological functioning, chronic mood regulation problems, and outbursts.

DYSREGULATION AS AN OVERARCHING CONCEPT

In everyday psychiatric vernacular, the term "dysregulated" is often used to describe youth who exhibit severe behavioral and affective outbursts. However, from a scientific perspective, the definition and measurement of such a clinical phenotype has been inconsistent, at best. This inconsistency hinders the ability to accurately characterize and effectively provide treatment to these youth who are among those most in need. Without a clear definition of the construct and methods for measuring it, we struggle to ascertain even basic information about this population. For example, prevalence rates of "dysregulated" children among outpatient psychiatry clinics can vary widely (ie, 7%–45%) depending on how the construct is measured.[1,2] Until the term dysregulation is operationalized, we will not be able to effectively explore deeper questions, such as what causes these disturbances, and how can they be rectified.

[a] Department of Psychiatry, University of Vermont, Vermont Center for Children, Youth and Families, Arnold 3421, 1 South Prospect Street, Burlington, VT 05401, USA; [b] Department of Psychiatry, University of Vermont, Vermont Center for Children, Youth and Families, Arnold 3421A, 1 South Prospect Street, Burlington, VT 05401, USA
* Corresponding author.
E-mail address: ralthoff@med.uvm.edu
Twitter: @childpsychvt (R.R.A.)

Child Adolesc Psychiatric Clin N Am 30 (2021) 321–333
https://doi.org/10.1016/j.chc.2020.10.004
1056-4993/21/© 2020 Elsevier Inc. All rights reserved.

From our perspective, dysregulation in children and adolescents is best conceptualized as a multifaceted construct that encompasses various distinct, yet related, processes involved with the management of emotions, behaviors, and cognitions.[3] Furthermore, we have argued that a comprehensive understanding of dysregulation in youth requires an examination of difficulties within each of these domains of regulation in addition to a broad, superordinate scale that combines functioning across all domains.[4] The complex and multifaceted nature of dysregulation produces similar measurement challenges as those that have been encountered in psychopathology research generally. Some central questions that arise include whether it should be measured dimensionally, categorically, or through some combination of the two approaches (eg, in a hierarchical model).[5–7] Furthermore, it is unclear whether measures should be based on self-report, parent-report, interviews, or mechanisms.[8–11] Because an exhaustive review of all previous efforts to describe the construct of dysregulation in children and adolescents is beyond the scope of this article, we concentrate primarily on the aspects of dysregulation that are most relevant to the other papers in this issue: emotional dysregulation. However, the measurement of emotional dysregulation overlaps with the measurement of other aberrant behaviors, such as aggression. Because aggressive behavior is a common manifestation of more severe emotional dysregulation, many of the instruments described here were either first developed as indices of aggression or have a strong aggression component. We caution the reader that aggression and emotional dysregulation are not the same construct and acknowledge that emotional dysregulation can manifest without any outward aggression. We concentrate here, however, on those clinically problematic and common situations where emotional dysregulation is associated with behavioral "outbursts."

It is attractive to attempt to categorize all of the ways to measure dysregulation by reading and characterizing the entire literature on the development of youth self-regulation. Having gone down that rabbit hole like Alice in Wonderland we have decided instead to concentrate only on emotional dysregulation, to avoid things becoming "curiouser and curiouser." Even with this focus on emotional dysregulation, after reviewing the scientific literature and consulting with experts in the field, we identified nearly 50 different methods for measuring the facets of emotion dysregulation without any clearly established gold standard. Therefore, we review a subset of these measures that represent the current measurement landscape. Although each measure on its own has unique strengths and weaknesses, we caution that the proliferation of measures of emotion dysregulation may also pose a threat to the integrity of the body of knowledge by limiting generalizability across studies. As such, we have organized measures of emotional dysregulation into broad measures of dysregulation and psychopathology and tools that are specific to emotional dysregulation and its behavioral manifestations, with the objective of helping guide researchers and clinicians in effectively assessing this complex construct.

BROAD MEASURES OF DYSREGULATION

Several measures of dysregulation have been derived from broadband measures of psychological functioning, which are often used as routine screening measures in mental health treatment settings.

Child Behavior Checklist Dysregulation Profile

The Child Behavior Checklist Dysregulation Profile (CBCL-DP) is a 113-item caregiver-report measure of a child's overall emotional, behavioral, and social functioning during the past 6 months. A set of empirically derived constructs from the CBCL have been used to measure dysregulation in children.[12] The CBCL was first used to measure dysregulation in the context of research on juvenile bipolar disorder, because it was identified that youth diagnosed with juvenile bipolar disorder were consistently rated by their caregivers as having elevations on the Anxious/Depressed, Attention Problems, and Aggressive Behavior subscales of the CBCL.[9] This profile, originally termed the CBCL-Juvenile Bipolar Disorder or CBCL-Mania profile, was further studied and later found to not be specific to youth with juvenile bipolar disorder or predictive of bipolar disorder in adulthood.[13–15] Rather, evidence suggested that this profile reflected a transdiagnostic construct of dysregulation that was associated with a multitude of psychological disorders, including substance use, anxiety and depression, and personality disorders.[6,16] As such, the profile has been rebranded as the CBCL-DP and has been widely used to examine problems with emotion regulation.[6,17,18] Most recently, attempts have been made to isolate the global effects of dysregulation from its specific components (ie, anxious/depressed, attention problems, aggressive behavior symptoms) using a bifactor model.[7] Results have showed that dysregulation predicts negative outcomes, such as suicidality, whereas specific symptom clusters do not. This suggests that broad difficulties with self-regulation may present unique clinical challenges. Furthermore, the CBCL-DP has been used to study biologic markers, cognitive function, and clinical outcomes.[19–21] Because the CBCL has excellent psychometric properties and is already widely used in research and clinical practice, it has immense potential to further the study of dysregulation without requiring clinicians or patients to spend time completing an additional measure.

Strengths and Difficulties Questionnaire–Dysregulation Profile

The Strengths and Difficulties Questionnaire (SDQ) is a 25-item, publicly available questionnaire included in the Development and Well-Being Assessment (DAWBA) family of instruments. This efficient screening tool has positively and negatively valanced items that are scored into five scales: (1) Emotional Symptoms, (2) Conduct Problems, (3) Hyperactivity/Inattention, (4) Peer Relationship Problems, and (5) Prosocial Behavior.[22] The SDQ and DAWBA have been used in many large data collection studies from which a five-item dysregulation scale (SDQ-DP) has been derived, which consists of two questions from the Emotional Symptoms and Conduct Problems scales and one question from the Hyperactivity/Inattention scale. Holtmann and colleagues[23] first validated this scale of the SDQ, demonstrating that it accurately corresponded with the characterizations based on CBCL-DP in 81% of cases. Subsequently, it has been demonstrated the SDQ-DP is associated with higher rates of severe psychopathology and the development of psychopathology later in life.[24,25] Both the five-item scale (and a 15-item scale, which includes all items of the Emotional Symptoms, Conduct Problems and Hyperactivity/Inattention Scales), have been examined in a bifactor model, and, like the CBCL-DP, reflects a broad syndrome of deficient self-regulation that is separable from challenges in each of its constituent domains.[26] This broad construct of dysregulation has been associated with nonsuicidal self-injury and higher post-traumatic stress symptoms in female adolescents following sexual assault.[27,28]

The CBCL-DP and SDQ-DP are advantageous because they allow clinicians to quickly assess a patient's overall level of dysregulation and permit researchers to examine the construct within a broadly phenotyped dataset. Nonetheless, these instruments have been criticized as simply measuring overall psychopathology rather than dysregulation specifically. Despite some evidence suggesting that dysregulation and a general psychopathology (or "p") factor are highly related but unique constructs, it could be reasonably argued that the overall propensity for expressing psychopathology may be related to one's ability to regulate emotion, and therefore, to some extent, the CBCL-DP and SDQ-DP scales index an individual's overall likelihood for psychopathology.[29] Therefore, we view the use of broad instruments, such as the CBCL-DP and SDQ-DP, as critical in the understanding of the progression of specific disorders if, for no other reason, than to control for a general propensity for any psychopathology in the development of a disorder. For instance, to understand the development of attention-deficit/hyperactivity disorder (ADHD) from a set of underlying risk factors, it might be best to control for a subject's overall propensity toward developing psychopathology by using one of these measures. McGough and coworkers[30] used this in examining electroencephalographic findings in ADHD, which revealed that all electroencephalographic findings between ADHD participants and control subjects were seen in a group that had elevations on the CBCL-DP.

Of note, several standardized diagnostic interviews (eg, K-SADS, DAWBA, CAPA, and DISC) have also been used to measure emotional dysregulation, which we have recently described elsewhere.[31] As such, we do not describe them again here.

MEASURES SPECIFIC TO EMOTION DYSREGULATION OR RELATED CONCEPTS

In contrast to broadband measures of dysregulation, several measures have been developed to specifically measure emotion (dys)regulation and related concepts, such as overt aggression, outbursts, and irritability.

Clinician-Administered Measures

The Overt Aggression Scale/Modified Overt Aggression Scale

The Overt Aggression Scale (OAS) is an observational measure designed to assess for the frequency and severity of different forms of violent behavior, including verbal aggression and physical aggression against objects, self, and others, within the context of an aggressive episode. The number of violent behaviors exhibited during the aggressive episode are summed, with acts of physical aggression toward self or others weighted more heavily than acts of aggression expressed verbally or toward objects. The original OAS measure was developed in children and adults being treated on an inpatient psychiatric unit and aggressive behaviors were observed and recorded by nurses and mental health aides.[32] A modified, retrospective version of this instrument was also developed to minimize the burden of frequent documenting and several other limitations encountered using the OAS.[33]

The Clinician Affective Reactivity Index

The Clinician Affective Reactivity Index (CL-ARI) is a semistructured interview designed to be conducted by a trained (master's level or higher) clinician with parents and children.[34] The interview contains 12 questions intended to obtain information about the frequency, duration, and severity of the child's temper outbursts and irritable mood and the child's level of functional impairment in school, family, and peer settings during the past week (a 3-day version is also available). Parents and children are asked the same questions separately and then the clinician rates each item on a six-point scale based on synthesis of reports and clinical judgment. Scores for the temper outburst, irritable mood, and impairments subscales are summed, such that higher scores reflect more severe impairment. The CL-ARI has good internal consistency ($\alpha = 0.89$) and adequate test-retest reliability over a 1-week period (intraclass correlation coefficient = .67) when tested in a subset of children with disruptive mood dysregulation disorder only. The CL-ARI total score showed moderate-to-high correlations with other clinician-rated measures of psychopathology (specifically, irritability, anxiety, and depression) and impairment, and moderate-to-low correlations with parent- and youth-report measures of psychopathology. Furthermore, the CL-ARI scores differed significantly between youth with disruptive mood dysregulation disorder, ADHD, and healthy control subjects.[34]

Children's Agitation Inventory

The Children's Agitation Inventory (CAI) is a 31-item checklist of behaviors identified by nursing staff as occurring during outbursts among children in an inpatient psychiatry unit. The measure systematically captures various aspects of aggression directed at property, others, and self; verbal expressions; mood and psychiatric symptoms; and level of cooperativeness. The presence of each of these behaviors is rated from the onset of the episode (eg, refusal of time out) and subsequently at 5, 15, 30, 45, 60, and 120 minutes thereafter. This measure was used previously on inpatient units and allows for direct observation of outbursts by trained clinical staff and provides insights into differences in the time required for children to regulate and the ways that the behaviors manifested may vary as a function of time.[35] However, the psychometric properties of the CAI have not yet been fully tested. Furthermore, the extent to which the behaviors described in an inpatient setting are generalizable to less severe outburst and outpatient populations is unclear. Finally, the CAI is a time-intensive measure, which may not be feasible in some settings.

Disappointing Gift Task

The Disappointing Gift Task is a widely used interactive assessment of children's affective and behavioral regulation.[36] Although several adaptions of this paradigm have been published,[37] in all cases, the experimenter presents the child with a wrapped gift containing an undesirable present (eg, a wood chip). Then, video recordings of the child's reactions on unwrapping the gift, including positive and negative facial expressions, vocalizations, body movements, and gaze direction are coded, with more negative reactions generally considered indicative of poorer emotion regulation ability.[37] The Disappointing Gift Task provides a more objective measure of differences in emotion regulation strategies than parent-report or child self-report measures by controlling some of the contextual factors. For research purposes, this instrument may be useful (although the coding is labor intensive) in characterizing emotional dysregulation especially in conjunction with other rating scales of emotion regulation; however, it is unlikely to be particularly useful in clinical practice.

Caregiver-Report Measures

Retrospective Modified Overt Aggression Scale

The Retrospective MOAS (R-MOAS) is an adaption of the OAS that is designed to capture the frequency of the same behaviors related to verbal aggression and physical aggression against objects and self. The measure contains 16 items that are rated in frequency during the past week on a Likert scale by a caregiver. Items related to more severely harmful behaviors are weighted, such that they contribute more to the individual's total R-MOAS score. In Blader and colleagues,[38] adequate psychometrics are reported, including internal consistency ($\alpha = 0.82$), convergent validity ($r = 0.52$ with CBCL Aggressive Behavior subscale), and divergent validity ($r = 0.22$ with CBCL Internalizing Problems scale). It has been used in children with ADHD, children treated with methylphenidate, adults with intellectual disabilities, and autism spectrum disorders (ASD).[38–41]

Outburst Monitoring Scale

The Outburst Monitoring Scale (OMS) is a brief, parent-report questionnaire that assesses various types of aggressive behaviors exhibited by the child during the past week. The measure was developed in accordance with the categories of aggressive behavior identified in the OAS with the objective of establishing a parent-report measure that could be used in conjunction with and directly compared with observational ratings obtained with these measures. Like the OAS, the OMS assesses acts of verbal aggression, and aggression toward property, self, and others; however, the OMS includes additional questions relating to aggression directed specifically toward adults and serious self-injurious behaviors. The frequency of each behavior is rated by parents on a five-point scale ranging from 0 (not at all) to 4 (three or more times in a typical day). Although the measure was developed and validated in a small sample (n = 23) that included only adolescents, initial findings showed promise, including moderate-to-high internal consistencies among subscales, significant correlations with other measures of disruptive behavior. Furthermore, OMS scores showed sensitivity to treatment effects when administered during the course of a small open-label trial of methylphenidate and quetiapine[42] and a clinical interview version has been used in the 5- to 9-year age group to measure outburst severity and performed well in distinguishing children with "frequent, impairing outbursts" from control subjects.[43]

Emotion Dysregulation Inventory

The Emotion Dysregulation Inventory (EDI) is a novel caregiver-report measure of problems with emotional control developed specifically for use among youth with ASD. The measure was designed in accordance with the National Institutes of Health Patient Reported Outcomes Measurement Information System (PROMIS) guidelines.[44] As such, item selection was based on literature review, conceptual models, and expert input and further refined based on trials for comprehension and readability among parents (see Mazefsky and colleagues[45] for detailed review of measurement development procedures) and item response theory calibrations.[44,45] The final version of the EDI consisted of 30 items related to different observable indicators of poor emotion regulation noted during the past 7 days. The degree to which each behavior presented a problem for the child was rated from "not at all" to "very severe" on a visual analog scale of a thermometer. The EDI items were tested and refined in a large sample of more than 1700 youth through which two factors of emotion dysregulation, reactivity and dysphoria, were derived.[45] Additionally, a seven-item short form for reactivity and six-item short form for dysphoria are available. The factor structure of the EDI was validated in a large sample of 1755 youth with ASD and was shown to be consistent across inpatient and community youth with ASD. In addition, the EDI correlates with other measures of emotion dysregulation, including the Aberrant Behavior Checklist and the CBCL. Test-retest reliability was stable over a 4-week period; however, the measure also showed sensitivity to treatment effects when administered on hospital admission and discharge.[45,46]

Emotion Regulation Checklist

The Emotion Regulation Checklist consists of 24 parent-rated items that assess how their child manages emotional experiences. It is scored on a four-point Likert scale and has been determined to yield two subscales: Lability/Negativity and Emotion Regulation. Psychometrics have shown internal consistency with Cronbach α between 0.80 and 0.96 for the two subscales and 0.89 for the entire scale.[45,47] It has been used commonly to examine the emotional responses of a child to traumatic stress and to measure physiologic responses to emotional situations.[48–50] The measure has an analogous teacher-report version and has been translated into various languages.[51]

Affective Reactivity Index

The Affective Reactivity Index (ARI) is a brief measure of the severity of a child's irritability symptoms during the past 6 months. The measure has analogous caregiver and youth self-report versions, which consist of six symptom items rated on a three-point scale of 0 (not true) and 2 (certainly true), which are summed to generate a total score. In addition, the ARI contains one impairment item about the degree to which the child's challenges with irritability cause problems in his or her life. The items in the ARI have been factor analytically demonstrated to reflect a single underlying irritability dimension and have analogous parent and youth self-report versions.[52] The ARI has been used and tested in several countries, including the United States, United Kingdom, and Brazil.[52,53] The ARI has strong psychometric properties, including internal consistencies ($\alpha = 0.88-0.92$), parent-child agreement ($r = 0.58$ in the United States and $r = 0.73$ in the United Kingdom), and longitudinal stability over a 1-year period ($r = 0.88$, for parent-report version only). Scores on the ARI were able to discriminate between children with different diagnoses, including bipolar disorder, severe mood dysregulation, and healthy controls. In addition, ARI scores converge with Emotional and Conduct Problem scores on the SDQ; however, in multivariate models, irritability scores on the ARI were the only predictor of Emotional Problems.[52] The psychometric properties of the ARI have been supported in a cross-cultural adaptation of the measure in Brazilian youth. Furthermore, item response theory analysis showed that the item "often loses temper" had the highest ability to discriminate youth in terms of their severity on the latent irritability trait.[53]

Youth Self-Report Measures

The Regulation of Emotions Questionnaire

The Regulation of Emotions Questionnaire (REQ) is a 19-item, self-report measure that assesses the frequency with which adolescents use various emotion regulation strategies. All items are rated on a Likert scale, ranging from 1 (never) to 5 (always). The measure assesses two dimensions of emotion regulation strategies (ie, functional vs dysfunctional and external vs internal resources), thereby yielding four factor-analytically derived subscales (ie, internal-functional, internal-dysfunctional, external-functional, external-dysfunctional). Although the REQ has not been widely used as a measure of emotion regulation, initial evidence has shown good psychometric properties, including internal consistency coefficients between 0.63 and 0.73, test-retest reliabilities of 0.70, significant positive correlations between higher levels of maladaptive emotion regulation strategies and parent-reported emotional-behavioral challenges and self-reported life dissatisfaction, and psychosomatic concerns.[54-56] Additionally, the items and factor structure of the REQ has been replicated across US and Turkish societies.[54,56]

Abbreviated Dysregulation Inventory

The Abbreviated Dysregulation Inventory is a 30-item, self-report questionnaire designed to assess emotional, cognitive, and behavioral aspects of psychological dysregulation among youth. Each item is rated on a Likert scale from 0 (never) to 3 (always true). Higher scores on the emotional and behavioral subscales indicate more dysregulation, whereas lower scores on the cognitive subscale indicate more dysregulation.[57] The Abbreviated Dysregulation Inventory was reportedly developed based on the items with the highest discriminant coefficients in the original Dysregulation Inventory (although the results of this analysis are not published). The three dimensions of this measure have been empirically validated through confirmatory factor analysis. In addition, the measure shows all subscales to have good internal consistency (α = 0.85–0.88) and good test-retest reliability over a 4-week period (r = 0.73–0.87). The emotional and behavioral dysregulation subscales were moderately associated with other measures of aggression, whereas the cognitive dysregulation subscale was not.[57]

Difficulties in Emotion Regulation Scale

The Difficulties in Emotion Regulation Scale (DERS) is a 36-item, self-report questionnaire that comprehensively assesses emotion regulation in six factor analytically derived domains (ie, nonacceptance of emotions, difficulties engaging in goal-directed behavior, impulse control difficulties, lack of emotional awareness, limited access to regulation strategies, and lack of emotional clarity). Each item is rated on a scale that reflects how often the item applies to the individual, ranging from 1 (almost never) to 5 (almost always). The initial adult version of the DERS demonstrated strong psychometric properties, including internal consistency (α = 0.93) and test-retest reliability (r = 0.88 over a period of 4–8 weeks).[58] Subscales of the DERS correlated in the expected direction with other measures of emotion regulation and experiential avoidance; however, only the awareness, access to strategies, and emotional clarity domains were associated with emotional expressivity. The DERS showed predictive validity of self-harm and history of partner abuse.[58] These psychometric properties and factor structure have replicated well in other samples, including adolescents, inpatient settings, and cross-cultural applications.[59–61] Most recently, a short-form of the DERS (DERS-SF) has been developed, which has retained the strong psychometric properties of the original measure.[62]

Emotion Regulation Questionnaire for Children and Adolescents

The Emotion Regulation Questionnaire for Children and Adolescents (ERQ-CA) is a 10-item, self-report questionnaire that assess children and adolescents' emotion regulation strategies commonly used in everyday life.[63] The ERQ-CA was created based largely on the previously developed Emotion Regulation Questionnaire for adults.[64] For each item, the youth rates the degree to which he/she agrees on a five-point scale from 1 (strongly disagree) to 5 (strongly agree). These items comprise two factor-analytically derived subscales of cognitive reappraisal and emotional suppression, each of which yield separate total scores. Psychometric testing of the ERQ-CA has shown good internal consistency (α = 0.69–0.85) and stability over a 12-month period (intraclass correlation coefficient = 0.37–0.57). The two factors of emotion regulation (cognitive reappraisal and emotional suppression) identified in the adult version of the survey were supported in the youth sample. The youth measure also demonstrated convergent validity, such that scores on the emotional suppression subscale were associated with higher levels of depression and neuroticism, and lower levels of extraversion, whereas higher scores on cognitive reappraisal subscale was associated with lower levels of depression and more extraversion.[63]

SUMMARY AND RECOMMENDATIONS

We have reviewed several broadband measures and measures specific to emotional dysregulation and its related manifestations. Although we were not able to exhaustively explore all measures (with apologies to authors whose measures may have been excluded), the current article was intended to provide a sense of the breadth of instruments available for measuring emotion dysregulation in children and adolescents. Although there were notable differences across the measures discussed (eg, informant, length, population, factor structures), nearly all measures of emotion dysregulation predicted serious negative sequelae. It is our hope that the American Academy of Child and Adolescent Psychiatry Presidential Initiative on Emotion Dysregulation can make progress toward establishing best practices for measuring the construct to streamline future research and work more expediently toward the development of treatments for patients with severe emotional dysregulation and outbursts.

Although each of the previously discussed instruments has its unique merits and limitations, current recommendations for the measurement of emotion dysregulation in new studies are to include (1) a broadband measure of psychological functioning (eg, the CBCL or SDQ); (2) a measure of chronic mood regulation problems (eg, the self- and/or parent-report ARI); and (3) characterization of the outbursts, although measures of outbursts themselves are admittedly the least well studied of this group and more work needs to be done to create a good measure of the structure of outbursts and the impairment they cause. For the moment, these instruments permit some examination of psychopathology broadly along with baseline irritability and behavior in the context of an outburst. We need a consistent approach by researchers and clinicians alike to gain a better grasp of the complex ways in which emotion regulation operates in and interacts with other aspects of children's' and adolescents' lives.

CLINICS CARE POINTS

- Many measures of emotion regulation difficulties in children and adolescent exist and appear to capture risk for adverse psychosocial outcomes.
- Assessment of emotion regulation problems may be enhanced by including a measure of broad psychological functioning together with measures of chronic mood problems and temper outbursts.
- Greater consistency in measurement approaches is needed in order to improve the coherency of the research body and its applicability to clinical care.

DISCLOSURE

Dr R. Althoff receives grant or research support from NIH and The Klingenstein Third Generation Foundation. He is a partner in WISER Systems, LLC and receives an honorarium as Associate Editor for the Journal of the American Academy of Child and Adolescent Psychiatry and for CME presentations for the MGH Psychiatry Academy. Ms M. Ametti has no disclosures or conflicts of interest to declare.

REFERENCES

1. Holtmann M, Goth K, Wöckel L, et al. CBCL-pediatric bipolar disorder phenotype: severe ADHD or bipolar disorder? J Neural Transm 2008;115(2):155–61.
2. Aitken M, Battaglia M, Marino C, et al. Clinical utility of the CBCL dysregulation profile in children with disruptive behavior. J Affect Disord 2019;253:87–95.
3. Althoff RR. Dysregulated children reconsidered. J Am Acad Child Adolesc Psychiatry 2010;49(4):302–5.
4. Ayer L, Althoff R, Ivanova M, et al. Child behavior checklist juvenile bipolar disorder (CBCL-JBD) and CBCL posttraumatic stress problems (CBCL-PTSP) scales

are measures of a single dysregulatory syndrome. J Child Psychol Psychiatry 2009;50(10):1291–300.

5. Biederman J, Petty CR, Day H, et al. Severity of the aggression/anxiety-depression/attention child behavior checklist profile discriminates between different levels of deficits in emotional regulation in youth with attention-deficit hyperactivity disorder. J Dev Behav Pediatr 2012;33(3):236–43.

6. Althoff RR, Verhulst FC, Rettew DC, et al. Adult outcomes of childhood dysregulation: a 14-year follow-up study. J Am Acad Child Adolesc Psychiatry 2010; 49(11):1105–16.

7. Deutz MH, Geeraerts SB, van Baar AL, et al. The dysregulation profile in middle childhood and adolescence across reporters: factor structure, measurement invariance, and links with self-harm and suicidal ideation. Eur child Adolesc Psychiatry 2016;25(4):431–42.

8. Althoff RR, Rettew DC, Ayer LA, et al. Cross-informant agreement of the dysregulation profile of the child behavior checklist. Psychiatry Res 2010;178(3):550–5.

9. Mick E, Biederman J, Pandina G, et al. A preliminary meta-analysis of the child behavior checklist in pediatric bipolar disorder. Biol Psychiatry 2003;53(11): 1021–7.

10. Rich BA, Brotman MA, Dickstein DP, et al. Deficits in attention to emotional stimuli distinguish youth with severe mood dysregulation from youth with bipolar disorder. J abnorm child Psychol 2010;38(5):695–706.

11. Bubenzer-Busch S, Herpertz-Dahlmann B, Kuzmanovic B, et al. Neural correlates of reactive aggression in children with attention-deficit/hyperactivity disorder and comorbid disruptive behaviour disorders. Acta psychiatr Scand 2016;133(4): 310–23.

12. Achenbach TM, Rescorla LA. Manual for the ASEBA school-age forms and profiles. Burlington (VT): University of Vermont Research Center for Children, Youth and Families; 2001.

13. Galanter CA, Carlson GA, Jensen PS, et al. Response to methylphenidate in children with attention deficit hyperactivity disorder and manic symptoms in the multimodal treatment study of children with attention deficit hyperactivity disorder titration trial. J Child Adolesc Psychopharmacol 2003;13(2):123–36.

14. Hudziak JJ, Althoff RR, Derks EM, et al. Prevalence and genetic architecture of child behavior checklist–juvenile bipolar disorder. Biol Psychiatry 2005;58(7): 562–8.

15. Youngstrom E, Meyers O, Demeter C, et al. Comparing diagnostic checklists for pediatric bipolar disorder in academic and community mental health settings. Bipolar Disord 2005;7(6):507–17.

16. Meyer SE, Carlson GA, Youngstrom E, et al. Long-term outcomes of youth who manifested the CBCL-pediatric bipolar disorder phenotype during childhood and/or adolescence. J Affect Disord 2009;113(3):227–35.

17. Surman CB, Biederman J, Spencer T, et al. Understanding deficient emotional self-regulation in adults with attention deficit hyperactivity disorder: a controlled study. Atten Defic Hyperact Disord 2013;5(3):273–81.

18. Frazier JA, Wood ME, Ware J, et al. Antecedents of the child behavior checklist-dysregulation profile in children born extremely preterm. J Am Acad Child Adolesc Psychiatry 2015;54(10):816–23.

19. Poustka L, Zohsel K, Blomeyer D, et al. Interacting effects of maternal responsiveness, infant regulatory problems and dopamine D4 receptor gene in the development of dysregulation during childhood: a longitudinal analysis. J Psychiatr Res 2015;70:83–90.

20. Basten M, van der Ende J, Tiemeier H, et al. Nonverbal intelligence in young children with dysregulation: the generation R study. Eur child Adolesc Psychiatry 2014;23(11):1061–70.

21. Masi G, Muratori P, Manfredi A, et al. Child behaviour checklist emotional dysregulation profiles in youth with disruptive behaviour disorders: clinical correlates and treatment implications. Psychiatry Res 2015;225(1):191–6.

22. Goodman A, Goodman R. Strengths and difficulties questionnaire as a dimensional measure of child mental health. J Am Acad Child Adolesc Psychiatry 2009;48(4):400–3.

23. Holtmann M, Becker A, Banaschewski T, et al. Psychometric validity of the strengths and difficulties questionnaire-dysregulation profile. Psychopathology 2011;44(1):53–9.

24. Carballo JJ, Serrano-Drozdowskyj E, Nieto RG, et al. Prevalence and correlates of psychopathology in children and adolescents evaluated with the strengths and difficulties questionnaire dysregulation profile in a clinical setting. Psychopathology 2014;47(5):303–11.

25. Kunze B, Wang B, Isensee C, et al. Gender associated developmental trajectories of SDQ-dysregulation profile and its predictors in children. Psychol Med 2018;48(3):404–15.

26. Deutz MHF, Shi Q, Vossen HGM, et al. Evaluation of the strengths and difficulties questionnaire-dysregulation profile (SDQ-DP). Psychol Assess 2018;30(9):1174–85.

27. Caro-Canizares I, Garcia-Nieto R, Diaz de Neira-Hernando M, et al. The SDQ dysregulation profile is associated with self-injurious thoughts and behaviors in adolescents evaluated at a clinical setting. Rev Psiquiatr Salud Ment 2019;12(4):242–50.

28. Villalta L, Khadr S, Chua K-C, et al. Complex post-traumatic stress symptoms in female adolescents: the role of emotion dysregulation in impairment and trauma exposure after an acute sexual assault. Eur J Psychotraumatol 2020;11(1):1710400.

29. Deutz MHF, Geeraerts SB, Belsky J, et al. General psychopathology and dysregulation profile in a longitudinal community sample: stability, antecedents and outcomes. Child Psychiatry Hum Dev 2020;51(1):114–26.

30. McGough JJ, McCracken JT, Cho AL, et al. A potential electroencephalography and cognitive biosignature for the child behavior checklist–dysregulation profile. J Am Acad Child Adolesc Psychiatry 2013;52(11):1173–82.

31. Ametti M, Althoff R. Measurement of irritability in children and adolescents. In: RA K, Brotman MA, Leibenluft E, editors. Irritability in pediatric psychopathology. New York: Oxford University Press; 2019.

32. Yudofsky SC, Silver JM, Jackson W, et al. The overt aggression scale for the objective rating of verbal and physical aggression. Am J Psychiatry 1986;143(1):35–9.

33. Sorgi P, Ratey JJ, Knoedler DW, et al. Rating aggression in the clinical setting: a retrospective adaptation of the overt aggression scale: preliminary results. J neuropsychiatry Clin neurosci 1991;3(2):S52–6.

34. Haller SP, Kircanski K, Stringaris A, et al. The clinician affective reactivity index: validity and reliability of a clinician-rated assessment of irritability. Behav Ther 2020;51(2):283–93.

35. Potegal M, Carlson GA, Margulies D, et al. The behavioral organization, temporal characteristics, and diagnostic concomitants of rage outbursts in child psychiatric inpatients. Curr Psychiatry Rep 2009;11(2):127–33.

36. Saarni C. An observational study of children's attempts to monitor their expressive behavior. Child Dev 1984;55(4):1504–13.

37. Patel R. Emotion Regulation and The Disappointing Gift Task: Implications for Understanding Children's Development. The Undergraduate Research Journal of Psychology at UCLA 2018;5:54–69.

38. Blader JC, Schooler NR, Jensen PS, et al. Adjunctive divalproex versus placebo for children with ADHD and aggression refractory to stimulant monotherapy. Am J Psychiatry 2009;166(12):1392–401.

39. Blader JC, Pliszka SR, Jensen PS, et al. Stimulant-responsive and stimulant-refractory aggressive behavior among children with ADHD. Pediatrics 2010; 126(4):e796–806.

40. Oliver PC, Crawford MJ, Rao B, et al. Modified overt aggression scale (MOAS) for people with intellectual disability and aggressive challenging behaviour: a reliability study. J Appl Res Intellect Disabil 2007;20(4):368–72.

41. Ratey JJ, Gutheil CM. The measurement of aggressive behavior: reflections on the use of the overt aggression scale and the modified overt aggression scale. J neuropsychiatry Clin neurosci 1991;3(2):S57–60.

42. Kronenberger WG, Giauque AL, Dunn DW. Development and validation of the outburst monitoring scale for children and adolescents. J child Adolesc Psychopharmacol 2007;17(4):511–26.

43. Roy AK, Klein RG, Angelosante A, et al. Clinical features of young children referred for impairing temper outbursts. J child Adolesc Psychopharmacol 2013;23(9):588–96.

44. Cella D, Yount S, Rothrock N, et al. The patient-reported outcomes measurement information system (PROMIS): progress of an NIH roadmap cooperative group during its first two years. Med Care 2007;45(5 Suppl 1):S3.

45. Mazefsky CA, Day TN, Siegel M, et al. Development of the emotion dysregulation inventory: a PROMISing method for creating sensitive and unbiased questionnaires for autism spectrum disorder. J Autism Dev Disord 2018;48(11):3736–46.

46. Mazefsky CA, Yu L, White SW, et al. The emotion dysregulation inventory: psychometric properties and item response theory calibration in an autism spectrum disorder sample. Autism Res 2016;11(6):928–41.

47. Shields A, Cicchetti D. Emotion regulation among school-age children: the development and validation of a new criterion Q-sort scale. Dev Psychol 1997; 33(6):906.

48. Choi JY, Oh KJ. Cumulative childhood trauma and psychological maladjustment of sexually abused children in Korea: mediating effects of emotion regulation. Child Abuse Negl 2014;38(2):296–303.

49. Muller RT, Vascotto NA, Konanur S, et al. Emotion regulation and psychopathology in a sample of maltreated children. J Child Adolesc Trauma 2013;6(1):25–40.

50. Kahle S, Miller JG, Lopez M, et al. Sympathetic recovery from anger is associated with emotion regulation. J Exp Child Psychol 2016;142:359–71.

51. Molina P, Sala MN, Zappulla C, et al. The emotion regulation checklist–Italian translation. Validation of parent and teacher versions. Eur J Dev Psychol 2014; 11(5):624–34.

52. Stringaris A, Goodman R, Ferdinando S, et al. The affective reactivity index: a concise irritability scale for clinical and research settings. J Child Psychol Psychiatry 2012;53(11):1109–17.

53. DeSousa DA, Stringaris A, Leibenluft E, et al. Cross-cultural adaptation and preliminary psychometric properties of the affective reactivity index in Brazilian

youth: implications for DSM-5 measured irritability. Trends Psychiatry Psychother 2013;35:171–80.

54. Yıldız MA, Duy B. The predictive role of emotion regulation strategies on depressive and psychosomatic symptoms in adolescents. Curr Psychol 2019;38(2): 387–96.

55. Phillips K, Power M. A new self-report measure of emotion regulation in adolescents: the regulation of emotions questionnaire. Clin Psychol Psychother 2007; 14(2):145–56.

56. Duy B, Yildiz MA. Ergenler için duygu düzenleme ölçeği'nin Türkçe'ye uyarlanması. Türk Psikolojik Danışma ve Rehberlik Dergisi 2014;5(41):23–35.

57. da Motta CDA, Rijo D, Vagos P, et al. The abbreviated dysregulation inventory: dimensionality and psychometric properties in Portuguese adolescents. J Child Fam Stud 2018;27(12):3806–15.

58. Gratz KL, Roemer L. Multidimensional assessment of emotion regulation and dysregulation: development, factor structure, and initial validation of the difficulties in emotion regulation scale. J Psychopathol Behav Assess 2004;26(1):41–54.

59. Neumann A, van Lier PA, Gratz KL, et al. Multidimensional assessment of emotion regulation difficulties in adolescents using the difficulties in emotion regulation scale. Assessment 2010;17(1):138–49.

60. Perez J, Venta A, Garnaat S, et al. The difficulties in emotion regulation scale: factor structure and association with nonsuicidal self-injury in adolescent inpatients. J Psychopathol Behav Assess 2012;34(3):393–404.

61. Sarıtaş-Atalar D, Gençöz T, Özen A. Confirmatory factor analyses of the difficulties in emotion regulation scale (DERS) in a Turkish adolescent sample. Eur J Psychol Assess 2015;31(1):12–9.

62. Kaufman EA, Xia M, Fosco G, et al. The difficulties in emotion regulation scale short form (DERS-SF): validation and replication in adolescent and adult samples. J Psychopathol Behav Assess 2016;38(3):443–55.

63. Gullone E, Taffe J. The emotion regulation questionnaire for children and adolescents (ERQ-CA): a psychometric evaluation. Psychol Assess 2012;24(2):409–17.

64. Gross JJ, John OP. Individual differences in two emotion regulation processes: implications for affect, relationships, and well-being. J Pers Soc Psychol 2003; 85(2):348.

Dysregulation by disorder/ condition

Understanding, Assessing, and Intervening with Emotion Dysregulation in Autism Spectrum Disorder

A Developmental Perspective

Jennifer Keluskar, PhD[a],*, Debra Reicher, PhD[a],
Amanda Gorecki, DO[a], Carla Mazefsky, PhD[b],
Judith A. Crowell, MD[c]

KEYWORDS

- Autism spectrum disorder • Emotion regulation • Child development

KEY POINTS

- Children and adolescents with autism spectrum disorder (ASD) present with social-communication deficits, restricted interests and repetitive behaviors, and sensory sensitivities, all of which render them more vulnerable to emotion dysregulation (ED).
- Associated features of ASD also contribute to ED, including difficulties with sleep and feeding, shifting attention appropriately, understanding and expressing emotions, and executive functions.
- It is essential to assess the timeline of ED throughout development, in addition to the duration, frequency, and intensity of outbursts. Understanding how the manifestation of ED may be reactionary and regulatory in nature can facilitate an accurate diagnosis and case conceptualization.
- Behavioral and cognitive behavioral interventions, parent training, medication management, and multidisciplinary collaboration have been used to treat ED in ASD. Gaining an understanding of how behavioral manifestations of ED are linked to the deficits associated with ASD, and promoting generalizability across settings are essential to treatment.

Funded by: SUNY.
[a] Stony Brook Outpatient, Department of Psychiatry, Renaissance School of Medicine, Stony Brook Hospital, Putnam Hall-South Campus, Stony Brook, NY 11794-8790, USA; [b] Department of Psychiatry, University of Pittsburgh, University of Pittsburgh School of Medicine, 3811 O'hara Street, Webster Hall Suite 300, Pittsburgh, PA 15213, USA; [c] Child and Adolescent Psychiatry, Stony Brook Outpatient, Department of Psychiatry, Stony Brook Hospital, Putnam Hall-South Campus, Stony Brook, NY 11794-8790, USA
* Corresponding author.
E-mail address: Jennifer.Keluskar@stonybrookmedicine.edu

Child Adolesc Psychiatric Clin N Am 30 (2021) 335–348
https://doi.org/10.1016/j.chc.2020.10.013   childpsych.theclinics.com
1056-4993/21/Published by Elsevier Inc.

Autism spectrum disorder (ASD) is characterized by 2 broad domains of impairment: (1) "Persistent deficits in social communication and social interactions across multiple contexts..." and (2) "restrictive, repetitive patterns of behavior, interests, or activities.[2]" Although irritability, tantrums, and self-injurious behaviors are not core symptoms, they cause considerable distress and impairment and add to the burden for people with ASD. Understanding the rates of such problems in children with ASD is complicated by the varied terminology used to identify the important symptoms and behaviors (see Daniel F. Connor and Leonard A. Doerfler's article, "The Many Faces (and Names) of Mood Dysregulation," elsewhere in this issue; and Lauren Spring and Gabrielle A. Carlson's article, "The Phenomenology of Outbursts," elsewhere in this issue). Thus, irritability, which has multiple definitions but generally refers to a propensity toward anger, has been co-opted by the Food and Drug Administration as a target behavior worthy of treatment.

Irritability is a consequence of poor emotion regulation (ER). Results of irritability and emotion dysregulation (ED) include impulsive aggression, tantrums, outbursts, agitation, and sometimes self-injurious behavior. Rates of ED in children with autism, when systematically ascertained, range from 50% to 80% (**Table 1**).

DEVELOPMENT OF EMOTION REGULATION IN AUTISM SPECTRUM DISORDER

Children with ASD have difficulties with both the reactive and regulatory components of ER from birth.[1] Core features of ASD, that is, deficits in social and communication functioning and rigid, repetitive behaviors, link to ED.[3,4] For example, social-communication deficits reduce a child's capacity to benefit from caregivers to serve as emotional regulators, in addition to interfering with the child's ability to understand and express emotions. Restricted and repetitive behaviors are most strongly associated with ED when other factors are held constant.[4] The deficits inherent in the ASD diagnosis can result in seemingly minor events precipitating intense, and prolonged outbursts.

Sensory Sensitivities

Abnormal sensitivity and responsiveness to stimuli (the reactive component of ER) is a core symptom of ASD, manifesting as either underresponsiveness and/or overresponsiveness, often present in the same individual. Some parents note underresponsive infants, recalling that their baby was "too good"; one mother stated that taking her infant son with her was like carrying her purse. This phenomenon is a pattern of early temperament in ASD characterized by decreased activity and notable passivity at approximately 6 months of age.[5] This pattern may be followed by extreme distress reactions, and many parents report infants' negativity and hypersensitivity as their first developmental concern. Indeed, siblings of children diagnosed with ASD (high-risk infants) who go on to develop ASD demonstrate greater negative affect by 12 to 24 months of age than those at low risk.

Dysregulated sensory experiences may result in behavioral outbursts, such as a child who experienced the tag on her shirt as feeling like a shard of glass or the toddler who turned the hot water so high that she had third degree burns before noticing pain. Sensory stimuli such as fluorescent lights, the hum of an air-conditioner, the sound of an automatic hand dryer, or toilet flush may be antecedents for irritability and even aggressive outbursts or self-injurious behavior that non-ASD individuals generally tolerate. In ASD samples, anxiety is also greater in individuals with more sensory dysfunction.[6]

Table 1
Rates of emotion dysregulation in youths with ASD

Sample	Assessment Tool	Percentage Endorsing, %	Reference
ASD Community (N = 1323; age range 6–17 y)	Emotion Dysregulation Inventory Reactivity Scale	56.8	Mazefsky et al,[41] 2018
ASD Inpatient (N = 432; age range 4–20 y)		22.3	
ASD Community (N = 1323; age range 6–17 y)	Emotion Dysregulation Inventory Dysphoria Scale	85	
ASD Inpatient (N = 432; age range 4–20 y)		61.1	
ASD; Clinic referred (N = 123; ages 5–21)	Child Behavior Checklist: Emotion Dysregulation Profile	82	Josji et al,[42] 2018
ADHD; Clinic referred (N = 123; ages 6–18)		53	
Community Controls (N = 123; ages 6–18)		2	
ASD Clinical (Autism Treatment Network) (N = 1584; ages 2–17 y)	Physical Aggression (Endorsement of hitting, biting, etc.)	53.7	Kanne & Mazurek,[42] 2018
2–4 y		54.6	
5–7 y		56.8	
8–10 y		50.9	
11–13 y		46.9	
14–17 y		48.3	

Abbreviations: ADHD, attention-deficit/hyperactivity disorder; ASD, autism spectrum disorder.

Eating, Sleeping, and Elimination

Bodily functions such as eating, elimination, and sleep often take significantly longer to be regulated in the ASD population and become vulnerabilities for ER deficits. There is a 2-way relationship between sleep and ER; although emotion impacts sleep, sleep plays a key role in regulating emotion.[7] A child who has been awake since 3:00 AM may experience ER difficulties throughout the day, resulting in tantrums and emotional lability. Sleep problems are approximately twice as common in young children with ASD compared with typically developing (TD) children.[8]

Individuals with ASD are approximately twice as likely as TD individuals to suffer from constipation and food selectivity.[9,10] The resulting discomfort may contribute to agitation and ER difficulty.[9]

Attention

Twelve-month-old infants later diagnosed with autism differ from other siblings and low-risk controls on a number of behavioral markers, including engagement of visual attention, prolonged latency to disengage visual attention, and a tendency to fixate on particular objects in the environment.[11]

"Sticky attention" refers to difficulty disengaging from stimuli and shifting attention appropriately. It has been highlighted as one of the earliest markers (7–14 months) of ASD.[12,13] Problems with attention disengagement interfere with social orienting,[14] and high-risk infants need a "bigger cue" to follow gaze, for example, full head movement rather than eye movement alone.[15] These impairments negatively impact self and others' soothing behaviors of re-orienting and distraction. Perseveration, an aspect of rigid and repetitive behavior, may lead to becoming "stuck" on distressing stimuli and rumination on distress.[16]

Although comorbidity rates of ADHD and ASD have been found to be high, even without formal ADHD, difficulties with attention are common. Children with ASD may have difficulty both attending to what is socially relevant and inflexibility in shifting attention.

Attachment

The bidirectional impact of the child's atypical affect and the caregiver response can have a cumulative effect over time. Parents of children with ASD are found to interact significantly less with their children than parents of TD children, and these reduced interactions are linked to the children's difficulty with ER.[17] Parents may have difficulty mirroring the emotions of ASD children, which also negatively impacts their ability to teach effective ER strategies. Parents can become distressed in response to the ASD child's atypical pattern of attachment and reciprocity.

Language and Knowledge of Emotions

Language and communication deficits in ASD interfere with the ability to express biological states (eg, hunger, pain), emotional states (eg, anger, sadness), and events/experiences that trigger emotions. For example, Delilah, a typically cooperative, nonverbal 11-year-old, developed an abrupt onset of severe ED over the course of several days. She was continually irritable and intermittently self-injurious (banging self on cheek) and aggressive toward others. When using her augmentative device, she pointed to a picture of teeth. Eventually, it was discovered that she had a tooth abscess. Following dental care, her dysregulation disappeared.

Individuals with ASD also have difficulty with facial recognition of emotions.[18] The prevalence of alexithymia (the inability to recognize or describe one's own emotions) in ASD is between 33.3% and 63.0%.[19] Intolerance of uncertainty, common in individuals with ASD, impairs interpretations of others' emotions, including hostile attributional bias,[20] and fuels secondary emotions of frustration or anger. For example, a 13-year-old boy with ASD often perceived neutral expressions as hostile and reacted with agitation and suspicion because he misperceived his mother as angry with him. When recognized, this was addressed in a series of therapy sessions using photographs of faces with different emotions.

Executive Functioning

Executive functions refer to the mental processes that play a role in initiating and organizing tasks, and in ER.[21] As noted, individuals with ASD have early difficulties with attention disengagement, for example, the "shift" function.[22] Shifting requires the ability to make transitions and flexibly problem-solve.[21]

Generativity, or the ability to generate multiple responses, is also impaired in individuals with ASD, and has implications for imaginative skills,[23] as well as leading the child to use the same approach repeatedly rather than to generate more effective solutions.[24] Such inflexibility and insistence on sameness may conflict with the demands of the outside world and be interpreted as oppositionality. This difficulty often escalates irritability and disruptive behavior, such as a child who hid under her desk, refusing to emerge, because she was not chosen for a game during recess as she had hoped.

With age, executive functions in children with ASD become increasingly divergent from those of peers, compounding the susceptibility to psychiatric distress in the middle-school-age years beyond that experienced by TD peers.[25] Deterioration in adaptive and metacognitive skills in adolescents with ASD can be conceptualized as a "second hit"[26]: neurologic and social-emotional vulnerabilities from earlier in childhood converge with hormonal changes and increasing demands, resulting in decline in functioning relative to the peer group. This can result in the onset of depression and anxiety during adolescence that must be interpreted within the context of ASD.

ASSESSING EMOTION DYSREGULATION

There are several approaches to assessing ER in ASD including parent interview, direct observation, and checklist/rating scales.

A careful history and observation should cover each of the developmental domains associated with ER. In addition to administration of instruments such as the Autism Diagnostic Observation Schedule-2 (ADOS-2)[27] and the Autism Diagnostic Interview –Revised (ADI-R),[28] routine questions should be asked about the child's mood, level of irritability, and the frequency, intensity, and duration of outbursts. An effort to understand the antecedents of ED should help in determining whether or not a comorbid diagnosis is warranted. In studies that did not take an understanding of ASD symptomatology into account, children had an average of 6+ additional diagnoses.[29,30] Thus, taking care to assess ED thoroughly may alleviate mistreatment, for example, polypharmacy or ineffective behavioral plans, because of the multiple diagnoses with which children with ASD can be labeled. Establishing a clear baseline and timeline can help parse out symptomatology, and clarify whether ED is a static or changing element, the latter often suggesting a true psychiatric comorbidity.[31]

Rating scales can provide support for observations and interviews. Existing scales vary in the definition of ER and thus there is some variability in the constructs measured. Rating scales often have norms allowing severity comparisons. Dimensional aspects also provide a way to monitor improvement. A sample of scales and descriptions is provided in **Table 2**. In addition, a functional behavior assessment systematically collects data to identify the function of a behavior so that a behavior plan can be tailored appropriately.

TREATMENT OF EMOTION DYSREGULATION IN AUTISM SPECTRUM DISORDER

ED and irritability resulting in aggressive behavior directed at self and/or others is a serious problem, and often a psychiatric emergency. Addressing these behaviors is essential to improving outcomes.

Table 2
Measures of emotion dysregulation for individuals with ASD

Measure	Description
The Emotion Dysregulation Inventory (EDI)[41]	Verbal and nonverbal youth; informant rating, for example, parent Assesses 2 factors: Reactivity (intense, rapidly escalating, and sustained negative emotional reactions) for example, trouble calming down, extreme reactions, reactions more severe than warranted, has outbursts, stays angry for longer than 5 min Dysphoria (anhedonia, sadness, and nervousness) Validated for ASD[42]: Items do not differ by gender, age, intellectual ability, or verbal ability
The Child Behavior Checklist (CBCL)-Dysregulation Profile[43]	Children and adolescents 6–18 y Broad measure of general dysregulation (not specific to ED) using elevated T scores in the Aggressive Behavior, Anxious/Depressed and Attention Problems Scales of CBCL Parent and teacher report Initially validated on neuro-typical children, can used meaningfully in children with ASD[4,44,45]
The Affective Reactivity Index (ARI)[46]	Children and adolescents; self and parent report Assesses irritability: Often loses temper, easily annoyed, stays angry a long time, angry most of the time, loses temper easily, irritability causes problems. Six symptom items and 1 impairment item
The Difficulties in Emotion Regulation Scale (DERS)[47]	Adults; self-report Four factors: Awareness and understanding of emotions, acceptance of emotions, controlling impulses and accessing effective emotion regulation strategies Used with adults with ASD,[47] less utility with children or individuals with cognitive and/or language impairment
The Aberrant Behavior Checklist (ABC)[48]	Developed as measure of treatment effects for pharmacologic intervention for the DD population Revised in 2017 with updated wording and expanded use including characterizing individuals with varying disorders Widely used in ASD treatment studies with results confirming validity as measure of behavior problems in ASD 58 items and 5 subscales measuring irritability, social withdrawal, stereotypic behavior, hyperactivity/noncompliance and inappropriate speech

Abbreviations: ASD, autism spectrum disorder; ED, emotion dysregulation; DD, developmental disability.

Behavioral Interventions

A functional behavior assessment is the process of gathering data to understand the function of behaviors such as social attention, access to preferred activities and sensory stimulation, or removal of demands or other unpleasant stimuli.[32] Once the function(s) are identified, a behavioral intervention plan is created using this information. Behavioral principles such as reinforcement, shaping, extinction, and time out from positive reinforcement can then be included in the plan.

In persons with ASD, the function of a behavior may not always be clear. ED may be the result of perceived interference with a stereotyped, intense interest. For example, Sam, a 14-year-old boy with ASD and a mild cognitive delay, presented at the clinic for his usual follow-up appointment. He usually played quietly in the office while the clinician spoke to his parents. On this day, entering the clinic, and unbeknownst to all, Sam spotted a sea creature–themed coloring book (his special interest) in the waiting room. He pushed his parents, appearing to want to leave. Efforts to stop him resulted in his throwing and breaking items and scratching the clinician when she tried to intervene. When it was clear he only wanted the sea creature book and got it, he calmed down.

Understanding the function of behavior is also essential so that behavior is not inadvertently reinforced, for example, using "time out" when the person is aggressing to avoid. A child sitting atop the slide and spitting at other children daily demonstrates this phenomenon. She was removed from the slide and remained on the playground for "time out" with her aide while the others returned to class. This persisted until it was understood that her spitting was to avoid transitioning back into class, which she had successfully achieved by sitting with her aide. It is important to teach alternative ways for the individual to meet the same function (ie, ask for a break as opposed to hitting him/herself). The use of functional communication training, which systematically teaches a functional equivalent to the maladaptive behavior, is an evidence-based practice.[33]

More specific therapeutic approaches, like cognitive behavioral therapy (CBT) can be used with individuals with ASD presenting with comorbid mood, anxiety, and/or behavioral difficulties and have at least average verbal capabilities. Targeting ED is a core feature of the treatment. However, individuals with ASD are more vulnerable to problems engaging in CBT[34].

Attwood and Scarpa[35] outlined effective modifications in CBT for individuals with ASD. Some constructs of dialectical behavior therapy[36] can also be useful in improving ER, such as the focus on the middle path, interpersonal effectiveness, and distress tolerance. Modifications include the use of visual supports, incorporating parent training (see the next section) and the child's special interests, such as creating a dinosaur board game with "extinction cards" representing cognitive distortions and "evolution cards" representing cognitive restructuring (**Boxes 1** and **2**).

Parent Training

Parents provide a vital connection among home, school, and therapy and are essential in assisting with the learning, maintenance, and generalization of ER. Parent training is designed to develop specific interventions to manage serious behavior problems that are often a result of poor ER in children with ASD. Some of the elements of parent-training interventions include positive reinforcement, clear requests, visual schedules, and planned ignoring that target specific parental concerns. Play time, direct instruction, videos, role play, coaching, and homework are often part of parent training programs. A meta-analysis concluded favorable effects of parent-training interventions

---

Box 1
Potential barriers to efficacy of traditional cognitive behavior therapy treatments in autism spectrum disorder

1. Greater internally driven responses may interfere with motivation toward behavioral change.

2. Greater or more intense baseline levels of negative affect/irritability may prohibit willingness to persist through treatment.

3. Hyperarousal makes it difficult to "sit with" discomfort of facing uncomfortable feelings.

4. Mood and anxiety symptoms may present differently than do the symptoms on which traditional interventions are based.

5. Trouble with information processing and perception can slow down mastery of cognitive concepts.

6. Cognitive factors: rigidity and perseveration can lead to treatment interfering behaviors.

Data from Mazefsky CA, Borue X, Day TN, et al. Emotion regulation patterns in adolescents with high-functioning autism spectrum disorder: Comparison to typically developing adolescents and association with psychiatric symptoms. Autism Res 2014;7(3):344-354.

---

on disruptive behavior.[36] Nevertheless, studies are necessary to increase the evidence base for parent-training interventions with the ASD population. Examples of parent-training programs targeting ER and disruptive behavior in ASD are provided in **Table 3**.

Box 2
Strategies for enhancing effectiveness of cognitive behavior therapy treatments in individuals with autism spectrum disorder (ASD)

1. A visual schedule of activities can be a useful tool for clarifying expectations, but with back and forth "dance" in which work and special interest breaks are interspersed. Incorporating special interests, such as creating a "Pokémon feelings chart," can also increase engagement.

2. Specific to emotion dysregulation, emotional self-awareness exercises are very useful, such as using numbers or colors on a feelings thermometer to rate the intensity of emotional experience. It is helpful to use terms such as "degree of discomfort" as a way to temporarily avoid emotion. The importance of emotional "balance" is emphasized.

3. Due to slower processing of verbal information, often exacerbated by fear of not knowing the "right" answer, they often automatically respond, "I don't know" to questions. It can be helpful to pause briefly before repeating or rewording a question or transitioning to a different topic.

4. Teaching perspective taking skills plays a role in helping individuals understand the rationale for others' behaviors and opens the door for understanding solutions that will aid self-soothing.

5. Incorporating work with parents is often critical even with older adolescents and young adults. It helps with generalization of skills and provides the parents with guidance on how their behaviors impact the course of treatment. For example, children with ASD may become excessively dependent on the parent to regulate their emotions, and parents may unintentionally reinforce a lack of independent, active coping strategies.

Data from Mazefsky CA, Borue X, Day TN, et al. Emotion regulation patterns in adolescents with high-functioning autism spectrum disorder: Comparison to typically developing adolescents and association with psychiatric symptoms. Autism Res 2014;7(3):344-354.

Table 3
Parent-training interventions for emotion regulation/disruptive behavior autism spectrum disorder (ASD)

Intervention	Description	Evidence Base Support
RUBI: Parent Training for Disruptive Behavior[48,49]	Eleven individual, weekly outpatient sessions, 7 supplemental sessions. Based on principles of ABA, manual for therapist and family, sessions consist of therapist script, activity sheet, fidelity checklists and video vignettes used to demonstrate common parenting errors and strategy implementation.	Parent training superior to parent education in decreasing disruptive behaviors in 3- to 7-year-olds. Sample size 180. Medication plus PT resulted in greater reduction of serious maladaptive behavior than MED alone, with a lower risperidone dose randomized control 124 children total.
Parent Child Interaction Therapy (PCIT)[50] Child-Directed Interaction Training (CDIT)[51]	Individual sessions typically 12–20, but not limited Teaches parents to use child-centered skills and avoid questions, commands, and criticism, using effective commands, contingent praise, warnings and developmentally appropriate time out-modified slightly for ASD (prohibited excessive talk intense interests, coached to redirect isolative play) Phase of PCIT: manualized 8-session individual approach	Wait list control group, 19 subjects No statistically significant reduction on parent reports of problem behaviors though child problem behaviors no longer in clinically significant range following PCIT. Thirty mother child dyads, randomized controlled trials (RCT) showed statistically significant improvements in child disruptive behavior
Triple P[52,53]	Individual and group components: 9 to 10 sessions Steps developed for families of children with developmental disabilities, theoretic roots in ABA and social learning theory. involves providing positive attention and managing behavior through parent practices that consider the function of behavior	RCT Significant improvements in parental reports of child behavior in 59 families, child 2–9 y RCT Significant short-term improvements on parent reported child behavior problems in 64 parents of children aged 2–9 y

(continued on next page)

Table 3
(continued)

Intervention	Description	Evidence Base Support
Compass for Hope (C-HOPE)[54]	Individual and group components-8-wk duration Discussion of unique and common characteristics of child, relaxation for parent, education on behavioral principles, development of individual behavior plan, teaching strategies to prevent disruptive behavior, environmental supports, review of plan and progress review/modification to plan.	Significant pre-post treatment gains found in child problem behavior 23 families receiving intervention vs 10 wait list control
Stress and Anger Management Program (STAMP)[55]	Nine-week group, cognitive behavioral therapy (CBT), emotion regulation treatment including parent-training component Developmental modification of Exploring Feelings CBT program Goal: generalize skills to settings outside clinic using modeling, home practice and instruction	23 children with waitlist control Child lability/negative affect decreased, emotion regulation not significantly changed
The Early Start Denver Model[56]	Individual and group settings Use of play to build positive relationships, teaching occurs during everyday activities	38 children with significant reduction of maladaptive behavior in 79% of the subjects on exit

Abbreviations: ABA, applied behavior analysis; PT, parent training; MED, medication.

Medication Management

Medication management in children with ASD is typically directed at improving behavioral and emotional control. Medication management may focus on treatment of irritability or target co-occurring anxiety, inattention/hyperactivity, or mood problems.

At present, only 2 medications, risperidone (for ages 5–17 years) and aripiprazole (for ages 6–17 years) have the approval of the Food and Drug Administration for treatment of irritability in ASD. A systematic review and meta-analysis examined efficacy and safety of pharmacologic treatments for irritability and aggression in children and adolescents with ASD.[37]

Thirteen placebo-controlled randomized controlled trials (RCTs) were reviewed assessing the efficacy of stimulants, alpha-agonists, and selective serotonin reuptake inhibitors (SSRI) for other symptoms though irritability using the ABC-I was also assessed.[37] Clonidine, methylphenidate, and tianeptine showed a moderate effect size in decreasing ABC-I. Guanfacine, an alpha-2 agonist, was helpful in managing ADHD symptoms in the ASD sample, but reduction of ABC-I was limited to effect size of 0.27.[38]

Three RCTs and 10 open label trials or retrospective chart reviews with "methodological weaknesses" examined SSRIs in individuals with ASD.[39] Citalopram, escitalopram, fluoxetine, fluvoxamine, and sertraline showed improvement in functioning related to anxiety or repetitive behaviors. Another review, however, did not find efficacy for SSRIs.[40] A more detailed discussion of psychopharmacologic interventions can be found in the article on "Psychopharmacology of Treating Explosive Behavior" by Carrie Vaudreuil, Abigail Farrell, and Janet Wozniak in Part II.

Multidisciplinary Collaboration

The importance of having ongoing contact with the treatment team, including teacher, speech and language pathologist, occupational therapist, behavior therapist, and any other treating professional is essential in effectively managing ED in children with ASD. This is particularly important to enhance consistency and maximize generalization of skills learned, which can often be challenging with this individuals.

SUMMARY

The ER deficits of children with ASD can be both reactive and regulatory and are often the reason for a family seeking treatment. Such difficulties can be conceptualized as sequelae of the core symptoms of ASD or an independent construct. Understanding the ER in ASD deficits from a developmental perspective provides a context for both assessment and intervention. A detailed assessment of the function and intensity of ER provides the basis for intervention. Evidence-based interventions include medication management, cognitive behavior therapy, behavioral interventions, and parent training. A comprehensive team approach involving all persons involved with the child is most likely to promote change and generalizability. Ongoing efforts to specifically tailor assessment and intervention to the ASD population are needed.

DISCLOSURE

The authors have nothing to disclose.

REFERENCES

1. Cibralic S, Kohlhoff J, Wallace N, et al. A systematic review of emotion regulation in children with Autism Spectrum Disorder. Res Autism Spectr Disord 2019.

2. American Psychiatric Association. Diagnostic and statistical manual of mental disorders. 5th Edition. Washington DC: American Psychiatric Publishing; 2013.

3. Mazefsky C, Herrington J, Siegel M, et al. The role of emotion regulation in autism spectrum disorder. J Am Acad Child Adolesc Psychiatry 2013;52:679–88.

4. Samson A, Phillips J, Parker K, et al. Emotion regulation and the core features of autism spectrum disorder. J Autism Dev Disord 2013;43.

5. Garon N, Zwaigenbaum L, Bryson S, et al. Temperament and its association with autism symptoms in a high risk population. J Abnorm Child Psychol 2016;44: 757–69.

6. Gillott A, Standen P. Levels of anxiety and sources of stress in adults with autism. J Intellect Disabil 2007;11:359–70.

7. Vandekerckhove M, Wang Y-L. Emotion, emotion regulation and sleep: an intimate relationship. Neuroscience 2018;5:1–22.

8. Reynolds A, Soke G, Sabourin K, et al. Sleep problems in 2-5 year-olds with autism spectrum disorder and other developmental delays. Pediatrics 2019; 143(3):e20180492.

9. Fulceri F, Morelli M, Santocchi E, et al. Gastrointestinal symptoms and behavioral problems in preschoolers with autism spectrum disorder. Dig Liver Dis 2016;48: 248–54.

10. Fattorusso A, Di Genova L, Battosta Dell'Isola G, et al. Autism spectrum dsorders and the gut microbiota. Nutrients 2019;11:521.

11. Zwaigenbaum L, Bryson S, Rogers T, et al. Behavioral manifestations of autism in the first year of life. Int J Dev Neurosci 2005;23:142–52.

12. Cheung C, Bedford R, Johnson M, et al. Visual search performance in infants associates with later ASD diagnosis. Developmental Cogn Neurosci 2018;29:4–10.

13. Elsabaugh M, Fernandes J, Webb S, et al, British Autism Study of Infant Siblings Team. Disengagement of visual attention in infancy is associated with emerging autism in toddlerhood. Biol Psychiatry 2013;74:189–94.

14. Mundy P, Sullivan L, Mastergeorge A. A parallel and distributed processing model of joint attention, social cognition, and autism. Autism Res 2009;2:2–21.

15. Thorup E, Gredeback P, Terje S, et al. Altered gaze following during live interaction in infants at risk for autism: an eye tracking study. Mol Autism 2016;7:12.

16. Keenan E, Gotham K, Lerner M. Hooked on a feeling: repetitive cognition and internalizing symptomatology in relation to autism spectrum symptomatology. Autism 2018;227.

17. Costa A, Steffgen G, Vogele C. The role of alexithymia in parent-child interaction and in the emotional lability of children with autism spectrum disorder. Autism Res 2019;12:458–68.

18. Harms M, Martin A, Wallace G. Facial emotion recognition in autism spectrum disorders: a review of behavioral and neuroimaging studies. Neuropsychol Rev 2010;20:290–322.

19. Kinnaird E, Steward C, Tchanturia K. Investigating alexithymia in autism: a systematic review and meta-analysis. Eur Psychiatry 2019;55:80–9.

20. Meyer J, Mundy P, Vaughan Van Hecke A, et al. Social attribution processes and comorbid psychiatric symptoms in children with Asperger syndrome. Autism 2006;10:383–402.

21. Gioia G, Isquith P, Guy S, et al. BRIEF 2 behavior rating inventory of executive function. In: Lutz FL, editor. Professional manual. 2nd edition. Lutz (FL): PAR; 2015. p. 1–5, 43.

22. Rodgers J, Herrema R, Honey E, et al. Towards a treatment of intolerance of uncertainty for autistic adults: a single case experimental study. J Autism Dev Disord 2018;48:2832–45.

23. Low J, Goddard E, Melser J. Generativity and imagination in autism spectrum disorder: evidence from individual differences in children's impossible entity drawings. Br J Dev Psychol 2009;27:425–44.

24. Granader Y, Wallace G, Kenworthy L. Characterizing the factor structure of parent reported executive function in autism spectrum disorder: the impact of cognitive inflexibility. J Autism Dev Disord 2014;44:3056–62.

25. Rosenthal M, Wallace G, Lawson R, et al. Impairments in real-world executive function increase from childhood to adolescence in autism spectrum disorders. Neuropsychology 2013;27:13–8.

26. Picci G, Scherf K. A two-hit model of autism: adolescence as the second hit. Clin Psychol Sci 2015;3:349–71.

27. Lord C, Rutter M, DiLavore P, et al. Autism diagnostic observation schedule. 2nd edition. Torrance (CA): Western Psychological Services; 2012.

28. Rutter M, LeCouteur A, Lord C. Autism diagnostic interview - revised. Los Angeles (CA): Western Psychological Services; 2008.

29. Mazefsky C, Oswald D, Day T, et al. ASD, a psychiatric disorder, or both? Psychiatric diagnoses in adolescents with high-functioning ASD. J Clin Child Adolesc Psychol 2012;41:516–23.

30. Joshi G, Petty C, Wozniak J, et al. The heavy burden of psychiatric comorbidity in youth with autism spectrum disorders: a large comparative study of a psychiatrically referred population. J Autism Dev Disord 2010;40:1361–70.

31. Collins H, Siegel M. Recognizing and treating comorbid psychiatric disorders in people with autism. Psychiatric Times 2019.

32. Iwata BA, Dorsey MF, Slifer KJ, et al. Toward a functional analysis of self injury. J Appl Behav Anal 1994;27:197.

33. Gerow S, Davis T, Radhakrishnan S, et al. Functional Communication Training: the strength of evidence across disabilities. Except Child 2018;85:86–103.

34. Mazefsky C, Borue X, Day T, et al. Emotion regulation patterns in adolescents with high functioning spectrum disorder: comparison to typically developing adolescents and association with psychiatric symptoms. Autism Res 2014;7:344–54.

35. Green SA, Wood JJ. Modifications of cognitive-behavioral therapy for children and adolescents with high-functioning ASD and their common difficulties. In: Scarpa A, White SW, Attwood T, editors. CBT for children & adolescents with high-functioning autism spectrum disorders. New York: The Guilford Press; 2013. p. 76–96.

36. Hartmann K, Urbano M, Manser K, et al. Modified dialectical behavior therapy to improve emotion regulation in autism spectrum disorders. In: Chaz E, Richardson, Wood R, editors. Autism Spectrum Disorders: New Research. New York: Nova Science Publishers, Inc; 2012. p. 41–72.

37. Fung L, Mahajan R, Nozzolillo A, et al. Pharmacologic treatment of severe irritability and problem behaviors in autism: a systematic review and meta-analysis. Pediatrics 2016;137:S124–35.

38. Scahill L, McCracken J, King B, et al. Extended-release guanfacine for hyperactivity in children with autism spectrum disorder. Am J Psychiatry 2015;172:1197–206.

39. Kolevzon A, Mathewson K, Hollander E. Selective serotonin reuptake inhibitors in autism: a review of efficacy and tolerability. J Clin Psychiatry 2006;67:407–14.

40. Williams K, Brignell A, Randall M, et al. Selective serotonin reuptake inhibitors (SSRIs) for autism spectrum disorders (ASD). Cochrane Database Syst Rev 2013;(8):CD004677.

41. Mazefsky C, Yu L, White S, et al. The Emotion Dysregulation Inventory: psychometric properties and item response theory calibration in an autism spectrum disorder sample. Autism Res 2018;11:928–41.

42. Josji G, Wozniak J, Fitzgerald M, et al. High risk for severe emotional dysregulation in psychiatrically referred youth with autism spectrum disorder: a controlled study. J Autism Dev Disord 2018;48:3101–15.

43. Kanne S, Mazurek M. Aggression in children and adolescents with ASD: prevalence and risk factors. J Autism Dev Disord 2011;41:926–37.

44. Achenbach TM, Rescorla LA. Manual for the ASEBA School-Age Forms & Profiles. Burlington (VT): University of Vermont; Research Center for Children, Youth, & Families; 2001. p. 27–41.

45. Keefer A, Singh V, Kalb L, et al. Investigating the factor structure of the child behavior checklist dysregulation profile in children and adolescents with autism spectrum disorder. Autism Res 2019;13(3):436–43.

46. Stringaris A, Goodman R, Ferdinando S, et al. The affective reactivity index: a concise irritability scale for clinical and research settings, Jd Psychiatry and Allied Disciplines. J Child Psychol Psychiatry 2012;53:1109–17.

47. Swain D, Scarpa A, White S, et al. Emotion dysregulaiton and anxiety in adults with ASD: does social motivation play a role? J Autism Dev Disord 2015;45:3971–7.

48. Aman M, Singh N, Stewart A, et al. The Aberrant Behavior Checklist: a behavior rating scale of the assessment of treatment effects. Am J Ment Defic 1985;89:485–91.

49. Bearss K, Burrell T, Stewart L, et al. Parent training in autism spectrum disorder: what's in a name? Clin Child Fam Psychol Rev 2015;18:170–82.

50. Soloman M, Ono M, Timmer S, et al. The effectiveness of parent child-interaction therapy for families of children on the autism spectrum. J Autism Dev Disord 2008;38:1767–76.

51. Ginn N, Clionsky L, Eyberg S, et al. Child-directed interaction training for young children with autism spectrum disorders: parent and child outcomes. J Clin Child Adolesc Psychol 2017;46:101–9.

52. Tellegen C, Sanders M. A randomized controlled trial evaluating a brief parenting program with children with ASD. Consult Clin Psychol 2014;82:1193–200.

53. Whittingham K, Sofronoff K, Sheffield J, et al. Stepping Stones Triple P: an RCT of a program with parents of a child with an autism spectrum disorder. J Abnorm Child Psychol 2009;37:469–80.

54. Kuravackel G, Ruble L, Reese R, et al. COMPASS for Hope: evaluating the effectiveness of a parent training and support program for children with ASD. J Autism Dev Disord 2018;48:404–16.

55. Factor RS, Swain DM, Antezana L, et al. Teaching emotion regulation to children with autism spectrum disorder: outcomes of the Stress and Anger Management Program (STAMP). Bull Menninger Clin 2019;83:235–58.

56. Fulton E, Eapen V, Crncec R, et al. ducing maladaptive behaviors in preschool-aged children with autism spectrum disorder using the early start Denver model. Front Pediatr 2014;40:1–10.

Attention-Deficit Hyperactivity Disorder and the Dysregulation of Emotion Generation and Emotional Expression

Joseph C. Blader, PhD

KEYWORDS

- Attention-deficit hyperactivity disorder • Disruptive mood dysregulation disorder
- Oppositional defiant disorder • Children • Adolescents • Aggressive behavior

KEY POINTS

- Individuals with attention-deficit/hyperactivity disorder (ADHD) frequently experience strong reactions to emotionally evocative situations. Difficulties modulating anger and other upsets have clinically significant behavioral consequences. Among youth with emotional dysregulation of this type, ADHD is ubiquitous.
- There are indications that those with ADHD may have anomalies in emotion generation or emotion expression that predispose to these problems. It is not established which of these processes is universally present among those with emotion dysregulation or if instead different combinations of them yield a variety of clinical phenotypes.
- Important conceptual issues concerns in this area include definitions of emotional regulation; whether in some individuals disinhibited, excessive expression of emotion does not necessarily indicate disturbed emotion generation; the time course of emotion and behavior among those with and without ADHD; and whether subgrouping on the basis of behavioral phenotyping, neurofunctional differences, or treatment response can improve clinical practice.
- Initial pharmacotherapy with agents that target ADHD offers, in most cases, an optimal balance of efficacy and adverse effect liabilities. Use of adjunctive medications is widespread but needs a stronger evidence base. Most current psychosocial treatments emphasize the reduction of family conflict and promoting improved distress tolerance and rule adherence; newer approaches that target emotional dysregulation processes directly are in various phases of development, refinement, and validation.

Department of Psychiatry and Behavioral Sciences, Joe R. and Teresa Lozano Long School of Medicine, University of Texas Health Science Center at San Antonio, 7703 Floyd Curl Drive, Mail Stop 7719, San Antonio, TX 78229, USA
E-mail address: blader@uthscsa.edu

Child Adolesc Psychiatric Clin N Am 30 (2021) 349–360
https://doi.org/10.1016/j.chc.2020.10.005
1056-4993/21/© 2020 Elsevier Inc. All rights reserved.

childpsych.theclinics.com

EMOTION DYSREGULATION AND ATTENTION-DEFICIT/HYPERACTIVITY DISORDER

Based on rates of co-occurrence, the association between emotional dysregulation and attention-deficit/hyperactivity disorder (ADHD) is strong. Approximately 30% to 45% of children with attention-deficit/hyperactivity disorder (ADHD) experience significant impairments because they are prone to anger, rageful outbursts, irritability, aggressive lashing out, or other indicators of dysregulated emotionality.[1–3] Conversely, ADHD is ubiquitous among youth who display such emotionally charged behaviors, with rates over 85% in clinical samples and slightly below in community.[4,5]

The association between emotion dysregulation and ADHD, based on the mechanisms that might explain their co-occurrence, however, remains unclear. Why are brittle frustration tolerance and easily triggered tempers so prevalent among those with ADHD? Do children with these difficulties have something more severe than ADHD? Are treatments for ADHD bound to be inadequate for them? Could they have a mood disorder that demands a different approach? Or maybe these behaviors do not indicate a problem with emotion processes per se; perhaps disinhibited people experience the same degree of emotional upset that others experience but they just show more intense, under-controlled behavior in reaction to it.

These questions are at the heart of child mental health practice. Every day clinicians encounter a volatile, dysregulated patient with ADHD whose behavior is deemed to be more than ADHD, who then starts antipsychotic treatment. Every day a child with lifelong brittle frustration tolerance is prescribed a behavior modification plan involving rewards and losses to shape behavior but instead proves to be another source of upset. Every day a pediatrician is deterred from treating the ADHD symptoms of a highly explosive patient because she is uncertain about the real underlying problem.

PHENOMENOLOGY AND IMPACT

The outbursts that signify emotion dysregulation often occur after provocations that age-mates usually handle with composure. Frequent upsets of this sort are disturbing to others and are not conducive to an enjoyable childhood or a satisfying image of self. ADHD alone adversely affects quality of life,[6] and severe emotion dysregulation further worsens risks for social rejection, academic failure, family distress, injury, mood and anxiety disorders, and early mortality.[7–10] It is the chief reason children receive antipsychotic medications and are psychiatrically hospitalized.[11–14]

Emotional dysregulation in this context differs from psychiatry's traditional framework for mood disturbances, such as major depression and bipolar disorder (BD). This framework emphasizes symptoms periodically coalescing and worsening to form an episode of illness. During that episode, symptoms are, in general, unremittingly present. Children with ADHD and emotional dysregulation usually show the inverse of this picture. They rarely have discrete well-defined episodes of disturbance. Instead, their susceptibility to affective dyscontrol is consistent over years.

In-between incidents of provoked rage and dyscontrol, when things seem to be going the child's preferred way, only a minority show significant mood problems. The *Diagnostic and Statistical Manual of Mental Disorders* (Fifth Edition) (*DSM-5*) introduced the diagnosis of disruptive mood dysregulation disorder (DMDD) within the depressive disorders rubric. It requires both intermittent rageful outbursts and a persistent mood disturbance (ie, irritable or angry mood most of the time). In clinical samples of children with ADHD and frequent but intermittent rage outbursts, up to 30% also have the persistent angry or irritable mood that DMDD requires.[15–17]

HISTORY AND DIAGNOSTIC PRACTICES

Well before ADHD became formalized as a diagnosis,[18] drastic displays of anger, reactive aggression, and weakly restrained emotional expression were known to be common among children with impulse control deficits. For most of the twentieth century, abnormally intense expressions of emotion were included in descriptions of children with ADHD symptoms.[19] Emotion-related problems essentially were defined out of ADHD, implicitly in the *Diagnostic and Statistical Manual of Mental Disorders* (Second Edition) and explicitly in the *Diagnostic and Statistical Manual of Mental Disorders* (Third Edition) (*DSM-III*), as the emphasis on the disorder shifted toward cognitive functions that could be measured in laboratory settings.[19] *DSM-III*'s introduction of oppositional defiant disorder (ODD) provided a rubric for some affective features frequent in ADHD; 4 of its 8 behavioral criteria involve affect.

In time, ADHD came to denote a problem of inattention and overactivity. ODD likewise evoked an image of bratty insubordination, a mere behavior problem, regarded by some observers as not quite a real psychiatric illness. In this context, concerns arose that ADHD and ODD underemphasized the affective disturbances so prominent among children with severe behavioral dyscontrol, many of whom required psychiatric hospitalization. To compensate, it became common to diagnose this presentation as a form of BD. This linkage, however, introduced its own set of difficulties, including vast inflation of BD's incidence among youth in the United States.[20,21] DMDD was designed partly to offer an alternative mood disorder rubric for these problems that avoided this and other concerns.

Currently, emotion regulation has developed as a focus of research interest in its own right.[22] Its relationship to ADHD is a significant area of emphasis,[23–25] as research has helped recast ADHD as a pervasive inadequacy of self-regulatory functions. Because only a minority of those with ADHD demonstrate severe emotion dysregulation, however, it is hard to maintain that emotional dysregulation is integral to ADHD or that ADHD is a sufficient cause for it. Understanding how problems in impulse control contribute to emotional disturbances for a subgroup of children is now a significant area of investigation.

CONCEPTUAL ISSUES
What Are the Elements of Emotion Regulation?

It is challenging to define emotion, which makes any definition of its dysregulation provisional at best. People differ in their susceptibility to certain mood states and in how these states affect observable behaviors. The regulatory processes that determine these differences remain uncertain. There likely is some process at work that enables a person to move on from an emotion-engaging event that is not resolved to the person's liking instead of dwelling on it. These processes are thought to involve some combination of explicit, effortful skills, such as reframing and reappraisal, and more automatic or implicit skills, such as habituation or distraction.[26–28]

Another framework for emotion-related processes distinguishes emotion generation from emotional expression. It resembles the common separation of bottom-up from top-down processes.[24] The next sections briefly review some ways both processes can be disrupted in ADHD.

Emotion Generation

Neurofunctional accounts of anger identify significant roles for responses to threat, reward, and frustration.[25,29] Some studies suggest that there are patterns of response that are specific to emotion dysregulation rather than generic to ADHD. For instance, measuring amygdalar reactivity to pictures of angry and neutral faces is a common paradigm to probe threat-related responses. Severely dysregulated youth with ADHD showed amygdala hypoactivity to neutral faces, whereas those with ADHD-only showed hyperactivation.[30] Within an ADHD sample, Hulvershorn and colleagues[31] found that higher amygdala-frontal connectivity correlated with irritability ratings, controlling for ADHD symptoms. Wiggins and colleagues[32] reported greater amygdala hypoactivation to intensely angry faces among more severely irritable subjects with DMDD whereas those with less irritability demonstrated hyperactivation; ADHD was present in 56% of DMDD participants but it was not tested as a moderator. Amygdala hypoactivation also is found in youth with callous-unemotional traits compared with those with ADHD but no conduct disturbance.[33]

Neural sensitivity to rewarding stimuli has been compared in several clinical and control groups. Aberrant connectivity[34] and hyporeactivity to reward anticipation[35] are reported among children with ADHD relative to controls. The experience of frustration is modeled in experimental settings by withholding a reward that a participant expects. Deveney and colleagues[36] found greater striatal deactivation during a frustrating event among severely mood dysregulated children than in a comparison group; 79% of this sample had ADHD. Low sensitivity to rewarding stimuli and greater shutdown of reward processing areas during frustration may predispose to irritability. These features might leave a person in a more dysphoric state because the hedonic setpoint requires more intense or frequent rewards than most daily life situations provide.

Stimulus novelty is thought be valued by those with ADHD, who may need more of it to maintain a level of optimal stimulation.[37,38] In ordinary situations that do not feature high stimulation, novelty, and interest, it suspected that the tedium and discomfort those with ADHD experience might contribute to their higher risk of irritability.[39]

Some cognitive deficits common in ADHD, such as working memory, may hamper adaptation to new reward contingencies. This in turn leads to perseveration of behaviors that are ineffective in obtaining desired outcomes and therefore amplifies frustration (reviewed by Leibenluft[25]).

Patients with ADHD also may have anomalous reactivity to alert signals in their environments. Studies of brain functional connectivity identify a salience network that comprises bilateral insular cortex and anterior cingulate gyrus.[40] In ADHD, this network[41] demonstrates abnormally high functional connectivity with the default mode and dorsal attention networks. These associations may contribute toward exaggerating the personal significance of events and promoting hyperreactivity. It is unclear whether this is a generic feature of ADHD or is more prominent among those with emotional dysregulation.

There also are effortful processes through which people regulate emotion generation. These include things to prevent unwanted emotional activation in the first place or to raise the likelihood of experiencing desirable events. People actively select, avoid, or modify situations and prepare for the emotional impact of events they can anticipate.[42,43] Planning and foresight often are underdeveloped or underutilized by youth and adults with ADHD. The low distress tolerance observed in ADHD might result in part from deficits in this form of anticipatory planning. For instance, a person may fail to anticipate and manage future disappointment when the chances for goal attainment are low.

Emotion Expression

It is easy to envision how poor response inhibition could lead to emotionally charged behavior that appears excessive. People often infer the strength of others' emotion from the intensity of behaviors that expresses it. Weak behavioral self-control that amplifies this expression of emotion leads to the perception that the underlying emotion is stronger compared with someone whose behavioral reaction is more muted. In cases of anger, individuals obviously are distressed, but a fair question is whether they have (1) a significant disturbance in mood or emotion generation per se or (2) difficulty curtailing behavioral outflow in responding to situations most people would find equally unpleasant but display less overt upset.

The issue relates in part to the imperfect correlation between 3 emotion response domains of subjective experience (including cognition), behavior, and autonomic physiology. Weak concordance between these response systems for several emotional states has been demonstrated in adults,[44,45] and similar results are seen for children and adolescents.[46,47] As an example of response system desynchrony, Faraone and colleagues[48] note, "it seems inappropriate to say that the stoic, unexpressive mourner is necessarily less grief-stricken than the sobbing one."

Viewing emotion dysregulation in ADHD as poorly modulated expression is parsimonious. The idea highlights ADHD's top-down self-regulatory deficits without having to invoke a separate disturbance of affect regulation. It also helps account for intense expressions of both positively and negatively valenced emotion observed in ADHD.[19,48] A majority of those with ADHD, however, do not display significant emotional dysregulation. Of course, few disorders have homogeneous presentations; the last part of this section takes up the issue of heterogeneity in ADHD's association with emotion dysregulation.

The Time Course of Emotion and Behavior

The traditional concepts of mood disorders in psychiatry emphasize pervasive, fairly stable symptoms during an episode. The concept of emotion dysregulation is broader, and, at least in child psychiatry, has focused more on emotional reactivity to situational provocations than on sustained disturbances in hedonic tone, outlook on the future and self, and so forth.

Emotion and behavior in these incidents have temporal features that include range, latency, rise time, peak intensity, and recovery time; collectively, they make the affective chronometry of a response.[49,50] These aspects of emotional reactivity have been examined mainly by developmental psychologists and emotion researchers, while largely neglected in clinical psychiatry literature. They seem highly relevant to ADHD, however, because impulsivity itself has a temporal dimension that may influence emotional reactions.

One framework applies these temporal characteristics separately to emotion generation and emotional expression.[48] In the context of ADHD, it proposes that emotional impulsivity is associated with (1) fast rise times for subjectively experienced emotion and (2) behavioral disinhibition that promotes more intense expression. Recovery time from a high peak intensity may take longer than in non-ADHD individuals but is nonetheless fairly complete. This pattern contrasts with reactivity in more pervasive mood disorders, in which baseline state already is negative and a noxious event intensifies dysphoria or anger, but behavioral expression is not generally as forceful as in ADHD with dysregulation.

Heterogeneity in Phenotypes and Mechanisms

The presence and degree of emotional dysregulation are characteristics that demonstrate wide variation among those with ADHD. Emerging data suggest that youth with ADHD and emotional dysregulation could represent a subgroup of ADHD with distinct neural and temperamental correlates.[2]

Within the group of patients with ADHD and emotion dysregulation, there might be diverse mechanisms that produce this apparently similar phenotype. Differences in the origins for emotion dysregulation may have implications for treatment. For instance, if a subgroup of children show emotion dysregulation because of weak response control, one prediction is that these symptoms should respond to stimulant treatment just as other aspects of behavioral inhibition improve with it.[24] This is the case for a large number of children with ADHD, emotional dysregulation, and aggressive behavior who receive structured titration of stimulant medication and brief psychosocial treatment; in 2 trials, more than half of children displayed remission of aggressive behavior and marked improvements in emotion-related symptoms, including anger and irritability.[15,51,52] Study participants whose aggressive behavior did not remit nevertheless did show improvement on core ADHD symptoms (hyperactivity, inattention, and impulsiveness). The nonremitters randomized to additional treatment with risperidone or divalproex sodium show greater improvements in aggression and mood than those randomized to placebo. Such results suggest heterogeneity in the processes that underlie emotion dysregulation in ADHD.

It is tempting, if simplifying, to suspect that patients with strong response to stimulant medications have a form of emotional dysregulation that reflects the general deficits in behavioral inhibition that characterize ADHD. The emotional dysregulation of stimulant nonresponders, in contrast, may indicate a pathogenesis in which emotion-generating, bottom-up processes are compromised.

ASSESSMENT CONSIDERATIONS FOR ATTENTION-DEFICIT/HYPERACTIVITY DISORDER AND EMOTION DYSREGULATION

This section highlights a few considerations in the clinical evaluation of patients with ADHD and features of emotion dysregulation. The discussion assumes the presence of ADHD has been validated and there are no major developmental concerns.

If parents are asked whether a child is "easily irritated" or "often irritable" the answer likely is yes, partly because they are focused on incidents of upsets, defiance, and inflexibility with seemingly minor provocations. This does not mean, however, that irritability or anger is the patient's prevailing mood, as is the case in a true mood disorder. It, therefore, is worthwhile to consider how dysregulation incidents differ from the child's baseline mood and the contexts in which they occur. Questions to pose include, "Does he seem pretty content when things seem to be going his way, or is he grouchy even at those times?" "When good things happen, how much does she seem to enjoy herself, or is she still negative or hard to please?" "Can you usually figure out what sets him off, or does he sometimes seem to become upset from out of nowhere?" "When she's starting to have a meltdown and you give into what she wants, does that change her mood, or does she still seem pretty mad for a long while?" and "If no one is doing anything to get on your nerves, do you still feel kind of annoyed or have a negative attitude that's hard to shake? Do you keep thinking about things that annoy you even if there's nobody bugging you at the moment?"

It also should be determined if low self-esteem, feelings of worthlessness, or self-harm preoccupation pervade even times of apparent calm. Weepiness, anhedonia, and sadness are less common in this patient group and may be more suggestive of depression.[15]

ADHD's comorbidity with anxiety disorders is high, and the trimorbidity involving these conditions with disruptive disorders also is significant. For these patients, outbursts that occur only in anxiety-provoking situations for the disorder (eg, at times of parental separation, when obsessive-compulsive disorder urges are blocked, and so forth) may be a complication of the anxiety disorder rather than more extensive emotion dysregulation. Similarly, outbursts around school or academic functioning—especially those that persists after treatment of attentional problems—can indicate a need to assess for a learning disorder.

It is not uncommon for more extreme outbursts to occur at home but not elsewhere. It is possible that this is more prevalent among children with anxiety, for whom feeling evaluated and judged by peers and other nonfamily may have an inhibitory effect.[53]

Although literature is limited, some trial data and clinical experience suggest that stimulant medications at times may cause or worsen irritability or dysphoria. Such affective toxicity often is a dose-dependent phenomenon, and some amphetamine-based products may pose higher risk.[54] Antidepressants may have a similar effect: an inpatient trial reported high rates of increased aggressive behavior among youth with severe mood dysregulation who received citalopram added to stimulant medication.[55] Clinicians also should be mindful of nonpsychiatric medications' potential to affect mood, including corticosteroids and certain antiepileptic drugs.[56]

Several rating scales for emotional dysregulation for use with children and adults are listed in a recent review by Faraone and colleagues.[48] It is important to obtain initial and follow-up data for both ADHD symptoms per se and for emotion-dysregulated difficulties. Some brief rating scales have separate subscales, along with norms, such as the 10-item Conners Global Index. Where gradations in aggressive behavior are the principal outcome, there are a few aggression-specific tools, including the Retrospective-Modified Overt Aggression Scale[51] and the Children's Aggression Scale.[57]

TREATMENT CONSIDERATIONS FOR ATTENTION-DEFICIT/HYPERACTIVITY DISORDER WITH EMOTION DYSREGULATION
Pharmacotherapy

There are few data on the treatment of youth with ADHD who also have emotional dysregulation. A somewhat larger literature is available for ADHD with impulsive aggression, which is relevant for the management of emotional dysregulation. Consensus guidelines[58,59] and clinical trial findings[51,52,55,60] indicate that medication that target ADHD symptoms usually should be first-line treatment. First, there is a good chance that stimulant treatment alone improves ADHD symptoms, aggressive behavior, emotional dysregulation, and irritability. Second, problems with inattentiveness, restlessness, and impulsiveness typically require treatment anyway because other medications that target impulsive aggression are less efficacious for these symptoms. Third, the adverse effect risk profile for stimulant treatment generally is favorable. Fourth, with improved impulse control, attention span, and frustration tolerance, psychosocial treatments are more likely to gain traction.

Approaches that include a titration protocol with prompt dose adjustments to improve response and tolerability avoid premature abandonment of a stimulant medication and are associated with high rates of remission of aggressive behavior and affect problems.[15,51] Situations that warrant greater caution in dosing and monitoring include patients with autism spectrum disorders and severe anxiety disorder who are susceptible to worsening of symptoms and, in the case of autism spectrum disorder, hyperacusis.

When impulsive aggressive behavior does not reduce adequately with optimized first-line stimulant medications, there are supportive data for the adjunctive use of the second-generation antipsychotic (SGA) risperidone and the anticonvulsant/mood stabilizer divalproex sodium. Risperidone's propensity for weight gain was observed in these studies, exceeding that for divalproex sodium. Other SGAs with few data include aripiprazole and quetiapine, although the latter's efficacy seems weak even in open trials. Lithium was reported efficacious relative to placebo in a trial for youth with BD,[61] and, along with early inpatient studies showing benefits for aggression, it also may help with emotional dysregulation that does not respond fully to stimulant treatment.

It is reasonable to suspect that if DMDD is classified as a depressive disorder, patients with emotional dysregulation might benefit from antidepressant treatment. An inpatient trial of adjunctive citalopram versus placebo for children with severe mood dysregulation who already had treatment with methylphenidate reported that 35% of those receiving citalopram were rated by staff as improved, compared with 6% receiving placebo.[55] Overall assessment of function was improved minimally over the trial's 8 weeks, with no difference between groups.

The noradrenergic α_2 receptor agonists guanfacine and clonidine are second-line treatments of ADHD and long have been used as stimulant adjuncts for disruptive behavior symptoms and aggressive behavior. What they contribute beyond optimized stimulant treatment among highly emotionally dysregulated patients is unclear. There is a significant need for stepped treatment trials that evaluate their efficacy in this context because their adverse effect liabilities are more favorable than SGAs or antiepileptic drugs. Of the 2, clonidine is by far the more potent antihypertensive; this property is related to its high affinity for an imidazoline receptor (I_1), which is irrelevant to its psychotropic activity. Guanfacine's binding profile shows greater specificity for α_2 receptors.

Psychosocial Treatments

Family-based behavioral treatments have been widely employed and studied in the treatment of disruptive behavior problems. They share emphases on (1) improving the parent-child relationship to become less conflictual and prone to mutual escalation over minor misbehavior, (2) improving communication and incentive to promote more cooperative behavior (giving directions constructively, praise and reward systems, and so forth), and (3) firm but nonhostile setting of limits and management of negative behaviors. Relative to no-treatment controls, large effect sizes for child behavioral improvement have been reported.[**] With very volatile and dysregulated children, these interventions need to be monitored so that strategies are implemented at home without exacerbating conflict. Some consequence-based approaches may prove inflammatory for children with minimal frustration tolerance. Contingencies and expectations have to be calibrated so that in a given situation the sequence of antecedent → target behavior (compliance and composure) → reward is highly likely.

Treatments that target emotion dysregulation specifically are being developed, refined, and studied gradually. Some are adaptations of dialectical behavior therapy, one of the leading interventions for borderline personality disorder.[62] Anger management approaches may be helpful, but many children with ADHD are deficient in the anticipatory self-awareness to monitor building rage and modulate their appraisals or responses in the brief interval that their short fuses afford. In these cases, initial treatment of ADHD might make these interventions more viable.

Anxiety disorder treatment capitalizes on the fact that exposure to stimuli long enough for arousal to extinguish reduces anxiety and apprehensiveness in subsequent exposures. Extending this reasoning to anger and emotion dysregulation, another approach is to work with families so that patients gradually are exposed to anger-provoking situations in a hierarchy that progresses from less provocative to more.[63]

SUMMARY

Growing appreciation that the self-regulatory deficits that characterize ADHD and other externalizing problems also have an impact on emotion-related functions has sparked interest on the interdependence of affective, executive control, and behavior-regulating processes. These processes may vary between patients. A similar trend in adult psychopathology has revealed the triangulation of affect dysregulation, executive control, and substance abuse,[64] whereas the specific neural functions that underly the relationship differ between subgroups of patients.

As analyses of processes that underlie emotional regulation become more sophisticated, they may reveal distinct pathogenic pathways for impulsive, emotionally dysregulated individuals. The robust response to first-line stimulant and behavioral treatments that many children with these difficulties experience, while other children remain highly impaired, hints at heterogeneity of mechanisms. Further refinement of these processes and capacity to identify them in routine clinical care will contribute to more judicious, safer, and effective care of one of the most vulnerable patient groups.

DISCLOSURE

The author has nothing to disclose.

REFERENCES

1. Liu L, Chen W, Vitoratou S, et al. Is emotional lability distinct from "angry/irritable mood," "negative affect," or other subdimensions of oppositional defiant disorder in children with ADHD? J Atten Disord 2019;23(8):859–68.
2. Karalunas SL, Fair D, Musser ED, et al. Subtyping attention-deficit/hyperactivity disorder using temperament dimensions: toward biologically based nosologic criteria. JAMA Psychiatry 2014;71(9):1015–24.
3. Stringaris A, Goodman R. Mood lability and psychopathology in youth. Psychol Med 2009;39(8):1237–45.
4. Axelson D, Findling RL, Fristad MA, et al. Examining the proposed disruptive mood dysregulation disorder diagnosis in children in the Longitudinal Assessment of Manic Symptoms study. J Clin Psychiatry 2012;73(10):1342–50.
5. Leibenluft E. Severe mood dysregulation, irritability, and the diagnostic boundaries of bipolar disorder in youths. Am J Psychiatry 2011;168(2):129–42.
6. Coghill D, Hodgkins P. Health-related quality of life of children with attention-deficit/hyperactivity disorder versus children with diabetes and healthy controls. Eur Child Adolesc Psychiatry 2016;25(3):261–71.
7. Copeland WE, Shanahan L, Egger H, et al. Adult diagnostic and functional outcomes of DSM-5 disruptive mood dysregulation disorder. Am J Psychiatry 2014;171(6):668–74.
8. Fang X, Massetti GM, Ouyang L, et al. Attention-deficit/hyperactivity disorder, conduct disorder, and young adult intimate partner violence. Arch Gen Psychiatry 2010;67(11):1179–86.
9. Dalsgaard S, Ostergaard SD, Leckman JF, et al. Mortality in children, adolescents, and adults with attention deficit hyperactivity disorder: a nationwide cohort study. Lancet 2015;385(9983):2190–6.
10. Bagwell CL, Molina BS, Pelham WE Jr, et al. Attention-deficit hyperactivity disorder and problems in peer relations: predictions from childhood to adolescence. J Am Acad Child Adolesc Psychiatry 2001;40(11):1285–92.

11. Birnbaum ML, Saito E, Gerhard T, et al. Pharmacoepidemiology of antipsychotic use in youth with ADHD: trends and clinical implications. Curr Psychiatry Rep 2013;15(8):1–13.
12. Olfson M, King M, Schoenbaum M. Treatment of young people with antipsychotic medications in the United States. JAMA Psychiatry 2015;72(9):867–74.
13. Blader JC. Acute inpatient care for psychiatric disorders in the United States, 1996 through 2007. Arch Gen Psychiatry 2011;68(12):1276–83.
14. Kreider AR, Matone M, Bellonci C, et al. Growth in the concurrent use af antipsychotics with other psychotropic medications in Medicaid-enrolled children. J Am Acad Child Adolesc Psychiatry 2014;53(9):960–70.
15. Blader JC, Pliszka SR, Kafantaris V, et al. Prevalence and treatment outcomes of persistent negative mood among children with attention-deficit/hyperactivity disorder and aggressive behavior. J Child Adolesc Psychopharmacol 2016;26(2):164–73.
16. Roy AK, Lopes V, Klein RG. Disruptive mood dysregulation disorder: a new diagnostic approach to chronic irritability in youth. Am J Psychiatry 2014;171(9):918–24.
17. Roy AK, Klein RG, Angelosante A, et al. Clinical features of young children referred for impairing temper outbursts. J Child Adolesc Psychopharmacol 2013;23(9):588–96.
18. American Psychiatric Association. Diagnostic and Statistical Manual of Mental Disorders. 3rd ed. Washington, DC: American Psychiatric Association; 1980.
19. Barkley RA. Emotional dysregulation is a core component of ADHD. In: Barkley RA, editor. Attention-deficit hyperactivity disorder: a handbook for diagnosis and treatment. 4th edition. New York: Guilford Press; 2015. p. 81–115.
20. Blader JC, Carlson GA. Increased rates of bipolar disorder diagnoses among U.S. child, adolescent, and adult inpatients, 1996-2004. Biol Psychiatry 2007;62(2):107–14.
21. Moreno C, Laje G, Blanco C, et al. National trends in the outpatient diagnosis and treatment of bipolar disorder in youth. Arch Gen Psychiatry 2007;64(9):1032–9.
22. Ochsner KN, Gross JJ. The neural bases of emotion and emotion regulation: a valuation perspective. In: Gross JJ, editor. Handbook of emotion regulation. 2nd edition. New York: Guilford; 2014. p. 23–43.
23. Graziano PA, Garcia A. Attention-deficit hyperactivity disorder and children's emotion dysregulation: a meta-analysis. Clin Psychol Rev 2016;46:106–23.
24. Shaw P, Stringaris A, Nigg J, et al. Emotion dysregulation in attention deficit hyperactivity disorder. Am J Psychiatry 2014;171(3):276–93.
25. Leibenluft E. Pediatric irritability: a systems neuroscience approach. Trends Cogn Sci 2017;21(4):277–89.
26. Gyurak A, Gross JJ, Etkin A. Explicit and implicit emotion regulation: a dual-process framework. Cogn Emot 2011;25(3):400–12.
27. Ochsner KN, Ray RR, Hughes B, et al. Bottom-up and top-down processes in emotion generation: common and distinct neural mechanisms. Psychol Sci 2009;20(11):1322–31.
28. Mauss IB, Levenson RW, McCarter L, et al. The tie that binds? Coherence among emotion experience, behavior, and physiology. Emotion 2005;5(2):175–90.
29. Blair RJR. Considering anger from a cognitive neuroscience perspective. Wiley Interdiscip Rev Cogn Sci 2012;3(1):65–74.
30. Brotman MA, Rich BA, Guyer AE, et al. Amygdala activation during emotion processing of neutral faces in children with severe mood dysregulation versus ADHD or bipolar disorder. Am J Psychiatry 2010;167(1):61–9.

31. Hulvershorn LA, Mennes M, Castellanos FX, et al. Abnormal amygdala functional connectivity associated with emotional lability in children with attention-deficit/ hyperactivity disorder. J Am Acad Child Adolesc Psychiatry 2014;53(3):351–61.

32. Wiggins JL, Brotman MA, Adleman NE, et al. Neural correlates of irritability in disruptive mood dysregulation and bipolar disorders. Am J Psychiatry 2016; 173(7):722–30.

33. Marsh AA, Finger EC, Mitchell DG, et al. Reduced amygdala response to fearful expressions in children and adolescents with callous-unemotional traits and disruptive behavior disorders. Am J Psychiatry 2008;165(6):712–20.

34. Costa Dias TG, Wilson VB, Bathula DR, et al. Reward circuit connectivity relates to delay discounting in children with attention-deficit/hyperactivity disorder. Eur Neuropsychopharmacol 2013;23(1):33–45.

35. Scheres A, Milham MP, Knutson B, et al. Ventral striatal hyporesponsiveness during reward anticipation in attention-deficit/hyperactivity disorder. Biol Psychiatry 2007;61(5):720–4.

36. Deveney CM, Connolly ME, Haring CT, et al. Neural mechanisms of frustration in chronically irritable children. Am J Psychiatry 2013;170(10):1186–94.

37. Antrop I, Roeyers H, Van Oost P, et al. Stimulation seeking and hyperactivity in children with ADHD. Attention deficit hyperactivity disorder. J Child Psychol Psychiatry 2000;41(2):225–31.

38. Zentall SS, Zentall TR. Optimal stimulation: a model of disordered activity and performance in normal and deviant children. Psychol Bull 1983;94(3):446–71.

39. Wilbertz G, Trueg A, Sonuga-Barke EJS, et al. Neural and psychophysiological markers of delay aversion in attention-deficit hyperactivity disorder. J Abnorm Psychol 2013;122(2):566–72.

40. Uddin LQ. Salience processing and insular cortical function and dysfunction. Nat Rev Neurosci 2015;16(1):55–61.

41. Sidlauskaite J, Sonuga-Barke E, Roeyers H, et al. Altered intrinsic organisation of brain networks implicated in attentional processes in adult attention-deficit/ hyperactivity disorder: a resting-state study of attention, default mode and salience network connectivity. Eur Arch Psychiatry Clin Neurosci 2016;266(4): 349–57.

42. Gross JJ, Barrett LF. Emotion generation and emotion regulation: one or two depends on your point of view. Emot Rev 2011;3(1):8–16.

43. Thompson RA. Emotion and emotion regulation: two sides of the developing coin. Emot Rev 2011;3(1):53–61.

44. Lanteigne DM, Flynn JJ, Eastabrook JM, et al. Discordant patterns among emotional experience, arousal, and expression in adolescence: relations with emotion regulation and internalizing problems. Can J Behav Sci 2014;46(1): 29–39.

45. Mauss IB, Robinson MD. Measures of emotion: a review. Cogn Emot 2009;23(2): 209–37.

46. Benoit Allen K, Allen B, Austin KE, et al. Synchrony–desynchrony in the tripartite model of fear: predicting treatment outcome in clinically phobic children. Behav Res Ther 2015;71(Supplement C):54–64.

47. Smith M, Hubbard JA, Laurenceau JP. Profiles of anger control in second-grade children: examination of self-report, observational, and physiological components. J Exp Child Psychol 2011;110(2):213–26.

48. Faraone SV, Rostain AL, Blader J, et al. Practitioner Review: emotional dysregulation in attention-deficit/hyperactivity disorder - implications for clinical recognition and intervention. J Child Psychol Psychiatry 2019;60(2):133–50.

49. Thompson RA. Emotion and self-regulation. In: Thompson RA, editor. Socioemotional development. Nebraska symposium on motivation 1988, vol. 36. Lincoln (NE): University of Nebraska Press; 1990. p. 367–467.

50. Lamm C, Granic I, Zelazo PD, et al. Magnitude and chronometry of neural mechanisms of emotion regulation in subtypes of aggressive children. Brain Cogn 2011;77(2):159–69.

51. Blader JC, Pliszka SR, Kafantaris V, et al. Stepped treatment for attention-deficit/ hyperactivity disorder and aggressive behavior: a randomized, controlled trial of adjunctive risperidone, divalproex sodium, or placebo after stimulant medication optimization. J Am Acad Child Adolesc Psychiatry 2020. https://doi.org/10.1016/j. jaac.2019.12.009.

52. Blader JC, Schooler NR, Jensen PS, et al. Adjunctive divalproex versus placebo for children with ADHD and aggression refractory to stimulant monotherapy. Am J Psychiatry 2009;166(12):1392–401.

53. Carlson GA, Blader JC. Diagnostic implications of informant disagreement for manic symptoms. J Child Adolesc Psychopharmacol 2011;21(5):399–405.

54. Stuckelman ZD, Mulqueen JM, Ferracioli-Oda E, et al. Risk of irritability with psychostimulant treatment in children with ADHD: a meta-analysis. J Clin Psychiatry 2017;78(6):e648–55.

55. Towbin K, Vidal-Ribas P, Brotman MA, et al. A double-blind randomized placebo-controlled trial of citalopram adjunctive to stimulant medication in youth with chronic severe irritability. J Am Acad Child Adolesc Psychiatry 2020;59(3): 350–61.

56. Halma E, de Louw AJ, Klinkenberg S, et al. Behavioral side-effects of levetiracetam in children with epilepsy: a systematic review. Seizure 2014;23(9):685–91.

57. Halperin JM, McKay KE, Newcorn JH. Development, reliability, and validity of the children's aggression scale-parent version. J Am Acad Child Adolesc Psychiatry 2002;41(3):245–52.

58. Pliszka SR, Crismon ML, Hughes CW, et al. The Texas Children's Medication Algorithm Project: revision of the algorithm for pharmacotherapy of attention-deficit/ hyperactivity disorder. J Am Acad Child Adolesc Psychiatry 2006;45(6):642–57.

59. Scotto Rosato N, Correll CU, Pappadopulos E, et al. Treatment of maladaptive aggression in youth: CERT Guidelines II. Treatments and ongoing management. Pediatrics 2012;129(6):e1577–86.

60. Aman MG, Bukstein OG, Gadow KD, et al. What does risperidone add to parent training and stimulant for severe aggression in child attention-deficit/hyperactivity disorder? J Am Acad Child Adolesc Psychiatry 2014;53(1):47–60.

61. Findling RL, Robb A, McNamara NK, et al. Lithium in the acute treatment of bipolar I disorder: a double-blind, placebo-controlled study. Pediatrics 2015;136(5): 885–94.

62. Perepletchikova F, Nathanson D, Axelrod S, et al. Randomized clinical trial of dialectical behavior therapy for pre-adolescent children with disruptive mood dysregulation disorder: feasibility and outcomes. J Am Acad Child Adolesc Psychiatry 2017;56(10):832–40.

63. Kircanski K, Craske MG, Averbeck BB, et al. Exposure therapy for pediatric irritability: theory and potential mechanisms. Behav Res Ther 2019;118:141–9.

64. Wilcox CE, Pommy JM, Adinoff B. Neural circuitry of impaired emotion regulation in substance use disorders. Am J Psychiatry 2016;173(4):344–61.

Ticked Off

Anger Outbursts and Aggressive Symptoms in Tourette Disorder

Marianna Ashurova, MD[a,b,*], Cathy Budman, MD[c,d],
Barbara J. Coffey, MD, MS[e]

KEYWORDS

- Rage in Tourette disorder • Explosive outbursts in Tourette disorder
- Aggressive symptoms in Tourette disorder
- Disruptive behaviors in Tourette disorder

KEY POINTS

- Explosive outbursts (rage) are common symptoms of impulsive aggression in Tourette disorder.
- Explosive outbursts often are associated with tic severity and psychiatric comorbidity in Tourette disorder.
- Explosive outbursts in Tourette disorder cause significant morbidity and require comprehensive evaluation with targeted treatments.

BACKGROUND

Aggression is a complex construct encompassing a range of different internal processes and external manifestations. The term, *aggression*, is applied to an array of different symptoms, behaviors, and experiences, some considered developmentally and/or socially appropriate, whereas others are regarded as maladaptive and pathologic. Aggressive symptoms and behavioral and emotional dysregulation are frequent reasons for referral to mental and behavioral health services and are among the core symptoms in several psychiatric disorders, including intermittent explosive disorder,

[a] Zucker Hillside Hospital, ACP Building Basement, 75-59 263rd Street, Glen Oaks, NY 11004, USA; [b] Child & Adolescent Psychiatry Consultation Liaison Service, Cohens Children's Medical Center, 268-01 76th Avenue, New Hyde Park, NY 11040, USA; [c] Long Island Center for Tourette, 1615 Northern Boulevard, Suite #306, Manhasset, NY 11030, USA; [d] Zucker School of Medicine, 500 Hofstra Boulevard, Hempstead, NY 11549, USA; [e] Department of Psychiatry and Behavioral Sciences, Child and Adolescent Psychiatry, Tourette Association Center of Excellence, University of Miami Miller School of Medicine, 1120 Northwest Fourteenth Street, Suite 1442, Miami, FL 33136, USA
* Corresponding author.
E-mail address: cbudmanmd@gmail.com

Child Adolesc Psychiatric Clin N Am 30 (2021) 361–373
https://doi.org/10.1016/j.chc.2020.10.006
1056-4993/21/© 2020 Elsevier Inc. All rights reserved.

childpsych.theclinics.com

disruptive mood dysregulation disorder, oppositional defiant disorder (ODD), and conduct disorder.[1,2] The most common aggressive symptoms in clinically referred children are impulsive in nature and associated with substantial functional impairment for the individual, family, and community.[3–8]

GENERAL FEATURES OF TOURETTE DISORDER

Tourette disorder (TD) is a neurodevelopmental disorder characterized by multiple repetitive movements (ie, motor tics) and at least 1 repetitive sound or vocalization (ie, phonic tic) that persist (not necessarily concurrently) for at least 1 year. Tics characteristically wax and wane in severity, change in type and in location, and are not due to any other underlying medical condition or substance.[2] Tics typically begin between 4 years and 6 years of age, peak in severity at 10 years to 12 years, and often decline during mid to late adolescence.[9] TD occurs worldwide and is reported more frequently in boys than girls.[10]

Whereas its exact prevalence is uncertain, TD is estimated to occur in approximately 0.52% to 0.77% of youth.[11] Chronic motor tic disorder, which may represent a milder form of TD, appears to be approximately twice as prevalent.[12,13] TD is one of the most heritable but heterogeneous neuropsychiatric disorders of childhood, resulting from a complex interplay between both genetic and environmental factors.[14,15]

Co-occurring psychiatric conditions are extremely common in TD and are associated with greater overall morbidity and lowered quality of life.[16–21] Approximately 90% of youth with TD have been reported to have at least 1 or more psychiatric conditions, including obsessive-compulsive disorder (OCD), attention-deficit/hyperactivity disorder (ADHD) (with 72.1% of people with TD having both OCD and ADHD), mood disorders (30%), non-OCD anxiety disorders (30%) and other impulse control problems, sleep disturbances, and school and social problems.[21–24]

ANGER OUTBURSTS AND AGGRESSIVE SYMPTOMS IN TOURETTE DISORDER

Excessive anger and aggressive symptoms have been reported in 25% to 70% of individuals with TD worldwide.[25–27] In an international survey of 3500 outpatients with TD, 37% reported a lifetime history of anger control problems and 25% experienced current anger problems.[28] Among youth with TD, such symptoms range in their intensity from persistent angry verbal protests and intense argumentativeness to more severe outbursts of verbal and/or physical aggression (ie, rage attacks or explosive outbursts). Aggressive behavior is grossly out of proportion to any stressor, is highly destructive to relationships and physical property, typically is directed at the primary caregiver, and may vary in duration from minutes to hours.[29–31] The aggressive outbursts occur mostly at home rather than in school or other settings.[25,31] Major life events (ie, being bullied, severe parental conflict, or parental divorce) that influence tic expression and severity also are linked with aggressive symptoms in TD.[32]

Common precipitants include failing to get one's way, experiencing unforeseen frustration or change in plans, and being reprimanded, criticized, or corrected.[25,30,33,34] Explosive outbursts also may be triggered by a cognitive or sensory urge or discomfort.[30] Family members typically experience these sudden angry behaviors as shocking and escalating with lightning speed; the angry responses are age-inappropriate, unpredictable, intense, excessive, and irrational.[30,31] Those experiencing these sudden fits of anger describe feeling "out of control" and are acutely distressed; most show signs of heightened physiologic arousal, including increased heart rate and psychomotor agitation. A majority of explosive outbursts are impulsive and reactive in nature; afterward, the individual often experiences feelings of increased physical/emotional calm and remorse. In contrast, predatory or proactive aggression is characterized by deliberate, planned, goal-directed behaviors with low autonomic arousal.[33,35,36]

DEVELOPMENTAL COURSE

The onset of anger control problems in TD typically occurs during early childhood and may persist into adolescence and adulthood.[37,38] These symptoms are a major cause of morbidity in TD and are associated with increased family stress and conflict; impaired social, academic, and occupational functioning; and increased rates of psychiatric hospitalization.[30,39–44]

The etiology of anger control problems in TD is multifactorial, stemming from a combination and synergism of biopsychosocial factors.[30,32,39,42] Tic complexity and severity, for example, correlate with worsening irritability and with a stronger association to vocal rather than motor tics.[18] Tic severity also contributes to worsening school performance, impaired social functioning, and reduced overall quality of life.[45–47] Psychiatric comorbidities, in particular ADHD, OCD, and mood disorders, are highly and significantly associated with aggression in TD as well.[26,27,29,33,34,48–50] Some studies comparing those with TD and OCD, with or without ADHD, found that ADHD is the main predictor of disruptive behaviors in TD as well as the main explanatory factor for lack of inhibitory control.[51–53] Untangling interactions among psychiatric comorbidities and their symptoms and evaluating functional consequences of different tic types, complexities, and severities, along with considering how varying psychosocial stresses have an impact over the course of development, pose major challenges for clinical assessment and management.

CLINICAL CORRELATES OF AGGRESSIVE SYMPTOMS IN TOURETTE DISORDER

Early studies that explored the phenomenology of explosive outbursts in clinically referred youth with TD reported an apparent association with underlying psychiatric comorbidity, in particular comorbid ADHD and/or OCD.[25,30,50,54–56]

In a study of 113 clinically referred youth with TD, ages 7 years to 17 years, 48 (43%) subjects with rage attacks were more likely to meet current *Diagnostic and Statistical Manual of Mental Disorders* (Fourth Edition) criteria for major depression, depression not otherwise specified, bipolar I disorder, ADHD, and ODD, and lifetime criteria for OCD and/or ODD than the 65 comparison subjects without explosive outbursts.[30] An investigation of 218 TD-affected individuals who participated in a genetic study (N = 104 from a nonclinical sample in Costa Rica, and N = 114 recruited from specialty US TD clinics) examined the prevalence and clinical correlates of explosive outbursts; 20% of all TD-affected individuals had explosive outbursts, with no significant differences in prevalence between the nonclinical and the clinical samples. In the overall sample, ADHD, greater tic severity, and lower age of tic onset were associated strongly with explosive outbursts. ADHD, male gender, and prenatal exposure to tobacco were significantly associated with explosive outbursts in the clinical sample, whereas lower age of onset and greater severity of tics were significantly associated with explosive outbursts in the nonclinical sample.[29]

A large study of 578 clinically referred individuals with TD showed a significant association between tic severity and current aggressive behaviors.[39] An association between tic severity and higher levels of irritability also was demonstrated in a clinical study of 101 patients with TD.[18] More recently, however, a clinical study of 47 youth ages 7 years to 17 years with TD from a tertiary pediatric Tourette clinic compared with a group of 32 healthy age-matched and sex-matched controls found no signifi-cant differences in aggression scores measured by the Overt Aggression Scale, and levels of aggression were not correlated with tic severity.[48] In this study, verbal aggression occurred in 70% of the youth with TD and was the most prevalent type. Although the probability of aggression in the TD cohort was increased by comorbid ADHD and OCD, only ADHD severity emerged as a significant predictor of aggression.[35]

NEUROBIOLOGICAL INFLUENCES ON AGGRESSIVE SYMPTOMS IN TOURETTE DISORDER

Studies in the non–TD-disordered population show that aggressive symptoms are influenced by interconnected circuitry that integrates activities associated with arousal, impulse control, motivation, memory, affect regulation, and sensory and so-cial processing.[57] Aggression dyscontrol may be the consequence of exaggerated ac-tivity in the subcortical circuits that mediate adaptive aggressive behaviors because they are triggered by endogenous or environmental cues at vulnerable time points or may be due to disturbed activity within multiple converging cortical and subcortical circuits; aggression also is shaped by social context and repeated environmental reward/reinforcement.[57,58] Evidence from neuroimaging studies of intermittent explo-sive disorder in adults suggests simultaneous hypofunction of the medial prefrontal cortex and hyperfunction of the amygdala.[59] Failure of top-down cognitive control may be common to TD, OCD, and ADHD.[60] Disturbances of circadian rhythms and ab-normalities of neurotransmission involving dopamine, serotonin, norepinephrine, and glutamate γ-aminobutyric acid neurotransmission in the prefrontal cortex as well as low testosterone and elevated cortisol have been associated with impulsive aggres-sion.[61–66] A recent study of 55 patients with TD and explosive outbursts using a multi-modal neuroimaging approach found structural changes in the right supplementary motor area as well as in the right hippocampus and in the left orbitofrontal cortex, sug-gesting lower connectivity within the sensorimotor cortico-basal ganglia network and aberrant connectivity pattern among the orbito-fontal cortex, amygdala, and hippocampus.[67]

EVALUATION AND DIFFERENTIAL DIAGNOSIS OF AGGRESSIVE SYMPTOMS IN TOURETTE DISORDER

Comprehensive evaluation of the individual with TD and aggressive symptoms is indi-cated. Multidisciplinary evaluation is helpful, given the clinical complexity of these cases. Detailed history should include developmental, medical, and behavioral symp-toms and conditions; family, social, and trauma background; alcohol and substance use; prescribed, over-the-counter medications and supplements; accidental or inten-tional toxic exposures; and psychosocial history and identifiable triggers[68] (**Table 1**).

Once underlying medical conditions and specific psychosocial triggers are excluded, the presence and severity of comorbid psychiatric disorders, in particular ADHD, OCD, and mood disorders, must be carefully explored. Other co-occurring psychiatric conditions should be considered, including autistic spectrum disorder, ODD, conduct disorder, posttraumatic stress disorder, borderline personality disor-der, specific learning disorders, and specific impulse control disorders.

Self-injurious behaviors (SIBs), such as pinching, slapping, biting, poking, and head-banging, that result in moderate to severe injury may occur up to 60% of all patients with TD and may be associated increased tic severity, copra phenomena, high levels of obsessiveness and hostility, OCD, ADHD, increased numbers and

Table 1
Differential diagnosis of aggression in youth with Tourette disorder

Causes	Clinical Examples
Metabolic encephalopathies	Hyperthyroidism, primary hyperparathyroidism
Infectious encephalopathies	Pediatric autoimmune neuropsychiatric symptoms
Autoimmune encephalopathies	Systemic lupus erythematosus (SLE), N-methyl-D-aspartate receptor encephalitis, Beçet syndrome
Traumatic brain injury	Postconcussive syndromes, head trauma
Seizure disorder	Partial complex seizures
Movement disorders	Wilson disease, Huntington disease
Acute intoxications/withdrawal states	Lead poisoning, alcohol intoxication, steroid abuse
Accidental/deliberate poisoning	Prescribed medication overdose
Medication side effects/interactions	Acute akathisia, antidepressant activation
Parasomnias	Night terrors
Physical pain	Injury secondary to tics
Sexual, physical, emotional abuse	Rape, trauma, bullying
Other psychosocial problems	Family conflict

severity of psychiatric disorders, episodic rages, affective dysregulation, and severe impulsivity.[39,69–73] Severe SIBs occur in only 5% of all TD cases.[74]

A recent clinical study of 165 consecutive patients ages 5 years to 50 years revealed a lifetime history of self-harming behaviors (SHBs) in 39.4%. In this sample, ADHD and OCD were found to be risk factors for lifetime SHBs, whereas only tic severity emerged as a statistically significant risk factor for current and lifetime SHBs in children. Anxiety and other psychiatric comorbidities, but not tic severity, was associated with SHBs in adults.[75]

ENVIRONMENTAL INFLUENCES ON AGGRESSIVE SYMPTOMS IN TOURETTE DISORDER

Usually the adverse impact on quality of life from TD is linked more closely with psychosocial and environmental factors than with tics themselves.[76,77] Many with TD and its co-occurring conditions struggle to attain competency and confidence in navigating normal age-appropriate development, family relationships, peer attachment, and academic and occupational performance.[78,79] A comparative study showed that parents of children with TD experience greater aggravation than parents of children without TD; parents who report being bothered by tics and rage symptoms are more likely to punish their children.[80,81] An authoritative parenting style with unrealistic expectations and minimal support negatively reinforces tics and aggression.[78] Conflict avoidance, failure to set appropriate expectations and limits, and family accommodation of OCD symptoms and/or tics also reinforce aggression.[82]

TREATMENT STRATEGIES FOR AGGRESSIVE SYMPTOMS IN TOURETTE DISORDER
Psychosocial Interventions

Considerable evidence supports the efficacy of behavioral interventions for reduction of aggressive behaviors in children with/without tic disorders. These include parent guidance or coaching therapy, teacher training, behavioral modification, and programs addressing skills deficits/issues within a patient-centered approach.[49,83]

Psychoeducation and parent and teacher training are particularly important, because parents, siblings, school staff, peers, health care providers, and others struggle to understand which behaviors of TD are deliberate or intentional and which are tics, compulsions, impulsive-compulsive symptoms, or medication side effects.[49,84,85] Diminished parental expectations for their children's self-control may be overgeneralized, leading to increased disruptive behaviors.[84] Family accommodation of tics and/or OCD symptoms occurs frequently and is associated with greater levels of overall functional impairment.[86,87] A study of children with OCD ages 6 years to 16 years showed that rage impaired quality of life out of proportion to OCD symptoms alone. This impairment was explained by family accommodation, resulting in either worsening rage and/or rage promoting increased familial accommodation.[87] In a study of youth ages 6 years to 18 years with tic disorder, 68% of parents of children with TD endorsed some form of tic accommodation during the month prior to study participation.[86] Family accommodation likely plays a significant role in fueling or contributing to explosive outbursts in TD as well.

How parental psychopathology and family expressed or repressed emotion also triggers and fuels explosive outbursts in youth with TD requires further consideration. Because TD is highly heritable, the likelihood that a parent(s) and/or siblings may suffer from tics and/or co-occurring psychiatric comorbidities is high. Therefore, screening, identifying, and treating psychopathology in family members of youth with TD with rage symptoms are imperative.

Providing the necessary parental psychoeducation and skills to better understand and manage their child's often puzzling behaviors has demonstrated significant efficacy for managing explosive outbursts in youth with TD. A randomized controlled trial of parent management training versus treatment as usual in youth with tic disorders and disruptive behaviors comparing 10 sessions of parent training (including psychoeducation about tics and co-occurring symptoms, limit and expectation setting, time-outs, and positive reinforcement) demonstrated a 51% decline in disruptive behaviors versus 19% in the treatment as usual, with a reported effect size of 0.96, comparable to that achieved by parent training for non–tic-associated ODD.[81] The clinical approach "brief trans-diagnostic parent training," also has demonstrated treatment efficacy for children with TD and aggression.[88]

Cognitive behavior therapy appears moderately effective in reducing anger and aggression in children without tic disorders and may have application for those with TD.[89–93] Treatment focuses on improving awareness of behavioral patterns and associated emotions and cognitions.[93]

A study that investigated anger control training (ACT) in adolescents with TD between the ages of 11 years and 16 years randomized subjects with ODD to receive either ACT or treatment as usual for 10 weeks. Among those who received ACT (ie, 10, 1-hour–long sessions that included managing anger, cognitive restructuring, and behavioral interventions), 52% demonstrated a reduction in disruptive behavior compared with an 11% reduction in the treatment as usual control group; these improvements were sustained at 3 month follow-up.[49] Additional studies are necessary to ascertain which treatment interventions are most useful at certain ages and with which particular clinical subtypes of TD.

Psychopharmacologic Interventions

Atypical antipsychotics

A majority of individuals with TD do not require pharmacologic intervention for tic suppression alone; however, depending on aggression severity, medication intervention may become more immediate, particularly when aggression occurs in more than 1 setting.[94] Atypical antipsychotics, such as aripiprazole and olanzapine, have been used to treated combined TD and aggression.[34,95] These agents, however, may cause serious adverse effects, including acute dystonic reactions, parkinsonism, akathisia, neuroleptic malignant syndrome, acute dystonia, weight gain, and risks for metabolic syndrome.[34,96] Nonetheless, when there is urgency to address severe, recurrent rage, particularly if accompanied by severe tics, unstable mood, severe OCD, and anxiety, use of atypical antipsychotics on at least an acute basis may be necessary.

Stimulants and α-agonists

ADHD is 1 of the 2 most frequently comorbid disorders with TD yet may be under-recognized and undertreated. This is of particular concern because many clinicians still avoid use of psychostimulants for treatment of combined ADHD and tics due to unwarranted concerns that these agents are contraindicated in TD. Treatment of TD with comorbid ADHD is particularly important, because both disruptive behavior and tic severity may be reduced when ADHD symptoms are treated.[97]

Tics, ADHD symptoms, and aggressive behaviors have been shown to improve by treatment with α-agonists (such as clonidine and guanfacine), psychostimulants, and targeted combined pharmacotherapy with an α-agonist and psychostimulant.[98–100] Youth with comorbid TD and ADHD who are treated with psychostimulants show an overall reduction of aggression and antisocial behavior.[35,101,102] Short-acting methyl-phenidate was found effective in treating oppositional behavior and peer aggression in children with ADHD and TD.[103] α-Agonists also have been reported to decrease irritability and aggression in conduct disorder co-occurring with TD.[104] When α-agonists are used to treat tics in TD without ADHD, treatment effect size is reduced.[98,99]

Selective serotonin inhibitors

Treatment of youth with OCD, tic, and rage attacks with serotonin reuptake inhibitors (SRIs) may be beneficial.[56] In an open-label study of paroxetine in 45 children with TD and explosive outbursts, 76% demonstrated reduced rage symptoms using an average dose of 33 mg/d. A majority of subjects met diagnostic criteria for OCD, ADHD, or both. However, 4 subjects experienced worsening of rage outbursts, and 1 subject experienced a hypomanic episode.[54] Using SRIs for treatment of rage requires close monitoring for adverse effects, such as activation, hypomania, and aggression.[105] Larger, randomized controlled studies are needed to confirm efficacy of these agents for treatment of explosive outbursts in TD.

SUMMARY

TD is a complex neurodevelopmental disorder characterized by multiple motor and phonic tics and is associated with high rates of psychiatric comorbidity. Symptoms of impulsive aggression commonly are encountered in the clinical setting, cause significant morbidity, and pose considerable diagnostic and treatment challenges. These symptoms usually are multifactorial in etiology and result from a complex interplay of illness severity and psychosocial factors, including tic severity, comorbid psychiatric disorders, and family accommodation of aggression. Treatment strategies require comprehensive evaluation and include both behavioral and pharmacologic interventions. More research is needed in this important area of scientific, clinical, and public health significance.

DISCLOSURE

The authors have nothing to disclose.

REFERENCES

1. Connor DF, Newcorn JH, Saylor KE, et al. Maladaptive aggression: with a focus on impulsive aggression in children and adolescents. J Child Adolesc Psychopharmacol 2019;29(8):576–91.
2. American Psychiatric Association. Diagnostic and statistical manual of mental disorders. 5th edition. Arlington,(VA): American Psychiatric Association; 2013.
3. Connor DF. On the challenge of maladaptive and impulsive aggression in the clinical treatment setting. J Child Adolesc Psychopharmacol 2016;26(1):2–3.
4. Kraft JT, Dalsgaard S, Obel C, et al. Prevalence and clinical correlates of tic disorders in a community sample of school-age children. Eur Child Adolesc Psychiatry 2012;21(1):5–13.
5. Mol Debes NMM, Hjalgrim H, Skov L. Validation of the presence of comorbidities in a Danish clinical cohort of children with Tourette syndrome. J Child Neurol 2008;23(9):1017–27.
6. Barratt ES, Stanford MS, Dowdy L, et al. Impulsive and premeditated aggression: a factor analysis of self- reported acts. Psychiatry Res 1999;86(2):163–73.
7. Cooper C, Robertson MM, Livingston G. Psychological morbidity and caregiver burden in parents of children with Tourette's disorder and psychiatric comorbidity. J Am Acad Child Adolesc Psychiatry 2003;42(11):1370–5.
8. Lee MY, Chen YC, Wang HS, et al. Parenting stress and related factors in parents of children with tourette syndrome. J Nurs Res 2007;15(3):165–74.
9. Leckman JF, King RA, Bloch MH. Clinical features of Tourette syndrome and tic disorders. J Obsessive Compuls Relat Disord 2014;3(4):372–9.
10. Cohen S, Leckman JF, Bloch MH. Clinical assessment of Tourette syndrome and tic disorders. Neurosci Biobehav Rev 2013;37(6):997–1007.
11. Scharf JM, Miller LL, Gauvin CA, et al. Population prevalence of Tourette syndrome: a systematic review and meta-analysis. Mov Disord 2015;30(2):221–8.
12. Knight T, Steeves T, Day L, et al. Prevalence of tic disorders: a systematic review and meta-analysis. Pediatr Neurol 2012;47(2):77–90.
13. Muller-Vahl KR, Sambrani T, Jakubovski E. Tic disorders revisited: introduction of the term "tic spectrum disorders. Eur Child Adolesc Psychiatry 2019;28(8): 1129–35.
14. Qi Y, Zheng Y, Li Z, et al. Genetic studies of tic disorders and Tourette syndrome. Psychiatric disorders. Methods Mol Biol 2011;547–71. https://doi.org/10.1007/978-1-4939-9554-7_32.
15. Mataix-Cols D, Isomura K, Perez-Vigil A, et al. Familial risks of Tourette syndrome and chronic tic disorders a population-based cohort study. JAMA Psychiatry 2015;72(8):787–93.
16. Eapen V, Cavanna AE, Robertson MM. Comorbidities, social impact, and quality of life in Tourette syndrome. Front Psychiatry 2016;7:5–10.
17. Gill CE, Kompoliti K. Clinical features of Tourette syndrome. J Child Neurol 2019. https://doi.org/10.1177/0883073819877335.
18. Cox JH, Cavanna AE. Irritability symptoms in Gilles de la Tourette syndrome. J Neuropsychiatry Clin Neurosci 2015;27(1):42–7.
19. Cavanna AE. The neuropsychiatry of Gilles de la Tourette syndrome: the etat de l'art. Rev Neurol (Paris) 2018;174(9):621–7.
20. Martino D, Ganos C, Pringsheim TM. Chapter Fifty-Three - Tourette syndrome and chronic tic disorders: the clinical spectrum beyond tics. In: International review of neurobiology, vol. 134. Calgary (Canada): Academic Press; 2017. p. 1461–90. https://doi.org/10.1016/bs.irn.2017.05.006.

21. Groth C, Mol Debes N, Rask CU, et al. Course of Tourette syndrome and comorbidities in a large prospective clinical study. J Am Acad Child Adolesc Psychiatry 2017;56(4):304–12.

22. Hirschtritt ME, Lee PC, Pauls DL, et al. Lifetime prevalence, age of risk, and genetic relationships of comorbid psychiatric disorders in Tourette syndrome. JAMA Psychiatry 2015;72(4):325–33.

23. Cavanna AE, Servo S, Monaco F, et al. The behavioral spectrum of Gilles de la Tourette syndrome. J Neuropsychiatry Clin Neurosci 2009;21(1):13–23.

24. Rizzo R, Gulisano M, Martino D, et al. Gilles de la Tourette syndrome, depression, depressive illness, and correlates in a child and adolescent population. J Child Adolesc Psychopharmacol 2017;27(3):243–9.

25. Kano Y, Ohta M, Nagai Y, et al. Rage attacks and aggressive symptoms in Japanese adolescents with Tourette syndrome, vol. 13. Arlington (TX): American Psychiatric Association Publishing; 2013. https://doi.org/10.1017/S1092852900016448.

26. Santangelo SL, Pauls DL, Goldstein JM, et al. Tourette's syndrome: what are the influences of gender and comorbid obsessive-compulsive disorder? J Am Acad Child Adolesc Psychiatry 1994;33(6):795–804.

27. Wand RR, Matazow GS, Shady GA, et al. Tourette syndrome: associated symptoms and most disabling features. Neurosci Biobehav Rev 1993;17(3):271–5.

28. Freeman RD, Fast DK, Burd L, et al. An international perspective on Tourette syndrome: selected findings from 3500 individuals in 22 countries. Dev Med Child Neurol 2000;42(7):436–47.

29. Chen K, Budman CL, Herrera LD, et al. Prevalence and clinical correlates of explosive outbursts in Tourette syndrome. Psychiatry Res 2013;205(3):269–75.

30. Budman CL, Rockmore L, Stokes J, et al. Clinical phenomenology of episodic rage in children with Tourette syndrome. J Psychosom Res 2003;55(1):59–65.

31. De Lange N, Olivier MAJ. Mothers' experiences of aggression in their Tourette's syndrome children. Int J Adv Counsell 2004;26(1):65–77.

32. Horesh N, Shmuel-Baruch S, Farbstein D, et al. Major and minor life events, personality and psychopathology in children with Tourette syndrome. Psychiatry Res 2018;260:1–9.

33. Budman CL, Feirman L. The relationship of Tourette's syndrome with its psychiatric comorbidities: is there an overlap? Psychiatr Ann 2001;31(9):541–8.

34. Budman C, Coffey BJ, Shechter R, et al. Aripiprazole in children and adolescents with Tourette disorder with and without explosive outbursts. J Child Adolesc Psychopharmacol 2008;18(5):509–15.

35. Connor DF, Glatt SJ, Lopez ID, et al. Psychopharmacology and aggression. I: a meta-analysis of stimulant effects on overt/covert aggression-related behaviors in ADHD. J Am Acad Child Adolesc Psychiatry 2002;41(3):253–61.

36. Connor DF. Aggression and antisocial behavior in children and adolescents: research and treatment. New York: Guilford Publications; 2004. p. 443–8.

37. Wright A, Rickards H, Cavanna AE. Impulse-control disorders in Gilles de la Tourette syndrome. J Neuropsychiatry Clin Neurosci 2012;24(1):16–27.

38. Frank MC, Piedad J, Rickards H, et al. The role of impulse control disorders in Tourette syndrome: an exploratory study. J Neurol Sci 2011;310(1–2):276–8.

39. Robertson MM, Cavanna AE, Eapen V. Gilles de la Tourette syndrome and disruptive behavior disorders: prevalence, associations, and explanation of the relationships. J Neuropsychiatry Clin Neurosci 2015;27(1):33–41.

40. Kadesjö B, Gillberg C. Tourette's disorder: epidemiology and comorbidity in primary school children. J Am Acad Child Adolesc Psychiatry 2000;39(5):548–55.

41. Sukhodolsky DG, Gladstone TR, Kaushal SA, et al. Tics and Tourette syndrome. In: Matson J, editor. Handbook of childhood psychopathology and developmental disabilities treatment. Cham (Switzerland): Springer; 2017. p. 241–56.

42. Kumar A, Trescher W, Byler D. Tourette syndrome and comorbid neuropsychiatric conditions. Curr Dev Disord Rep 2016;3(4):217–21.

43. Dooley JM, Brna PM, Gordon KE. Parent perceptions of symptom severity in Tourette's syndrome. Arch Dis Child 1999;81(5):440–1.

44. Coffey BJ, Park KS. Behavioral and emotional aspects of Tourette syndrome. Neurol Clin 1997;15(2):277–89.

45. Zhu Y, Leung KM, Liu PZ, et al. Comorbid behavioural problems in Tourette's syndrome are positively correlated with the severity of tic symptoms. Aust N Z J Psychiatry 2006;40(1):67–73.

46. Quast LF, Rosenthal LD, Cushman GK, et al. Relations between tic severity, emotion regulation, and social outcomes in youth with Tourette syndrome. Child Psychiatry Hum Dev 2019. https://doi.org/10.1007/s10578-019-00948-8.

47. Zinner SH, Conelea CA, Glew GM, et al. Peer victimization in youth with Tourette syndrome and other chronic tic disorders. Child Psychiatry Hum Dev 2012; 43(1):124–36.

48. Benaroya-Milshtein N, Shmuel-Baruch S, Apter A, et al. Aggressive symptoms in children with tic disorders. Eur Child Adolesc Psychiatry 2019;6–8. https://doi.org/10.1007/s00787-019-01386-6.

49. Sukhodolsky DG, Vitulano LA, Carroll DH, et al. Randomized trial of anger control training for adolescents with Tourette's syndrome and disruptive behavior, vol. 48. Arlington (TX): American Psychiatric Association Publishing; 2009. https://doi.org/10.1097/CHI.0b013e3181985050.

50. Stephens RJ, Sandor P. Aggressive behaviour in children with Tourette syndrome and comorbid attention-deficit hyperactivity disorder and obsessive-compulsive disorder. Can J Psychiatry 1999;44(10):1036–42.

51. Lathif N, Chishty ME, Awe D. Tourette Syndrome and violence: is there a link? Eur Psychiatry 2015;30(Retz 2009):1237.

52. Morand-Beaulieu S, Grot S, Lavoie J, et al. The puzzling question of inhibitory control in Tourette syndrome: a meta-analysis. Neurosci Biobehav Rev 2017; 80:240–62.

53. Sukhodolsky DG, Scahill L, Zhang H, et al. Disruptive behavior in children with Tourette's syndrome: association with ADHD comorbidity, tic severity, and functional impairment. J Am Acad Child Adolesc Psychiatry 2003;42(1):98–105.

54. Bruun RD, Budman CL. Paroxetine treatment of episodic rages associated with Tourette's disorder. J Clin Psychiatry 1998;59(11):581–4.

55. Budman CL, Bruun RD, Park KS, et al. Explosive outbursts in children with Tourette's disorder. J Am Acad Child Adolesc Psychiatry 2000;39(10):1270–6.

56. Budman CL. Treatment of aggression in Tourette syndrome. Adv Neurol 2006; 99:222–624.

57. Flanigan ME, Russo SJ. Recent advances in the study of aggression. Neuropsychopharmacology 2019;44(2):241–4.

58. Covington HE, Newman EL, Leonard MZ, et al. Translational models of adaptive and excessive fighting: an emerging role for neural circuits in pathological aggression. F1000Res 2019;8. https://doi.org/10.12688/f1000research.18883.1.

59. Fanning JR, Keedy S, Berman ME, et al. Neural correlates of aggressive behavior in real time: a review of fMRI studies of laboratory reactive aggression. Curr Behav Neurosci Rep 2017;4(2):138–50.

60. Hirschtritt ME, Darrow SM, Illmann C, et al. Genetic and phenotypic overlap of specific obsessive-compulsive and attention-deficit/hyperactive subtypes with Tourette syndrome. Psychol Med 2018;48(2):279–93.

61. Rosell DR, Siever LJ. The neurobiology of aggression and violence. CNS Spectr 2015;20(3):254–79.

62. Weisbrot DM, Ettinger AB. Aggression and violence in mood disorders. Child Adolesc Psychiatr Clin N Am 2002;11(3):649–71.

63. Siegel A, Victoroff J. Understanding human aggression: new insights from neuroscience. Int J Law Psychiatry 2009;32(4):209–15.

64. Shiina A. Neurobiological basis of reactive aggression: a review. Int J Forensic Sci Pathol 2015;3(3):94–8.

65. Tricklebank MD, Petrinovic MM. Chapter nine - Serotonin and aggression. In: Tricklebank MD, Daly E, editors. The serotonin system: history, neuropharmacology, and pathology. London: Academic Press; 2019. p. 155–80. https://doi.org/10.1016/B978-0-12-813323-1.00009-8.

66. Hood S, Amir S. Biological clocks and rhythms of anger and aggression. Front Behav Neurosci 2018;12:1–12.

67. Atkinson-Clement C, Sofia F, Fernandez-Egea E, et al. Structural and functional abnormalities within sensori-motor and limbic networks underpin intermittent explosive symptoms in Tourette disorder. J Psychiatr Res 2020;125:1–6.

68. Heyneman EK. The aggressive child. Child Adolesc Psychiatr Clin N Am 2003; 12(4):667–77.

69. Sambrani T, Jakubovski E, Muller-Vahl KR. New insights into clinical characteristics of Gilles de la Tourette syndrome: findings in 1032 patients from a single German center. Front Neurosci 2016;10(415). https://doi.org/10.3389/fnins.2016.00415.

70. Kano Y, Ohta M, Nagai Y, et al. Association between Tourette syndrome and comorbidities in Japan. Brain Dev 2010;32(3):201–7.

71. Mathews CA, Waller J, Glidden DV, et al. Self injurous behaviour in Tourette syndrome: correlates with impulsivity and impulse control. J Neurol Neurosurg Psychiatry 2004;75(8):1149–55.

72. Robertson MM, Stern JS. Gilles de la Tourette syndrome: symptomatic treatment based on evidence. Eur Child Adolesc Psychiatry 2000;9(Suppl. 1):60–75.

73. Robertson MM, Lees AJ. Self-injurious behaviour and the Gilles de la Tourette syndrome: a clinical study and review of the literature. Psychol Med 1989; 19(3):611–25.

74. Cheung MYC, Shahed J, Jankovic J. Malignant Tourette syndrome. Mov Disord 2007;22(12):1743–50.

75. Szejko N, Jakubczyk A, Janik P. Prevalence and clinical correlates of self-harm behaviors in Gilles de la Tourette syndrome. Front Psychiatry 2019;10:1–9.

76. Espil FM, Capriotti MR, Conelea CA, et al. The role of parental perceptions of tic frequency and intensity in predicting tic-related functional impairment in youth with chronic tic disorders. Child Psychiatry Hum Dev 2014;45(6):657–65.

77. McGuire JF, Arnold E, Park JM, et al. Living with tics: reduced impairment and improved quality of life for youth with chronic tic disorders. Psychiatry Res 2015; 225(3):571–9.

78. Juncos J, Chilakamarri J. Treatment of non-motor symptoms in Tourette syndrome. In: Reich S, Factor S, editors. Therapy of movement disorders. Cham (Switzerland): Humana; 2019. p. 267–72. https://doi.org/10.1007/978-3-319-97897-0_61.

79. O'Hare D, Helmes E, Eapen V, et al. The impact of tic severity, comorbidity and peer attachment on quality of life outcomes and functioning in Tourette's syndrome: parental perspectives. Child Psychiatry Hum Dev 2016;47(4):563–73.

80. Robinson LR, Bitsko RH, Schieve LA, et al. Tourette syndrome, parenting aggravation, and the contribution of co-occurring conditions among a nationally representative sample. Disabil Health J 2013;6(1):26–35.

81. Scahill L, Sukhodolsky DG, Bearss K, et al. Randomized trial of parent management training in children with tic disorders and disruptive behavior. J Child Neurol 2006;21(8):650–6.

82. Nadeau JM, Hieneman M. Managing avoidance and accommodation of tics and related behaviors. In: McGuire J, Murphy T, Piacentini J, et al, editors. The clinician's guide to treatment and management of youth with Tourette syndrome and tic disorders. Academic Press; 2018. p. 177–200. https://doi.org/10.1016/b978-0-12-811980-8.00009-1.

83. Carlson GA, Chua J, Pan K, et al. Behavior modification is associated with reduced psychotropic medication use in children with aggression in inpatient rreatment: a retrospective cohort study. J Am Acad Child Adolesc Psychiatry 2019. https://doi.org/10.1016/j.jaac.2019.07.940.

84. Walkup JT. The psychiatry of Tourette syndrome. CNS Spectr 1999;4(2):54–61. https://doi.org/10.1017/S109285290001138X.

85. Monahan M, Agazzi H, Jordan-Arthur B. The implementation of parent–child interaction therapy for the treatment of Tourette syndrome and disruptive behavior. Clin Case Stud 2018;17(1):38–54.

86. Storch EA, Johnco C, McGuire JF, et al. An initial study of family accommodation in children and adolescents with chronic tic disorders. Eur Child Adolesc Psychiatry 2017;26(1):99–109.

87. Storch E, Jones A, Lack C, et al. Rage attacks in pediatric obsessive-compulsive disorder: phenomenology and clinical correlates. J Acad Child Psychiatry 2012;51:582–92.

88. Specht MW, Edwards KR, Perry-Parish C, et al. Brief trans-diagnostic parent training: a strengths-based, parent-centered treatment for youth with Tourette syndrome. In: The clinician's guide to treatment and management of youth with Tourette syndrome and tic disorders. Academic Press; 2018. p. 225–53.

89. Sukhodolsky DG, Smith SD, McCauley SA, et al. Behavioral interventions for anger, irritability, and aggression in children and adolescents. J Child Adolesc Psychopharmacol 2016;26(1):58–64.

90. Verdellen C, Van De Griendt J, Hartmann A, et al, ESSTS Guidelines Group. European clinical guidelines for Tourette syndrome and other tic disorders. Part III: behavioural and psychosocial interventions. Eur Child Adolesc Psychiatry 2011;20(4):197–207.

91. Wilhelm S, Peterson AL, Piacentini J, et al. Randomized trial of behavior therapy for adults with Tourette syndrome. Arch Gen Psychiatry 2012;69(8):795–803.

92. Tudor ME, Bertschinger E, Piasecka J, et al. Cognitive behavioral therapy for anger and aggression in a child with Tourette's syndrome. Clin Case Stud 2018;17(4):220–32.

93. Leclerc J, O'Connor KP, Forget J, et al. Behavioral program for managing explosive outbursts in children with Tourette syndrome. J Dev Phys Disabil 2011;23(1):33–47.

94. Black KJ, Black ER, Greene DJ, et al. Provisional tic disorder: what to tell parents when their child first starts ticcing. F1000Res 2016;5:1–18.

95. Stephens RJ, Bassel C, Sandor P. Olanzapine in the treatment of aggression and tics in children with Tourette's syndrome - a pilot study, vol. 14. Arlington (TX): American Psychiatric Association Publishing; 2013. https://doi.org/10.1089/1044546041648959.

96. Shapiro AK, Shapiro E, Wayne H. Treatment of Tourette's syndrome with haloperidol, review of 34 Cases. Arch Gen Psychiatry 1973;28(1):93-7.

97. Osland ST, Steeves TD, Pringsheim T. Pharmacological treatment for attention deficit hyperactivity disorder (ADHD) in children with comorbid tic disorders. Cochrane Database Syst Rev 2018;(6):CD007990. https://doi.org/10.1002/14651858.CD007990.

98. Egolf A, Coffey BJ. Current pharmacotherapeutic approaches for the treatment of Tourette syndrome. Drugs Today 2014;50(2):159-79.

99. Murphy TK, Fernandez TV, Coffey BJ, et al. Extended-release guanfacine does not show a large effect on tic severity in children with chronic tic disorders. J Child Adolesc Psychopharmacol 2017;27(9):762-70.

100. The Tourette's Syndrome Study Group. Treatment of ADHD in children with tics: a randomized controlled trial. Neurology 2002;58(4):527-36.

101. Pliszka S. Practice parameter for the assessment and treatment of children and adolescents with attention-deficit/hyperactivity disorder. J Am Acad Child Adolesc Psychiatry 2007;46(7):894-921.

102. Srour M, Lesperance P, Richer F, et al. Psychopharmacology of tic disorders, vol. 17. Arlington (TX): American Psychiatric Association Publishing; 2013.

103. Gadow KD, Nolan EE. Methylphenidate and comorbid anxiety disorder in children with both chronic multiple tic disorder and ADHD, vol. 15. Arlington (TX): American Psychiatric Association Publishing; 2013. https://doi.org/10.1177/1087054709356405.

104. Pisano S, Masi G. Recommendations for the pharmacological management of irritability and aggression in conduct disorder patients. Expert Opin Pharmacother 2019;1-3. https://doi.org/10.1080/14656566.2019.1685498.

105. Riddle MA, King RA, Hardin MT, et al. Behavioral side effects of fluoxetine in children and adolescents. J Child Adolesc Psychopharmacol 1990;1(3):193-8.

Emotional Dysregulation
A Trauma-Informed Approach

Brooks R. Keeshin, MD[a,b,*], Beverly J. Bryant, MD[c],
Elizabeth R. Gargaro, MD[a,b]

KEYWORDS

- Trauma • Posttraumatic stress disorder • Emotional dysregulation
- Evidence-based trauma-focused treatment • Trauma-informed care
- Inpatient treatment • Residential treatment

KEY POINTS

- Traumatic experiences, subsequent traumatic stress, and other trauma reactions are highly associated with youth who experience emotional dysregulation.
- Evidence-based, trauma-focused treatments should be considered for youth with emotional dysregulation and a clear history of both trauma exposures and resultant traumatic stress.
- Trauma-informed approaches to severely emotionally dysregulated youth, including youth in in-patient and residential settings, can improve emotional and behavioral outbursts while maintaining the safety of the milieu.
- Knowledge about trauma exposures and symptoms is an important consideration when prescribing psychopharmacologic interventions in severely emotionally dysregulated youth.

BACKGROUND

All trauma is inherently complex. This is the first of 12 core principles of childhood trauma as described by the National Child Traumatic Stress Network.[1] There is no limit to the types of trauma that a child can experience, either in quality, severity, or frequency. Emotional dysregulation is highly correlated with prior childhood trauma, where challenges in emotional regulation among trauma-exposed youth are observed at up to twice the rate of controls, and there is significant overrepresentation of trauma among youth identified with severe emotional dysregulation (see Dvir and colleagues[2]

[a] Department of Pediatrics, University of Utah, 295 Chipeta Way, Salt Lake City, UT 84108, USA;
[b] Department of Psychiatry, University of Utah, 5021 Chipeta Way, Salt Lake City, UT 84108, USA; [c] Department of Psychiatry, University of Texas Health Science Center at Tyler, 11937 US-271, Tyler, TX 75708, USA
* Corresponding author. University of Utah Department of Pediatrics, Center for Safe and Healthy Families, 81 N Mario Capecchi Drive, Salt Lake City, UT 84113, USA
E-mail address: Brooks.Keeshin@hsc.utah.edu

Child Adolesc Psychiatric Clin N Am 30 (2021) 375–387
https://doi.org/10.1016/j.chc.2020.10.007
1056-4993/21/© 2020 Elsevier Inc. All rights reserved.

for review). Although many emotionally dysregulated, trauma-exposed youth experience some or all of the symptoms consistent with posttraumatic stress disorder (PTSD), the connection linking trauma and emotional dysregulation likely is complex, including disruptions in development and attachment, which also likely contribute to the association between trauma and emotional dysregulation.[2]

The trouble with Mark

Mark is an 8-year-old white boy, brought into an outpatient clinic by his foster mother because he grabbed a knife and threatened another foster child in the home. Mark has a history of "abuse" but no one has details because he is in his fifth foster home with his third child protective services worker. Mark is sent for evaluation in the emergency department and is admitted to a children's psychiatric inpatient unit. There, he continues to be aggressive, requiring seclusion and as-needed intramuscular (IM) neuroleptic medication to manage behaviors. In spite of multiple episodes of seclusion and as-needed injections, Mark's behavior does not improve. The inpatient staff feel threatened, worn out, and angry. When asked if there is a pattern to the explosive behavior, they report that it is a result of Mark "not getting his way" or being told "no." When asked to wait his turn by a favorite staff member, Mark becomes angry and begins crying, yelling, and disrupting the activity. After the area is cleared for safety, the child psychiatrist on the unit approaches Mark and urges him to take some deep breaths and talk about what is on his mind. He is asked what it means when someone he likes tells him "no" and he relates that it means that the person is saying that they do not like him, like other kids more, and will never care about what he wants. The favorite staff member reassures Mark that he still likes him. Mark is able to de-escalate to the point where he can recognize his cognitive distortions. Later, in a calmer state, Mark agrees to "practice" being told "no" and makes a safety agreement. Eventually, Mark is able to recognize that feelings of rejection remind him of violence and abandonment experienced earlier in life.

Often, the question is not whether or not trauma is present in emotionally dysregulated youth, but instead how trauma has an impact on the development, conceptualization, treatment planning, and treatment monitoring of youth with severe emotional dysregulation and aggression. What does it mean to provide care with a "trauma lens" or, better yet, how can an environmental/trauma framework be weighed/balanced with a more biological/behavioral approach?

The goal of this article is to identify how trauma can be incorporated into the treatment of severely emotionally dysregulated youth in a thoughtful and coherent manner. Principles of trauma-informed care (TIC) are discussed and examined, looking at trauma-focused versus trauma-informed approaches to treatment. Additionally, inpatient and residential settings are examined and psychopharmacologic interventions in severely emotionally dysregulated youth considered.

TRAUMA-INFORMED CARE

TIC is an overarching framework emphasizing the impact of trauma when working with individuals and families.[3] TIC has 4 tenets to consider when providing care within a strength-based approach: realizing, recognizing, responding to trauma, and resisting retraumatization.[4] TIC realizes how frequently trauma occurs among the children and families served, especially in high-risk populations. It ensures that systems apply trauma-informed principles universally, given the ubiquitous nature of trauma. In order to incorporate trauma into evaluation and case formulation, traumatic experiences and symptoms must be recognized. Responding requires that consideration of the impact of trauma continues to be a part of treatment planning and monitoring. Finally, awareness of potential retraumatization among traumatized children and families is the first step in mitigating that possibility in clinical interactions.[4] These 4 Rs identify the critical importance of understanding the prevalence and impact of trauma on individuals, families, and communities and ensuring that interventions actively treat trauma and prevent unnecessary trauma-related distress.

Trauma-focused treatments may be useful in youth with significant emotional dysregulation. Aggression, emotional lability and irritability often are overwhelming symptoms that are functionally impairing and challenging to address. Younger children with significant emotional dysregulation are at risk of being kicked out of daycare/school, and their reactivity causes poor social development with peers and strained relationships at home. Likewise, emotionally dysregulated older children and adolescents can be aggressive or violent or they can engage in high-risk behaviors, impairing their function and putting their safety and the safety of others at risk.

Trauma-focused interventions, however, are not a panacea for all trauma-exposed children with emotional dysregulation. It is important for clinicians not to oversell what can be gained with the use of outpatient, evidence-based trauma therapies. For some children and adolescents, evidence-based trauma treatments can address emotional dysregulation effectively. For others, trauma-informed, but not necessarily trauma-focused, approaches are necessary for severe emotional dysregulation that has generalized beyond traditional trauma-related cues and symptoms.

Evidence-based trauma treatments like trauma-focused cognitive-behavior therapy (TF-CBT) are designed to decrease symptoms associated with traumatic stress.[5] Numerous randomized controlled trials demonstrate the effectiveness of trauma-focused therapies in ameliorating symptoms.[6] Although trauma-focused therapies improve a variety of trauma-related or trauma-exacerbated symptoms (eg, behavioral problems and depression), most of these therapies target posttraumatic stress symptoms.[7] Several points are worth considering that may connect trauma, trauma reminders, and emotional dysregulation.

Temporal association

A simple yet critical place to start is to identify a complete inventory of potentially traumatic exposures. Tools, such as the UCLA PTSD Reaction Index for the *Diagnostic and Statistical Manual of Mental Disorders* (Fifth Edition), can be helpful to identify the specific trauma exposures as well as age(s) when the exposures occurred.[8] Identifying temporal proximity between specific traumas and the development or exacerbation of emotional dysregulation is strong evidence that trauma-focused work should be considered at some point during the treatment process.

Presence of symptoms of traumatic stress

When considering trauma-focused therapy for patients with emotional dysregulation, it is critical to account for current or recent symptoms of traumatic stress. Due to the heterogeneity and breadth of trauma symptoms, it can be difficult to appreciate trauma symptoms during traditional clinical interviews. For example, intrusive symptoms of traumatic events occur in many forms. Memories, nightmares, dissociative flashbacks, and physiologic or psychological distress can be precipitated by trauma reminders. Absent a standardized approach, it can be challenging to elicit how a youth re-experiences trauma. Avoidance, another core feature of traumatic stress, often exacerbates this diagnostic challenge. Patients often do not fill in the gaps when the interviewer asks a question that is close, but not directly related to, the patient's own experience of intrusive symptoms. Additionally, corroborative information about symptoms from parents or caregivers may elucidate the breadth of trauma symptoms. Similar to traumatic events, standardized symptom reports, such as the self and parent report versions of the UCLA PTSD Reaction Index, help systematically identify traumatic stress.[8]

Identification of trauma reminders

Exploring the presence of trauma reminders is pertinent, especially to the trauma-informed assessment of the emotionally dysregulated youth.[9] Behavioral outbursts in emotional dysregulation can be overwhelming, and it is easy to focus on the event itself and safety implications. Equally important, however, is identifying triggers for the explosive episode, because these may include specific trauma reminders. Trauma reminders often are innocuous, everyday events that leave no particular impression on outside observers. Furthermore, youth may have little or no conscious insight about the connection between reminders and emotional outbursts. Proximity between the reminder and explosive event is helpful, although some children may experience a trauma reminder that begins an upslope of emotional dysregulation, taking minutes to hours before becoming clinically apparent.

Clarification of syndromic overlap

Histories of depression, anxiety, and attention-deficit/hyperactivity disorder commonly are seen among those with concerns for emotional dysregulation. It is important to remember that although previous providers may have assigned these diagnoses, traumatic stress may be a better explanation of the constellation of symptoms and behaviors.[10] Certainly, comorbidity is possible, but re-examination of prior diagnoses may be warranted, especially when youth have failed to adequately respond to evidence-based first-line pharmacologic treatments of common pediatric disorders. When trying to connect trauma to emotional dysregulation, it is critical to identify the temporal association of trauma and trauma reminders, while evaluating the presence of traumatic stress symptoms and clarifying historical comorbidity versus syndromic overlap. Furthermore, in stable patients, with emotional dysregulation associated with trauma, a trauma-focused outpatient treatment, such as TF-CBT, could be the definitive treatment to address the constellation of posttraumatic symptoms, including emotional dysregulation.[7]

Presence of moderating and promotive factors

As described by Layne and colleagues[11] and demonstrated in **Fig. 1**, promotive, inhibitory and facilitative factors all are worth identifying when assessing the causative impact of traumatic experiences on poor outcomes, including emotional dysregulation. Promotive factors are independently associated with positive outcomes, for example, in a child with a healthy range of emotions who appropriately responds to positive and negative thoughts and experiences. The absence or presence of promotive factors is critical in establishing a child's functional emotional baseline and contextualizes function prior to the traumatic event(s). Furthermore, even with strong promotive factors, an increase in inhibitory factors or a decrease in facilitative factors has an impact on the effect of promotive factors on emotional regulation.

Protective and vulnerability factors also have an impact on the potential severity and chronicity of the child's trauma response.[11] In children with access to strong protective factors, even significant traumatic experiences may have little to no impact. Conversely, a child with significant vulnerabilities may be at risk for extreme trauma responses, including severe emotional dysregulation.

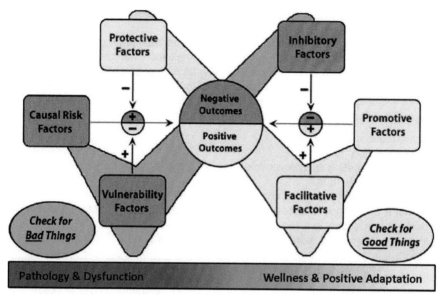

Fig. 1. Double Checks Heuristic identifies both moderating and promotive factors when considering reactions to trauma. (*From* Layne CM, Kaplow JB, Youngstrom EA. Applying Evidence-Based Assessment to Childhood Trauma and Bereavement: Concepts, Principles, and Practices. In: Landolt MA, Cloitre M, Schnyder U, editors. Evidence-Based Treatments for Trauma Related Disorders in Children and Adolescents. Cham: Springer, Cham; 2017; with permission.)

For patients with severe dysregulation who are not suitable for evidence-based trauma-focused therapy, the initial effort may need to address moderating and promotive factors. For example, a teenager with a history of sexual abuse and explosive self-injurious behavior is a danger to herself and others. Limiting access to items that she can use to harm herself (limiting a vulnerability), identifying adults she can access as part of a safety plan during future times of distress (increasing a protective factor), and improving the parent-child relationship (facilitating a promotive factor) may be the most important first steps in a trauma-informed intervention. At some point, directly addressing the trauma likely is necessary for optimal recovery.

INPATIENT MANAGEMENT OF EMOTIONAL DYSREGULATION

Due to the pervasive impact of trauma, people in need of trauma-specific services represent the majority of adults seeking public mental health services.[12] The extent of trauma exposure among children and adolescents admitted for aggression is less well documented. Bryant and colleagues[13] studied inpatient encounters with the clinical diagnosis of disruptive mood dysregulation disorder (DMDD) over a 5-year period. Patients with DMDD and a history of child maltreatment had significantly longer hospital stays, higher use of neuroleptics, and higher use of seclusion and restraint, even in the absence of a comorbid diagnosis of PTSD.[13] Patients with a high degree of trauma exposure have a lowered ability to tolerate distress.[14] Important changes in the trajectory of brain development in trauma-exposed children lead to heightened activity of the amygdala and reduced language processing during perceived threat.[15,16] Saxe and colleagues[15] describe outbursts after exposure to trauma reminders as "survival-in-the moment" states, leading some to suggest the idea of "universal trauma precautions" on inpatient units.[12]

The use of coercive techniques, including seclusion, restraints, physical holds, and intramuscular injections, although inevitable in some circumstances, can be traumatic for staff and patients alike. With shortened lengths of stay and limited information on admission, many inpatient staff are unaware of the trauma histories of patients. The National Association of State Mental Health Program Directors outlined 6 factors considered critical to successful implementation of TIC[12]:

1. Active leadership support
2. Data collection
3. Rigorous debriefing and prevention-focused analysis of events resulting in the use of coercive techniques
4. Trauma-informed education and skill development of the staff
5. Use of a wide range of assessments, including identification of strengths and capacity for emotional self-regulation
6. Involvement and inclusion of consumers at all levels of care

Yet, adoption of TIC on inpatient units sometimes can backfire. Evaluating admissions of aggressive children to an inpatient unit over 10 years, Carlson and colleagues[17] described an increased use of seclusion, restraints, physical holds, and unscheduled medications when a behavior modification program was abandoned in favor of a de-escalation approach. When emphasis shifted to de-escalation and/or distraction of a raging child rather than behavioral intervention, a "small army of adults" was needed to deal with outbursts, and verbally focused engagement often worsened agitation. This is consistent with the neurobiology of childhood maltreatment, which suggests use of language-based interventions in the presence of rage is unlikely to be effective. During such times, the emphasis must be on the safety of both the child and the environment.

TIC and behavior modification need not be mutually exclusive concepts when dealing with aggressive, emotionally dysregulated outbursts. Children exposed to trauma may engage in behaviors that appear to be attention-seeking or even manipulative. Also, a behavior may start out as one and end up as the other. Enforcing behavioral consequences in the heat of the moment, however, can escalate the acuity rather than reduce it. Saxe and colleagues[15] describe a treatment grid based on trauma systems therapy (TST) (**Fig. 2**). In TST, both the child and the environment are assessed in terms of stability. As an enhancement tool, color codes have been added to make it easier to communicate the components of the grid on the inpatient unit.

Children are categorized as regulated (green), emotionally dysregulated (yellow), or emotionally and behaviorally dysregulated (red). Similarly, a child's environment is assessed as stable (green), distressed (yellow), or threatening (red). The environment may be composed of the family, caregiver, or the staff of the inpatient unit. Depending on where the situation maps out on the grid, an appropriate level of intervention is chosen.

Treatment interventions then are divided into 3 categories:

1. Safety-focused (red): primary goal is to define and establish safety parameters for the child and environment and to develop a plan for managing "survival in the moment" states.
2. Regulation-focused (yellow): primary goal is to teach the child coping skills (as discussed previously).

Relational/Social/Physical Environment of Child

		Stable	Distressed	Threatening
Self-Regulation Capacity	Regulated	Beyond Trauma	Beyond Trauma	Safety Focused
	Emotionally Dysregulated	Self-regulation focused	Self-regulation focused	Safety Focused
	Emotionally & Behaviorally Dysregulated	Self-regulation focused	Safety Focused	Safety Focused

Fig. 2. Trauma Systems Therapy Treatment Grid. (*Adapted from* Saxe GN, Ellis BH, Brown AD. Trauma Systems Therapy for Children and Teens, 2nd edition. New York: The Guilford Press; 2016; with permission.)

3. Beyond trauma (green): primary goal is to help the child and family gain perspective on the trauma so that it is no longer the defining characteristic of the child's identity.

Clearly, when an out-of-control environment poses acute risk (red), treatment interventions must focus on safety. Additionally, significant emotional and behavioral dysregulation in a distressed environment warrants a safety focused response. In the heat of the moment, using a "lot of language," either for reassurance or for behavioral consequences, is ineffective and possibly deleterious. Efforts to incentivize adaptive behavior must be introduced while the child is receptive. In TST, the importance of the 2 Es (environment past and environment present), the 3 As (awareness, affect, and actions), and the 4 Rs (regulating, revving, re-experiencing, and reconstituting) are emphasized. TST advocates a "moment-by-moment" assessment of dysregulated behavior and identification of the "earliest detectable warning signals." When triggers are identified, the Managing Emotions Guide in TST contains a list of specific interventions. Coping skills (such as relaxation and affect management) and emotional identification skills (such as naming and expressing feelings) decrease the likelihood of future severe emotional outbursts.

When working with families exposed to trauma, it is useful to be aware of the Karpman Drama Triangle.[15] Trauma-exposed family members sometimes assume rigid roles: victim, persecutor, or rescuer. Stress has an impact on these roles. Fulkerson provides an example: a child feels "victimized" when taken to the hospital for out-of-control behavior. In the hospital, continued dangerous behaviors require intervention by the hospital staff (such as seclusion/restraint).[18] As a result, staff shift into the "persecutor" role previously held by the family that initially brought the child into the hospital. The family steps in to "rescue" the child, complaining to hospital administration and signing the child out against medical advice. It is easy for resentment to build in such situations and for staff members to feel hopeless and frustrated because they "can't seem to do anything right."

Inpatient treatment has a limited but important role in the overall treatment of traumatized children. Specific efforts to identify trauma and its connection to emotional dysregulation are the first steps. An easy-to-apply schematic (red/yellow/green) allows for quick and consistent engagement. This can be taught to parents and transferred to the home and school environment post discharge. If possible, these patients and their families need to be discharged to a trauma-informed system of care, including wraparound services, such as home-based care. The availability of such services varies widely, however, as does communication between inpatient and outpatient providers, leading to high readmission rates.[15]

TRAUMA-INFORMED RESIDENTIAL TREATMENT

Residential treatment is considered one of the most restrictive and expensive levels of care. Increasingly, severe psychological or behavioral dysfunction is required for placement in this level of care.[19] It often is considered the end of the road for youth unresponsive to lower levels of care or with little improvement despite multiple inpatient admissions.

Although community-based treatment tends to focus on individual, family, and/or group therapy, and inpatient treatment provides crisis stabilization, the therapeutic impact of residential programs is achieved largely through extended exposure to the structured, therapeutic milieu. The milieu provides opportunities and challenges to build success in aspects of daily life.[20] The emphasis on milieu as well as the severity of patients is a barrier to implementation of many trauma-focused therapies.[19] Trauma-informed practice should be the goal, especially given trauma-exposed prevalence rates greater than 70% (Zelechoski and colleagues[21]) and poor outcomes among these youth.[22]

High prevalence of complex trauma in residential care is not surprising given the impacts of childhood trauma on brain structure and development. Exposure to trauma during critical periods has an impact on areas of the brain responsible for emotion regulation, learning, and memory.[23] These deficits can lead to problematic behaviors and impairments in psychosocial functioning. Trauma exposure is associated with increased psychosocial dysfunction across a broad range of domains, including behavioral problems, substance abuse, running away, academic problems, attachment problems, self-injury, suicidality, and criminal activity.[22]

Caring and supportive environments are protective factors against the neurobiological effects of trauma through coregulation of stress responsivity. This relational security potentially can reduce the effect of trauma and prevent the development of long-term maladaptive behavior.[23] Residential treatment offers the opportunity to provide the care, structure, and skills children/youth need to successfully address severe deficits in psychosocial function and learn to develop relational security.

Residential treatment, however, may cue trauma reminders or cause additional trauma. Any events/behavior that disrupt/threaten the safety of the milieu pose a barrier to therapeutic growth, particularly in traumatized youth. Aggression may create feelings of insecurity and foster negative communication on the unit.[19,20] Seclusions/restraints jeopardize feelings of safety. As discussed previously, there are deleterious effects of seclusions/restraints, including retraumatization.

In their 2017 systematic review, Bryson and colleagues[23] identified key components to successful implementation of TIC in the psychiatric inpatient or residential setting. They defined successful implementation as reduction/cessation of seclusions/restraints and other "coercive measures."[21,23] Among the broad range of interventions evaluated, 5 key components stood out, which overlap substantially with the components identified by Muskett.[12] Key features in successful implementation were senior leadership commitment, sufficient staff support, inclusion of patients/families in the process, aligning policy and programming with trauma-informed principles, and using data to motivate change. Their review concluded that both targeted approaches and broad interventions could be effective in reducing coercive measures. Data regarding long-term outcomes/efficacy of trauma-informed interventions in the residential setting are limited; however, broad frameworks and specific therapeutic models have shown promise. A full discussion of every residential model is beyond the scope of this article; thus, 1 framework and 1 model are highlighted as examples of residential trauma-informed practices.

The attachment, regulation, and competency (ARC) framework has been implemented in a diverse range of populations and settings. This framework focuses on 3 domains (attachment, regulation, and competency) because they frequently are disrupted among traumatized youth. Within these 3 domains are 9 core building blocks of interventions and a tenth component—trauma experience integration. This last component shifts the framework from trauma informed to trauma focused. Hodgdon and colleagues[24] described the process and outcome of implementation of this framework in 2 different residential programs, noting that the framework "offers a guiding structure for providers working with trauma-impacted youth and their caregivers, while allowing significant flexibility in application."[23]

The flexibility of the ARC framework prompted 1 program with prior training in dialectical behavioral therapy (DBT) to emphasize additional training for milieu staff regarding DBT skills and their connection to the components of ARC.[24] Many residential programs are DBT based/informed given its effectiveness with severe emotional dysregulation, chronic suicidality, self-harm, and aggression. DBT also has been adapted in juvenile justice settings.[25,26] The ability to integrate trauma-informed and behavioral principles is a particular advantage of the ARC framework in residential care.

Literature from the juvenile justice system (JJS) also sheds light on trauma-informed interventions for severe emotional/behavioral dysregulation/aggression. Approximately 70% of JJS involved youth have at least 1 diagnosable mental health condition. Additionally, more than 80% of youths involved with JJS have 1 or more traumatic exposures. Historically, JJS involvement also has been a common source of retraumatization. A shift toward more TIC has produced data regarding trauma-informed behavioral interventions for aggressive youth in JJS residential settings.

Trauma Affect Regulation: Guide for Education and Therapy (TARGET) is a therapeutic model with efficacy in reducing aggressive behavior in adolescents, including an adaptation in the residential setting and a randomized trial within JJS placements.[27] This model is listed in the California Evidence Based Clearinghouse for Child Welfare as having promising research evidence regarding its efficacy in treating trauma.

TARGET is a manualized intervention that teaches a "seven step sequence of skills for processing and managing trauma-related reactions to current stressful experiences."[27] TARGET can be implemented by staff with no mental health training, making it versatile in application. It has gender-specific interventions and is applicable to both adolescents and adults. The 7 skills create the acronym FREEDOM and incorporate self-regulation, trauma processing, and strengths-based reintegration. The acronym stands for focusing (with its subskill of slow down, orient, and self-check [SOS]); recognizing triggers; emotion self-check; evaluate cognitions; define/deliberate goals; options for action; and make a contribution.[25] In juvenile detention facilities, each TARGET session during the first 7 days in detention was associated with 54% fewer disciplinary incidents and 72 fewer minutes of disciplinary seclusion during an approximately 2-week average stay.[27]

PSYCHOTROPIC MEDICATION CONSIDERATIONS

Children with a history of trauma, especially maltreatment, have higher rates of psychotropic medication use.[28–30] Behavioral issues, emotional dysregulation, and aggression tend to be primary drivers of pharmacologic treatment. Not all trauma-exposed children have PTSD, and some children may have identifiable, distinguishable disorders for which pharmacotherapy may be indicated. The following pharmacologic principles that are specific to trauma in children, however, should be considered:

1. There are no psychotropic medications with strong evidence in the treatment of PTSD in youth. Selective serotonin reuptake inhibitors have negative randomized controlled trials in youth with PTSD.
2. When untreated, both traumatic stress symptoms and chronic insomnia can exacerbate emotional dysregulation. If insomnia is a concern, improving sleep can be a logical first step.
3. Trauma can mimic other disorders, increasing the importance of diagnostic clarity, particularly when youth do not respond to first-line treatments.
4. Ecophenotypic variation suggests individuals with a history of trauma and symptoms that meet criteria for a disorder may be biologically distinct and, therefore, do not respond as predicted to traditional pharmacologic interventions.[31]
5. Trauma and adversity are associated with increased risk of obesity, diabetes, and cardiovascular disease. Traumatized youth are more likely to be involved in multiple systems, subject to frequent transitions and disjointed and interrupted care. These factors should be taken into account when considering medications with a high-risk side-effect profile or when prescribing a complex psychotropic regimen.[32]
6. Second-generation antipsychotics may be particularly challenging to use in this population. Outside of as-needed use in an inpatient setting, there is minimal evidence to support off-label use of these medications for trauma-associated emotional dysregulation/aggression.[33]
7. In children with persistent and functionally impairing emotional dysregulation despite complex polypharmacy, a systematic approach to deprescribing may be necessary to identify medications that are exacerbating or perpetuating symptoms as well as medications that are helpful.

SUMMARY

Emotional dysregulation has emerged as a transdiagnostic symptom of psychiatric illness in children and is a common reason for seeking treatment, especially among traumatized youth. Unappreciated trauma exposures and symptoms, syndromic overlap, ecophenotypic variation, moderating factors, and safety concerns make it difficult to determine the extent to which trauma has an impact on the diagnosis and management of the emotionally dysregulated child. Psychopharmacologic interventions often are only partially effective. Without recognition of the trauma component, children can enter a revolving door of inpatient hospitalizations and a pattern of escalating medication use, resulting in multiple medications at high doses. It is easy for caregivers and members of the treatment team to feel defeated. Without the specific skills needed to identify and manage these cases, it is easy to give up or resort to chemical restraint.

In Mark's case, his aggressive outbursts had worn out the inpatient staff, his teachers, and his foster parents. His behavior was seen as willful and manipulative, occurring "when he didn't get his way." Ultimately, a moment-by-moment analysis of the behavior revealed a trauma component. Recognizing where he fell on the TST grid allowed for de-escalation. When Mark had moved from red to yellow and finally to green, he was able to learn from the experience and develop a plan for exposure therapy to being told "no." Mark had to be taught different skills at each level but ultimately needed to face the cognitive distortions that told him that "no" is the same thing as being abandoned and maltreated.

Evidence-based treatments for PTSD symptoms do exist and skills-based training for the child, the caregiver, and the treatment team is imperative. Trauma-focused interventions need to be used to treat trauma-related symptoms. All children respond to trauma-informed, safe, and predictable environments where they can experience success. Trauma-informed behavioral interventions include limited language when the child is not receptive and positive reinforcement when possible. This allows children to experience success with positive adaptive behaviors. Caregivers, whether they are parents, teachers, foster families, or nursing staff, can become victims themselves when emotional dysregulation results in violence. Seclusion and restraint are a last resort but not uncommon in the treatment of these children. Structured, trauma-informed behavioral programming in higher levels of care can reduce the need for coercive measures while allowing the child to build skills. The world does not adapt to traumatized children. The job of clinicians is to recognize and teach these children and their families to adapt to it. Trauma-informed training and treatment are essential parts of the healing and recovery process for traumatized children and their families.

DISCLOSURES

Dr B.R. Keeshin has received research support from Substance Abuse and Mental Health Services Administration and the Utah Department of Health. Dr B. Bryant has received support from the Child Access to Mental Health and Psychiatry grant supported by the Health Resources and Services Administration. Dr E. Gargaro receives support from the Utah Department of Health.

REFERENCES

1. NCTSN Core Curriculum on Childhood Trauma Task Force. The 12 core concepts: concepts for understanding traumatic stress responses in children and families. Core curriculum on childhood trauma. Los Angeles (CA) and Durham (NC): UCLA-Duke University National Center for Child Traumatic Stress; 2012.
2. Dvir Y, Ford JD, Hill M, et al. Childhood maltreatment, emotional dysregulation, and psychiatric comorbidities. Harv Rev Psychiatry 2014;22(3):149–61.
3. Substance Abuse and Mental Health Services Administration. SAMHSA's concept of trauma and guidance for a trauma-informed approach. HHS Publication No. 14-4884. Rockville (MD): Substance Abuse and Mental Health Services Administration; 2014.
4. Substance Abuse and Mental Health Services Administration. Trauma-informed care in behavioral health services. Treatment improvement protocol (TIP) Series 57. HHS publication No. 13-4801. Rockville (MD): Substance Abuse and Mental Health Services Administration; 2014.
5. Cohen JA, Bukstein O, Walter H, et al. Practice parameter for the assessment and treatment of children and adolescents with posttraumatic stress disorder. J Am Acad Child Adolesc Psychiatry 2010;49(4):414–30.
6. Morina N, Koerssen R, Pollet TV. Interventions for children and adolescents with posttraumatic stress disorder: a meta-analysis of comparative outcome studies. Clin Psychol Rev 2016;47:41–54.

7. Cohen JA, Berliner L, Mannarino A. Trauma focused CBT for children with co-occurring trauma and behavior problems. Child Abuse Negl 2010;34(4):215–24.

8. Kaplow JB, Rolon-Arroyo B, Layne CM, et al. Validation of the UCLA PTSD reaction index for DSM-5: a developmentally-informed assessment tool for youth. J Am Acad Child Adolesc Psychiatry 2020;59(1):186–94.

9. Pynoos RS, Steinberg AM, Piacentini JC. A developmental psychopathology model of childhood traumatic stress and intersection with anxiety disorders. Biol Psychiatry 1999;46(11):1542–54.

10. Keeshin B, Forkey HC, Fouras G, et al. Children exposed to maltreatment: assessment and the role of psychotropic medication. Pediatrics 2020;145(2):e20193751.

11. Layne CM, Kaplow JB, Youngstrom EA. Applying evidence-based assessment to childhood trauma and bereavement: concepts, principles, and practices. In: Landolt MA, Cloitre M, Schnyder U, editors. Evidence-based treatments for trauma related disorders in children and adolescents, vol. 22. Cham (Switzerland): Springer; 2017. p. 67–96. Complete. Cham.

12. Muskett C. Trauma-informed care in inpatient mental health settings: a review of the literature. Int J Ment Health Nurs 2014;23(1):51–9.

13. Bryant B, Bear M, Rowlett J. DMDD patients with and without a history of childhood abuse and/or neglect: comparison of hospital length of stay, use of antipsychotics, and restraints. J Am Acad Child Adolesc Psychiatry 2018;57(10):S182.

14. Berenz EC, Vujanovic AA, Rappaport L, et al. Childhood trauma and distress tolerance in a trauma-exposed acute-care psychiatric inpatient sample. Psychol Trauma 2018;10(3):368–75.

15. Saxe GN, Ellis BH, Brown AB, editors. Trauma Systems Therapy for Traumatized Children and Teens, 2nd edition. New York: Guilford Press; 2015.

16. Teicher MH, Samson JA, Anderson CM, et al. The effects of childhood maltreatment on brain structure, function and connectivity. Nat Rev Neurosci 2016;17(10):652–66.

17. Carlson GA, Chua J, Pan K, et al. Behavior modification is associated with reduced psychotropic medication use in children with aggression in inpatient treatment: a retrospective cohort study. J Am Acad Child Adolesc Psychiatry 2019;59(5):632–41.e4.

18. Fulkerson M. Integrating the Karpman Drama Triangle With Choice Theory and Reality Therapy. International Journal of Reality Therapy 2003;23(1):12–4.

19. Hodgdon HB, Liebman R, Martin L, et al. The effects of trauma type and executive dysfunction on symptom expression of polyvictimized youth in residential care. J Trauma Stress 2018;31(2):255–64.

20. Knorth EJ, Klomp M, Van den Bergh PM, et al. Aggressive adolescents in residential care: a selective review of treatment requirements and models. Adolescence 2007;42(167):461–85.

21. Zelechoski AD, Sharma R, Beserra K, et al. Traumatized youth in residential treatment settings: prevalence, clinical presentation, treatment, and policy implications. J Fam Viol 2013;28(7):639–52.

22. Briggs EC, Greeson JKP, Layne CM, et al. Trauma exposure, psychosocial functioning, and treatment needs of youth in residential care: preliminary findings from the NCTSN core data set. J Child Adolesc Trauma 2012;5(1):1–15.

23. Bryson SA, Gauvin E, Jamieson A, et al. What are effective strategies for implementing trauma-informed care in youth inpatient psychiatric and residential treatment settings? A realist systematic review. Int J Ment Health Syst 2017;11(1):36.

24. Hodgdon HB, Kinniburgh K, Gabowitz D, et al. Development and implementation of trauma-informed programming in youth residential treatment centers using the ARC framework. J Fam Viol 2013;28(7):679–92.
25. Shelton D, Kesten K, Zhang W, et al. Impact of a dialectic behavior therapy-corrections modified (DBT-CM) upon behaviorally challenged incarcerated male adolescents. J Child Adolesc Psychiatr Nurs 2011;24(2):105–13.
26. Trupin EW, Stewart DG, Beach B, et al. Effectiveness of a Dialectal Behavior Therapy Program for Incarcerated Female Juvenile Offenders. Child and Adolescent Mental Health 2002;7(3):121-27.
27. Ford JD, Steinberg KL, Hawke J, et al. Randomized trial comparison of emotion regulation and relational psychotherapies for PTSD with girls involved in delinquency. J Clin Child Adolesc Psychol 2012;41(1):27–37.
28. Raghavan R, Brown DS, Allaire BT, et al. Medicaid expenditures on psychotropic medications for maltreated children: a study of 36 States. Psychiatr Serv 2014; 65(12):1445–51.
29. Leslie LK, Raghavan R, Zhsang J, et al. Rates of psychotropic medication use over time among youth in child welfare/child protective services. J Child Adolesc Psychopharmacol 2010;20(2):135–43.
30. Keeshin BR, Strawn JR, Luebbe AM, et al. Hospitalized youth and child abuse: a systematic examination of psychiatric morbidity and clinical severity. Child Abuse Negl 2014;38(1):76–83.
31. Teicher MH, Samson JA. Childhood maltreatment and psychopathology: a case for ecophenotypic variants as clinically and neurobiologically distinct subtypes. Am J Psychiatry 2013;170(10):1114–33.
32. Felitti VJ, Anda RF, Nordenberg D, et al. Relationship of childhood abuse and household dysfunction to many of the leading causes of death in adults. The adverse childhood experiences (ACE) Study. Am J Prev Med 1998;14(4):245–58.
33. Miller L, Riddle MA, Pruitt D, et al. Antipsychotic treatment patterns and aggressive behavior among adolescents in residential facilities. J Behav Health Serv Res 2013;40(1):97–110.

Dysregulation and Suicide in Children and Adolescents

Tami D. Benton, MD[a,b],*, Eli Muhrer, MD[a], Jason D. Jones, PhD[a,b], Jason Lewis, PhD[a,b]

KEYWORDS

- Dysregulation • Mood and behavioral dysregulation • Irritability • Suicidality
- Pediatric • Adolescent

KEY POINTS

- Emotional and behavioral dysregulation have been associated consistently with suicidal behaviors across diagnostic groups and are independent risk factors acutely and chronically.
- Interventions for suicidal youth should focus on dysregulation using evidence-based psychotherapies as well as treating the underlying psychiatric disorder pharmacologically.
- Instruments are available to measure dysregulation across diagnostic groups and can be used to monitor change when treating suicidal youth.

EPIDEMIOLOGY

Suicide is a public health crisis. It is the second leading cause of death among adolescents ages 15 years to 19 years old. In addition, nationwide, 7% of youth reported 1 or more suicide attempts in the past year.[1] Generally, suicide attempts and plans increase linearly from age 12 years to age 15 years and then more slowly through age 17. Recent data, however, show that among African American youths, suicide rates are significantly higher than those of white youths from ages 5 years to 12 years old (incident rate ratio [IRR] 1.82), whereas from ages 13 years to 17 years old, the suicide rate was approximately 50% lower among African American youths compared with white youths (IRR 0.51).[2]

According to data from the most recent nationwide Youth Risk Behavior Survey,[1] 17% of high school students (grades 9–12) reported seriously considering suicide in the past year. These rates were higher among female students than male students (22% vs 12%, respectively), a little higher among white students than African

[a] Department of Child and Adolescent Psychiatry and Behavioral Sciences, Children's Hospital of Philadelphia, 3440 Market Street, Suite 410, Philadelphia, PA 19104, USA; [b] Department of Psychiatry, University of Pennsylvania, Perelman School of Medicine, Philadelphia, PA, USA
* Corresponding author. Department of Child and Adolescent Psychiatry and Behavioral Sciences, Children's Hospital of Philadelphia, 3440 Market Street, Suite 410, Philadelphia, PA 19104.
E-mail address: bentont@email.chop.edu

Child Adolesc Psychiatric Clin N Am 30 (2021) 389–399
https://doi.org/10.1016/j.chc.2020.10.008
1056-4993/21/© 2020 Elsevier Inc. All rights reserved.

American students (17% vs 15%, respectively), and much higher among sexual minority youth than non-sexual minority youth (48% vs 13%, respectively). Rates for minority youth and girls are showing an upward trend compared with prior years.[3] Sexual minority youth have higher rates of suicidal ideation and attempts as well. From a developmental perspective, suicidal ideation is rare before 10 years of age, increases slowly through age 12 years, and then accelerates through age 17 years.[1]

In studies examining children with specific mental disorders who died by suicide, Trigylidas and colleagues[4] found that depression was present in 40.8%, oppositional defiant disorder (ODD)/conduct disorder in 20.1%, attention-deficit/hyperactivity disorder in 20.6%, and bipolar disorder in 16.3%. Taken together, as in other studies,[5] 90% of those who died by suicide had a diagnosis. Although these diagnoses are disparate, deficits in emotional regulation cut across all of them. Emotion dysregulation, in general, has been found to have high cross-sectional correlations with measures of variability in suicidal ideation[5] and prospective associations with suicidal behavior,[6,7] confounding diagnostic clarity for clinicians and hindering the development of therapeutic targets for researchers.

EMOTIONAL DYSREGULATION AND SUICIDALITY

First, as discussed elsewhere in this issue, there are several overlapping concepts in the study of emotion dysregulation. This article focuses on dysregulation, anger, and irritability. In terms of emotion dysregulation, there are a plethora of studies of the Child Behavior Checklist (CBCL)[8]–Dysregulation Profile (DP).[9] Empirically derived, it captures a mixed phenotype of severe dysregulation, including significant elevations on CBCL subscales of anxiety/depression, attention deficit/hyperactivity, and aggression. In other words, it reflects a combination of mood and behavior, internalizing and externalizing symptoms and behaviors. The CBCL-DP is linked to adolescent self-harm, suicidal ideation, and suicidal behaviors[1,2] compared with the general population.[10–15] Furthermore, both youth-reported and parent-reported dysregulation were uniquely and positively related to adolescent-reported self-harm and suicidal ideation.[16] Finally, studies of community samples show that young adults with a higher CBCL-DP score in childhood were at increased risk for suicidality in adulthood.[17]

Studies of clinical samples using standardized measures for dysregulation show similar findings. For instance, among inpatient youth seeking treatment of mood symptoms, those with the CBCL-DP profile presented more severe suicidal ideation compared with those without the profile.[18] Another study of youths hospitalized for elevated suicide risk[17] found that prior weeks' levels of affective and behavioral dysregulation were significant predictors of the subsequent weeks' level of suicidal ideation. Furthermore, prior weeks' suicidal ideation significantly predicted subsequent elevation of affective sensitivity, negative affective intensity, and behavioral dysregulation. Finally, affect sensitivity and peer invalidation were significant positive, prospective predictors of elevated suicidal ideation over the subsequent 6 months. Taken together, these findings suggest that affective and behavioral dysregulation are related to suicidal ideation level both acutely (over weeks) and chronically (over months). Both trait-level of dysregulation using the CBCL-DP and current degree (state) of dysregulation are important when predicting individual risk of suicidality.

Anger dysregulation is an important variable to measure in suicidality. The State-Trait Anger Expression Inventory (STAXI) was used in 180 adolescent inpatients, and higher trait and expressed anger were associated with increased likelihood of suicide attempts over 13 years after discharge.[19] For girls, trait anger and the inward/outward expression of anger mediated the risk for suicide attempts only when associated with major depression. Diagnosis for boys[19] was not contributory. Relatedly, both male students and female students at high risk for suicidal ideation were found to have higher rates of state and trait anger on the STAXI.[20] Finally, in a study comparing affective, behavioral, and cognitive functioning in adolescents with a single suicide attempt versus multiple suicide attempts, the multiple attempts group was more likely to be diagnosed with a mood disorder, with greater depression severity and more anger on the STAXI. Even after controlling for mood disorder, however, anger, affect dysregulation, and serious self-mutilation remained significant, underscoring the need to target both anger and affect dysregulation to reduce risk of future suicidal behavior.[21]

Haskevitch and colleagues used the Difficulties in Emotion Regulation Scale (DERS)[22] along with psychiatric diagnosis to study suicidal ideation and attempts in a sample of 547 adolescent inpatients.[23] Besides finding that suicidal ideation was uniquely and significantly associated with lifetime attempts, the limited ability to access emotion regulation strategies was associated most significantly with suicidal ideation in the past year compared with other aspects of dysregulation. Also, teens who attempted suicide before hospitalization reported higher levels of affect dysregulation than those with suicidal ideation only, and a history of past suicide attempts was positively correlated with a greater degree of affect dysregulation.[23] Finally, multiple attempters, relative to single attempters, not only report more severe negative affective states but also exhibit more affect dysregulation and disorders characterized by poor impulse control.[21]

The authors suggest that suicidal behavior and attempts may be similar to self-injurious behaviors in that they may be efforts to reduce intolerable emotionally states and have reinforcing properties. Specifically, in the immediate short term, suicidal behavior initially may reduce emotional arousal, whereas in the long term ultimately may increase negative affect and itself becomes another stressor.[24]

Although the negative affect emphasized in emotional dysregulation often is sadness or anxiety, recent research has shown that irritability plays a significant role in suicide and suicidal behaviors. For instance, psychological autopsies of children and young adolescents showed that emotional and behavioral disturbances associated with irritability frequently are reported during the period preceding the suicidal act.[25-27] Prospectively, Orri and colleagues[28] followed approximately 1400 youths longitudinally from birth to age 17. They used teacher assessments of irritability (defined by tantrums and aggressive responding to provocation) yearly from ages 6 years to 12 years and children self-reported measures of mental health symptoms and suicidality from ages 13 years to 17 years. Four irritability/aggression trajectories were identified: consistently low, rising, declining, and persistent. Children with either a rising or persistent irritability trajectory (vs a low trajectory) were at higher suicidal risk. A rising irritability trajectory across childhood, however, represented a direct risk for suicidality, whereas persistent irritability/aggression appeared to be a distal marker of suicidality acting through more proximal depressive symptoms, which accounted for most (73%) of the association.

Benarous and colleagues[29] completed a systematic review designed to understand the relationship between irritability and suicide. They equated anger with irritability and found reports of school-based adolescent samples where teens with self-reported high levels of anger were more likely to report suicidal ideation.[20,30,31] Similarly, Goldston[32] found that level of anger differentiated those with a single suicide attempt from those with multiple suicide attempts in hospitalized teens. In another sample of inpatient adolescents, the level of self-reported (but not parent-reported) irritability was positively associated with suicidal ideation after controlling for gender, depressive symptoms, substance use disorder, disruptive behavior disorder, and posttraumatic stress disorders.[33] On the other hand, in a 35-year follow-up study, irritability as rated by parents appeared to be a significant risk factor for their child's suicidal behavior in adulthood beyond associations with other adult disorders.[34]

Irritability includes an affective and behavioral dimension. Extensive research describes the association between internalizing symptoms and suicidality. What is less clear is the role that externalizing disorders and behaviors play in suicidality in youths. For instance, Aebi and colleagues[35] found that in a sample of 158 detained male adolescents diagnosed with ODD, those with the irritable (vs defiant/headstrong) subtype of ODD were at risk for anxiety, reoffending, and suicidal behavior. Myers and colleagues[36] found that angry mood among youth ages 7 years to 17 years old with a major depressive episode predicted later suicidality in a 3-year follow-up study. Thus, it is not only the experience of sadness that predicts suicidal behavior but also the presence of angry mood.

Taken together, these findings suggest that irritability and, more generally, emotional dysregulation play a key role in suicidal behavior in youth. Consequently, treatment must target all aspects of emotional dysregulation because this factor cuts across both externalizing and internalizing disorders.

ASSESSMENT OF SUICIDE AND EMOTIONAL DYSREGULATION

The biopsychosocial model posits that genetic, experiential, psychological, clinical, sociologic, and environmental factors contribute to an individual's suicide risk. A comprehensive review of suicide assessment can be found in the reviews by Turecki and colleagues[5] and by Turecki and Brent.[37] Any 1 or all of these factors may interact at a given time to precipitate suicidality for an individual; however, their relative contributions to risk for suicide are mediated by factors unique to that individual. Specific to this review, however, the assessment of emotional dysregulation, a transdiagnostic symptom complex common to disorders that decrease restraint and increases distress, is essential when assessing suicide risk among youth.[38] Fortunately, robust research demonstrates that emotional dysregulation can be measured across psychiatric disorders.

The CBCL-DP, described previously, is one of the most commonly used tools for the assessment of dysregulation and measurement of change over time[39] and has been associated strongly with adolescent self-harm and suicidal behaviors.[16] In addition, a DP can be generated from the items of the widely used Strengths and Difficulties Questionnaire[40] (SDQ). Like the CBCL-DP, which incorporates internalizing and externalizing dimensions, the SDQ-DP is based on items that assess hyperactivity/inattention and emotional and conduct/aggression problems.[41] The SDQ-DP and CBCL-DP are highly correlated[42] and the SDQ-DP also is associated with suicidal behaviors.[43]

The DERS[22] is a widely used self-report measure that asks respondents to rate emotion regulation difficulties when distressed. The original form is a 36-item self-report measure of 6 facets of emotion dysregulation:

1. Lack of emotional awareness
2. Lack of emotional clarity
3. Impulse control difficulties
4. Difficulty engaging in goal-directed behavior
5. Nonacceptance of certain emotional responses
6. Lack of access regulation strategies[23]

As discussed previously, the DERS has been associated with suicidality.

Several investigators are using ecological momentary assessment to gather data in real time about mood and behaviors using mobile electronic devices, such as phones or watches.[44] Ecological momentary assessment generates many data points over time and enables researchers to conduct fine-grained analyses of the associations between emotional dynamics and suicidal ideation and attempts. For example, 1 study assessed mood multiple times a day for 6 days and found that affective instability was associated with more frequent and more severe suicidal ideation.[45] These instruments can be useful when assessing risks for suicidal behaviors and for identifying targets for psychotherapeutic interventions.

TREATMENT

Given the associations between adolescent emotional dysregulation and suicidal ideation and behavior, discussed previously, there is a need for effective clinical interventions targeting emotional dysregulation and suicidal behaviors specifically. Currently, there are several interventions developed for the symptoms of emotional regulation in adult samples that are utilized for suicidal children and adolescents. These interventions consist of individually focused interventions, family-focused interventions, and multicomponent interventions, each of which is discussed later. Unfortunately, compared with adults, evidence on the use of psychosocial treatments in children and adolescents is relatively weak.[46] Recent reviews suggest that dialectical behavior therapy for adolescents (DBT-A) is the only intervention meeting criteria as a well-established treatment of reducing suicidal ideation and behavior in adolescents. When different types of cognitive behavior therapy (CBT) are grouped together, this type of intervention has positive findings as well. Lastly, recent evidence further suggests that family components are promising.[47]

INDIVIDUALLY FOCUSED INTERVENTIONS

Although CBT is an effective treatment of unipolar depression, it does not address suicidality adequately. To rectify this, a form of CBT for suicide prevention (CBT-SP)[48] has been developed for adolescents, which integrates traditional components of CBT for depression, such as cognitive restructuring and behavioral activation, with elements of DBT-A. The latter specifically targets emotion regulation. The underlying theory of suicidal behavior as conceptualized by CBT-SP is based on vulnerabilities in emotion dysregulation and cognitive rigidity.[49] CBT-SP aims to target emotion dysregulation through teaching several different emotion regulation skills, including relaxation, mindfulness, emotion identification, and hope building. There also is an emphasis on family engagement and collaboration, an important addition given the role of family factors in the development of dysregulation. The overall goal of this treatment is to reduce the risk of suicidal behaviors in adolescents, which is accomplished partly by addressing emotion dysregulation directly, a core vulnerability for suicide. At this time, however, evidence for the effectiveness of CBT-SP is promising, albeit limited[50]

Interpersonal psychotherapy for adolescents[51] is another individually focused therapeutic intervention that has shown promise in the treatment of youth presenting with self-injurious thoughts and behaviors.[52] Although interpersonal psychotherapy for adolescents focuses on the impact of adolescents' interpersonal relationships on psychiatric symptoms and overall functioning, there also is an emphasis on emotion regulation skills, including emotion education and awareness, mood monitoring, and feeling expression.

FAMILY-FOCUSED INTERVENTIONS FOR DYSREGULATION AND SUICIDE

Considering the familial origins of dysregulation,[53,54] it is not surprising that several psychosocial interventions for suicide either target or include a component focused on family relationships.[55] For example, attachment-based family therapy,[56] rooted in attachment theory, teaches interpersonal and affect regulation skills, with the goal of improving the quality of parent-adolescent relationships and fostering a sense of relational security in which the adolescent can confront and manage difficult issues.[57,58]

Another example of a family-focused intervention is Safe Alternatives for Teens and Youths (SAFETY), a cognitive-behavioral family treatment that is informed by DBT and designed to prevent suicide attempts. SAFETY aims to bolster protective interpersonal supports within the family and teach skills that promote safer behaviors and reactions in the face of stressors. SAFETY has been shown to reduce suicidal ideation and prevent suicide attempts among high-risk youth.[59,60]

Several other family-focused interventions and interventions that combine individual treatment with a family component have been developed and tested, and the efficacy of these programs has been reviewed in detail elsewhere.[55,61,62] It is noteworthy that across meta-analyses and reviews of interventions for youth suicidal behaviors, investigators have concluded that those interventions that include a family component show the strongest evidence of efficacy in the treatment of suicidality and self-harm.[61–63] Thus, the evidence strongly suggest that a consideration of family context is an essential feature of interventions aimed at treatment of youth suicidality and dysregulation.

DIALECTICAL BEHAVIOR THERAPY FOR ADOLESCENTS

DBT-A is a multicomponent cognitive behavior treatment that targets emotional, interpersonal, and behavioral dysregulation that leads to self-injurious thoughts and behaviors.[24] Standard adult DBT was adapted for use with adolescents experiencing deficits in core features of adolescent development, namely mood stability, interpersonal relationships, behavioral control, and identity development.[64] DBT-A itself consists of multiple components: individual therapy, skills training group, phone coaching, and ancillary treatment (ie, psychopharmacotherapy, school meetings/phone contact, family therapy, and telephone consultation for family members on using skills). The skills training group is focused primarily on teaching skills for enhancing emotion regulation in addition to other constructs, such as distress tolerance and interpersonal effectiveness. Additionally, keeping in mind the research demonstrating the efficacy of family components in interventions targeting adolescent suicidality, the skills training group adapted for DBT-A is multifamily, consisting of a group of adolescents and their parents.

According to DBT theory, the behaviors of suicidal and self-injurious individuals stem from a combination of a biological sensitivity and environmental factors.[24] The biological vulnerability consists of all aspects of emotion dysregulation, including emotional sensitivity, defined as easily triggered emotions, emotional reactivity, the experience of more intense emotions, and a slow return to emotional baseline. This theory, as applied to adolescents, further posits that in instances in which there is a mismatch between parents/caretakers and teens with respect to these 3 aspects of emotional sensitivity, an invalidating environment is created, leading to emotional dysregulation and other problems in the core features of adolescent development.[64]

PHARMACOLOGIC INTERVENTIONS FOR DYSREGULATION AND SUICIDE

There are no pharmacologic treatments of emotional dysregulation in the context of suicidal ideation or behaviors, although there are many trials focused on treating the behavioral manifestations of dysregulation, such as anger dysregulation or irritability, aggression, and suicidality. Pharmacologic treatments should be guided by evidence-based clinical practice guidelines developed for treating underlying psychiatric disorders when dysregulation is identified.[65,66]

DISCUSSION

A growing body of evidence shows that there exists a strong relationship between emotional dysregulation and suicidality for youth; it cuts across psychiatric diagnostic groups, increases risk for self-harm, and can be measured as a stable symptom profile across diagnostic groups. Currently, structured assessment of emotional dysregulation or externalizing behaviors are not consistently integrated in the suicide risk assessment for children and adolescents and should be. Although suicide risk assessments of clinical, environmental, social, and biological risk factors have provided information that demonstrates increase risk for suicide at some point in time, they have not been specific in predicting who actually will attempt or complete suicide.

Diagnosis mediates an individual's response to a stressor and may increase the impact of a particular stressor or reduce the restraint needed to resist suicidal impulses when stressed but does not predict who will make a suicide attempt. An individual's ability to tolerate negative affect and exhibit skills and strategies to regulate their behaviors in response to negative feelings determines their response to suicidal thoughts. This is consistent with extant literature that has found that emotion dysregulation contributes to suicidality, even after accounting for psychiatric disorders.

Ongoing treatment studies show effective and promising interventions for emotional dysregulation for children and adolescents. DBT-A has demonstrated effectiveness for emotional dysregulation and suicidal behaviors among adolescents. CBT and family-based interventions are identified as promising interventions for youth as well. Although pharmacotherapy trials are ongoing, there are no medications approved for the treatment of emotional dysregulation or suicide among youths.

What is increasingly clear is that emotion dysregulation plays a role in moving individuals from suicidal thoughts to action. Future research must integrate the assessment of this transdiagnostic construct in clinical studies to advance recognition and develop focused interventions to reverse the trend of suicide in children and adolescents.

DISCLOSURE

The authors have nothing to disclose.

REFERENCES

1. Kann L, McManus T, Harris WA, et al. Youth risk behavior surveillance —United States, 2017. MMWR Surveill Summ 2018;67(8):479.
2. Bridge JA, Horowitz LM, Fontanella CA, et al. Age-related racial disparity in suicide rates among US Youths from 2001 through 2015. JAMA Pediatr 2018;172(7):697–9.
3. Lindsey MA, Sheftall AH, Xiao Y, et al. Trends of suicidal behaviors among high school students in the United States: 1991–2017. Pediatrics 2019;144(5):e20191187.

4. Trigylidas TE, Reynolds EM, Teshome G, et al. Paediatric suicide in the USA: analysis of the National child death case reporting system. Inj Prev 2016;22(4):268.

5. Turecki G, Brent DA, Gunnell D, et al. Suicide and suicide risk. Nat Rev Dis Primers 2019;5(1):74.

6. Yen S, Shea MT, Sanislow CA, et al. Personality traits as prospective predictors of suicide attempts. Acta Psychiatr Scand 2009;120(3):222–9.

7. Yen S, Shea MT, Sanislow CA, et al. Borderline personality disorder criteria associated with prospectively observed suicidal behavior. Avicenna J Phytomed 2004;161(7):1296–8.

8. Achenbach T, Edelbrock C. The child behavior checklist manual. Burlington (VT): The University of Vermont; 1991. Published online.

9. Althoff RR, Rettew DC, Ayer LA, et al. Cross-informant agreement of the dysregulation profile of the child behavior checklist. Psychiatry Res 2010;178(3):550–5.

10. Ayer L, Althoff R, Ivanova M, et al. Child behavior checklist juvenile bipolar disorder (CBCL-JBD) and CBCL posttraumatic stress problems (CBCL-PTSP) scales are measures of a single dysregulatory syndrome. J Child Psychol Psychiatry 2009;50(10):1291–300.

11. Meyer SE, Carlson GA, Youngstrom E, et al. Long-term outcomes of youth who manifested the CBCL-pediatric bipolar disorder phenotype during childhood and/or adolescence. J Affect Disord 2009;113(3):227–35.

12. Holtmann M, Goth K, Wöckel L, et al. CBCL-pediatric bipolar disorder phenotype: severe ADHD or bipolar disorder? J Neural Transm 2008;115(2):155–61.

13. Hudziak JJ, Althoff RR, Derks EM, et al. Prevalence and genetic architecture of child behavior checklist–juvenile bipolar disorder. Biol Psychiatry 2005;58(7): 562–8.

14. Althoff RR, Rettew DC, Faraone SV, et al. Latent class analysis shows strong heritability of the child behavior checklist–juvenile bipolar phenotype. Biol Psychiatry 2006;60(9):903–11.

15. Volk HE, Todd RD. Does the child behavior checklist juvenile bipolar disorder phenotype identify bipolar disorder? Biol Psychiatry 2007;62(2):115–20.

16. Deutz MHF, Geeraerts SB, van Baar AL, et al. The dysregulation profile in middle childhood and adolescence across reporters: factor structure, measurement invariance, and links with self-harm and suicidal ideation. Eur Child Adolesc Psychiatry 2016;25(4):431–42.

17. Selby EA, Yen S, Spirito A. Time varying prediction of thoughts of death and suicidal ideation in adolescents: weekly ratings over 6-month Follow-Up. J Clin Child Adolesc Psychol 2013;42(4):481–95.

18. Mbekou V, Gignac M, MacNeil S, et al. The CBCL dysregulated profile: an indicator of pediatric bipolar disorder or of psychopathology severity? J Affect Disord 2014;155:299–302.

19. Daniel SS, Goldston DB, Erkanli A, et al. Trait anger, anger expression, and suicide attempts among adolescents and young adults: a prospective study. J Clin Child Adolesc Psychol 2009;38(5):661–71.

20. Park Y-J, Ryu H, Han KS, et al. Anger, anger expression, and suicidal ideation in Korean adolescents. Arch Psychiatr Nurs 2010;24(3):168–77.

21. Esposito C, Spirito A, Boergers J, et al. Affective, behavioral, and cognitive functioning in adolescents with multiple suicide attempts. Suicide Life Threat Behav 2003;33(4):389–99.

22. Gratz KL, Roemer L. Multidimensional assessment of emotion regulation and dysregulation: development, factor structure, and initial validation of the difficulties in emotion regulation scale. J Psychopathol Behav Assess 2004;26(1):41–54.

23. Zlotnik C, Donaldson D, Spirito A, et al. Affect regulation and suicide attempts in adolescent inpatients. J Am Acad Child Adolesc Psychiatry 1997;36(6):793–8.
24. Linehan M. Special treatment strategies. Suicidal behavior strategies. Cognitive-behavioral treatment of borderline personality disorder. 1993.
25. Shaffer D. Suicide in childhood and early adolescence. J Child Psychol Psychiatry 1974;15(4):275–91.
26. Hoberman HM, Garfinkel BD. Completed suicide in children and adolescents. J Am Acad Child Adolesc Psychiatry 1988;27(6):689–95.
27. Freuchen A, Kjelsberg E, Lundervold AJ, et al. Differences between children and adolescents who commit suicide and their peers: a psychological autopsy of suicide victims compared to accident victims and a community sample. Child Adolesc Psychiatry Ment Health 2012;6(1):1.
28. Orri M, Galera C, Turecki G, et al. Pathways of association between childhood irritability and adolescent suicidality. J Am Acad Child Adolesc Psychiatry 2019; 58(1):99–107.e3.
29. Benarous X, Consoli A, Cohen D, et al. Suicidal behaviors and irritability in children and adolescents: a systematic review of the nature and mechanisms of the association. Eur Child Adolesc Psychiatry 2019;28(5):667–83.
30. Zhang P, Roberts RE, Liu Z, et al. Hostility, physical aggression and trait anger as predictors for suicidal behavior in Chinese adolescents: a school-based study. PLoS One 2012;7(2):e31044.
31. Sigfusdottir ID, Asgeirsdottir BB, Gudjonsson GH, et al. Suicidal ideations and attempts among adolescents subjected to childhood sexual abuse and family conflict/violence: the mediating role of anger and depressed mood. J Adolesc 2013; 36(6):1227–36.
32. Kelley A, Goldston DB, Brunstetter R, et al. First-time suicide attempters, repeat attempters, and previous attempters on an adolescent inpatient psychiatry unit. J Am Acad Child Adolesc Psychiatry 1996;35(5):631–9.
33. Frazier EA, Liu RT, Massing-Schaffer M, et al. Adolescent but not parent report of irritability is related to suicidal ideation in psychiatrically hospitalized adolescents. Arch Suicide Res 2016;20(2):280–9.
34. Pickles A, Aglan A, Collishaw S, et al. Predictors of suicidality across the life span: the Isle of Wight study. Psychol Med 2010;40(9):1453–66.
35. Aebi M, Barra S, Bessler C, et al. Oppositional defiant disorder dimensions and subtypes among detained male adolescent offenders. J Child Psychol Psychiatry 2016;57(6):729–36.
36. Myers K, McCauley E, Calderon R, et al. The 3-year longitudinal course of suicidality and predictive factors for subsequent suicidality in youths with major depressive disorder. J Am Acad Child Adolesc Psychiatry 1991;30(5):804–10.
37. Turecki G, Brent DA. Suicide and suicidal behaviour. Lancet 2016;387(10024): 1227–39. https://doi.org/10.1016/S0140-6736(15)00234-2.
38. Nock MK, Green JG, Hwang I, et al. Prevalence, correlates, and treatment of lifetime suicidal behavior among adolescents: results from the national comorbidity survey replication adolescent supplement. JAMA Psychiatry 2013;70(3):300–10.
39. Biederman J, Petty CR, Day H, et al. Severity of the aggression/anxiety-depression/attention (AAA) CBCL profile discriminates between different levels of deficits in emotional regulation in youth with ADHD. J Dev Behav Pediatr 2012;33(3):236.
40. Goodman R. The strengths and difficulties questionnaire: a research note. J Child Psychol Psychiatry 1997;38(5):581–6.

41. Deutz MHF, Shi Q, Vossen HGM, et al. Evaluation of the strengths and difficulties questionnaire-dysregulation profile (SDQ-DP). Psychol Assess 2018;30(9): 1174–85.

42. Holtmann M, Buchmann AF, Esser G, et al. The child behavior checklist-dysregulation profile predicts substance use, suicidality, and functional impairment: a longitudinal analysis. J Child Psychol Psychiatry 2011;52(2):139–47.

43. Caro-Cañizares I, García-Nieto R, Díaz de Neira-Hernando M, et al. The SDQ dysregulation profile is associated with self-injurious thoughts and behaviors in adolescents evaluated at a clinical setting. Rev Psiquiatr Salud Ment 2019; 12(4):242–50.

44. Ecological momentary assessment of mood disorders and mood dysregulation. - PubMed - NCBI. Available at: https://www.ncbi.nlm.nih.gov/pubmed/19947781. Accessed March 2, 2020.

45. Palmier-Claus JE, Taylor PJ, Gooding P, et al. Affective variability predicts suicidal ideation in individuals at ultra-high risk of developing psychosis: an experience sampling study. Br J Clin Psychol 2012;51(1):72–83.

46. Calati R, Courtet P. Is psychotherapy effective for reducing suicide attempt and non-suicidal self-injury rates? Meta-analysis and meta-regression of literature data. J Psychiatr Res 2016;79:8–20.

47. Iyengar U, Snowden N, Asarnow JR, et al. A further look at therapeutic interventions for suicide attempts and self-harm in adolescents: an updated systematic review of randomized controlled trials. Front Psychiatry 2018;9:583.

48. Stanley B, Brown G, Brent DA, et al. Cognitive-behavioral therapy for suicide prevention (CBT-SP): treatment model, feasibility, and acceptability. J Am Acad Child Adolesc Psychiatry 2009;48(10):1005–13.

49. Bryan CJ. Cognitive behavioral therapy for suicide prevention (CBT-SP): implications for meeting standard of care expectations with suicidal patients. Behav Sci Law 2019;37(3):247–58.

50. Alavi A, Sharifi B, Ghanizadeh A, et al. Effectiveness of cognitive-behavioral therapy in decreasing suicidal ideation and hopelessness of the adolescents with previous suicidal attempts. Iran J Pediatr 2013;23(4):467–72.

51. Mufson L, Moreau D, Weissman MM, et al. Modification of interpersonal psychotherapy with depressed adolescents (IPT-A): phase I and II studies. J Am Acad Child Adolesc Psychiatry 1994;33(5):695–705.

52. Tang T-C, Jou S-H, Ko C-H, et al. Randomized study of school-based intensive interpersonal psychotherapy for depressed adolescents with suicidal risk and parasuicide behaviors. Psychiatry Clin Neurosci 2009;63(4):463–70.

53. Cassidy J. Emotion regulation: influences of attachment relationships. Monogr Soc Res Child Dev 1994;59(2–3):228–49.

54. Morris AS, Silk JS, Steinberg L, et al. The role of the family context in the development of emotion regulation. Soc Dev 2007;16(2):361–88.

55. Frey LM, Hunt QA. Treatment for suicidal thoughts and behavior: a review of family-based interventions. J Marital Fam Ther 2018;44(1):107–24.

56. Diamond GS, Diamond GM, Levy SA. Attachment-based family therapy for depressed adolescents. Published online 2014.

57. Diamond GS, Wintersteen MB, Brown GK, et al. Attachment-based family therapy for adolescents with suicidal ideation: a randomized controlled trial. J Am Acad Child Adolesc Psychiatry 2010;49(2):122–31.

58. Diamond GS, Kobak RR, Krauthamer Ewing ES, et al. A randomized controlled trial: attachment-based family and nondirective supportive treatments for youth who are suicidal. J Am Acad Child Adolesc Psychiatry 2019;58(7):721–31.

59. Asarnow JR, Berk M, Hughes JL, et al. The SAFETY program: a treatment-development trial of a cognitive-behavioral family treatment for adolescent suicide attempters. J Clin Child Adolesc Psychol 2015;44(1):194–203.

60. Asarnow JR, Hughes JL, Babeva KN, et al. Cognitive-behavioral family treatment for suicide attempt prevention: a randomized controlled trial. J Am Acad Child Adolesc Psychiatry 2017;56(6):506–14.

61. Glenn CR, Esposito EC, Porter AC, et al. Evidence base update of psychosocial treatments for self-injurious thoughts and behaviors in youth. J Clin Child Adolesc Psychol 2019;48(3):357–92.

62. Ougrin D, Tranah T, Stahl D, et al. Therapeutic interventions for suicide attempts and self-harm in adolescents: systematic review and meta-analysis. J Am Acad Child Adolesc Psychiatry 2015;54(2):97–107.e2.

63. Asarnow JR, Mehlum L. Practitioner Review: treatment for suicidal and self-harming adolescents – advances in suicide prevention care. J Child Psychol Psychiatry 2019;60(10):1046–54.

64. Miller AL, Rathus JH, DuBose AP, et al. Dialectical behavior therapy for adolescents. Dialectical behavior therapy in clinical practice: applications across disorders and settings. 2007:245–63.

65. Jensen PS, Youngstrom E, Steiner H, et al. Consensus report on impulsive aggression as a symptom across diagnostic categories in child psychiatry: implications for medication studies. J Am Acad Child Adolesc Psychiatry 2007;46(3):309–22.

66. Pappadopulos E, Macintyre JC, Crismon ML, et al. Treatment recommendations for the use of antipsychotics for aggressive youth (TRAAY). Part II. J Am Acad Child Adolesc Psychiatry 2003;42(2):145–61.

How and Why Are Irritability and Depression Linked?

Pablo Vidal-Ribas, PhD[a],*, Argyris Stringaris, MD, PhD, FRCPsych[b]

KEYWORDS

- Irritability • Depression • Risk • Genetics • Prospective

KEY POINTS

- Based on its time course and the baseline mood of the child, current psychiatric nosology differentiates between episodic irritability and chronic irritability.
- Episodic irritability is cross-sectionally linked to depression as a cardinal symptom in young people, but the manifestation of irritability alone is rare.
- Chronic irritability is longitudinally linked to depression rather than other disorders, such as bipolar disorder, attention-deficit/hyperactivity disorder, and conduct disorder.
- The association between irritability and depression can be explained by shared genetic and environmental risk factors, including higher rates of family history of depression, specific childhood temperaments and personality styles, and negative parenting styles.

INTRODUCTION

Irritability, defined as an increased proneness to anger relative to peers at the same developmental level,[1] is linked to depression. This review first distinguishes between two types of irritability, namely episodic irritability and chronic irritability. Then, the two types of associations between irritability and depression, namely cross-sectional and longitudinal, are described. Next, possible explanations are provided, mechanistic pathways that might explain the link between irritability and depression proposed, and treatment implications of such association discussed. Finally, future directions to investigate unanswered questions in the link between irritability and depression are discussed.

[a] Social and Behavioral Science Branch, Eunice Kennedy Shriver National Institute of Child Health and Human Development, National Institutes of Health, 6710 Rockledge Drive, Building 6710B, Room 3153A, Bethesda, MD 20892, USA; [b] Section of Clinical and Computational Psychiatry, National Institute of Mental Health, National Institutes of Health, 9000 Rockville Pike, Building 15k, Room 208, Bethesda, MD 20892, USA
* Corresponding author.
E-mail address: pablo.vidal-ribasbelil@nih.gov
Twitter: @PabloVidalRibas (P.V.-R.); @argStringaris (A.S.)

Child Adolesc Psychiatric Clin N Am 30 (2021) 401–414
https://doi.org/10.1016/j.chc.2020.10.009
1056-4993/21/Published by Elsevier Inc.

childpsych.theclinics.com

DURATION MATTERS: EPISODIC IRRITABILITY VERSUS CHRONIC IRRITABILITY

Current psychiatric nosology makes a distinction between chronic irritability and episodic irritability[2] (**Fig. 1**). Chronic irritability is persistent, does not constitute a change from a child's baseline mood, refers to how a child usually appears, and characterizes disruptive mood dysregulation disorder (DMDD) and its precursor, severe mood dysregulation (SMD),[3] in which clinically significant irritability must have been present for at least 12 months. Chronic irritability also is seen in oppositional defiant disorder (ODD), in which symptoms must have been present for at least 6 months.[4] By contrast, in depression or bipolar disorder (BD), irritability is episodic; it is a distinct change in relation to a child's baseline mood, analogous to the manic elevation in mood or the depressive decline.[4]

The *Diagnostic and Statistical Manual of Mental Disorders* (Fifth Edition) (*DSM-5*) also distinguishes between tonic irritability and phasic irritability.[5] Tonic irritability refers to persistently angry, grumpy, or grouchy mood, usually lasting days and weeks. Phasic irritability refers to behavioral outbursts of intense anger, manifested by brief or protracted verbal or physical aggression. Irritable or angry mood (ie, tonic irritability) and severe recurrent temper outbursts (ie, phasic irritability) occurring frequently over this grumpy mood are the core symptoms of DMDD. These irritability components, however, also can be seen during episodes of depression or mania coursing with irritability (see **Fig. 1**). The implications of tonic irritability versus phasic irritability are discussed later.

CROSS-SECTIONAL LINKS BETWEEN EPISODIC IRRITABILITY AND DEPRESSION

The study of irritability as part of depression has existed for a long time. For example, anger was considered an important symptom in Burton's concept of melancholia,[6] and, for Freud, melancholia involved self-directed hostility originally directed toward

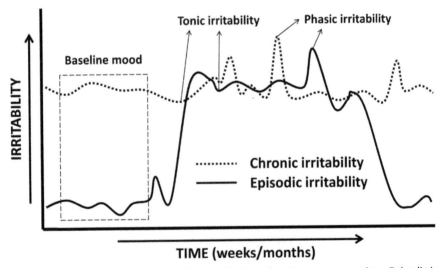

Fig. 1. Simplified depiction of types of irritability based on its course over time. Episodic irritability (*solid line*) constitutes a change from a child's baseline mood and is present in depression and BD. Chronic irritability (*dotted line*) is persistent over time and is the hallmark of DMDD and ODD. Although phasic irritability (ie, temper outbursts) and tonic irritability (ie, irritable or angry mood between outbursts) usually have been studied in the context of chronic irritability, these also take place in the context of episodic irritability as long as the mood episode lasts.

others, underlying what he described as "the undoubtedly pleasurable self-torture of melancholy."[7]

Today, the *DSM-5* considers irritability as a cardinal symptom of depression only in children and adolescents but not in adults.[4] Because irritability is part of a depression episode, it is therefore considered episodic irritability; that is, it should be a change from a patient's baseline mood noticeable by parents, teachers, or friends. That said, presenting only irritable mood—in contrast to low mood or anhedonia—as a cardinal symptom in pediatric depression seems rare.[8]

LONGITUDINAL LINKS BETWEEN CHRONIC IRRITABILITY AND DEPRESSION

Besides depression, irritability is listed as a cardinal or associated symptom of several behavioral and emotional psychiatric disorders in the *DSM-5*, particularly in children and adolescents.[4] However, in a recent meta-analysis that included articles in which dimensional irritability or categorical (ie, DMDD) irritability was a predictor of any future psychiatric outcome, irritability was found to be a significant predictor of depression, anxiety, and ODD but not of BD, conduct disorder (CD), attention-deficit/hyperactivity disorder (ADHD), or substance abuse[9] (**Fig. 2**). The association between chronic irritability and future depression has been demonstrated using the two main conceptualizations of chronic irritability, named the irritability dimension of ODD and DMDD.

According to the *DSM-5*, ODD is characterized by persistent irritable mood, and argumentative and defiant behaviors against authority figures, such as parents or teachers (APA, 2013). In previous versions of the *Diagnostical and Statistical Manual of Mental Disorders*, if ODD was accompanied by serious antisocial behaviors and violence, then a diagnosis of CD superseded that of ODD. Research has shown, however, that ODD not only is a precursor of CD[10] and ADHD[11] but also of emotional disorders, such as depression and anxiety.[10] There are at least two distinct symptoms dimensions: irritability/negative affectivity dimension and headstrong/oppositional behavior dimension.[12] Stringaris and Goodman[13] showed that irritability in ODD—defined as touchiness, easy annoyance, and anger—was a significantly stronger predictor of depression and generalized anxiety disorder than of conduct problems, whereas the headstrong dimension of ODD was predictive of ADHD and CD. These distinct longitudinal associations across ODD dimensions have also been replicated in other samples and countries[14,15] and by latent class analysis.[16]

Chronic irritability of the type seen in DMDD has also been associated with the development of depression and other emotional problems.[2,17,18] In the Children in the Community study in the United States, children and adolescents with chronic irritability followedup to late adolescence and young adulthood showed higher odds of developing depression and anxiety 10 years and 20 years later, even when considering the presence of depression at baseline.[2,18] In another prospective population-based study, children and adolescents with DMDD were more likely to suffer from depression and anxiety disorders in young adulthood than those without a psychiatric disorder or those with a psychiatric disorder other than DMDD. However, no differences were found in rates of antisocial behaviors or substance abuse disorders.[17] These findings have also been replicated in young children.[19]

THREE POTENTIAL MODELS TO EXPLAIN THE LINK BETWEEN IRRITABILITY AND DEPRESSION

Three plausible models have been suggested to describe the potential mechanisms that explain the association between irritability and depression. The one with the

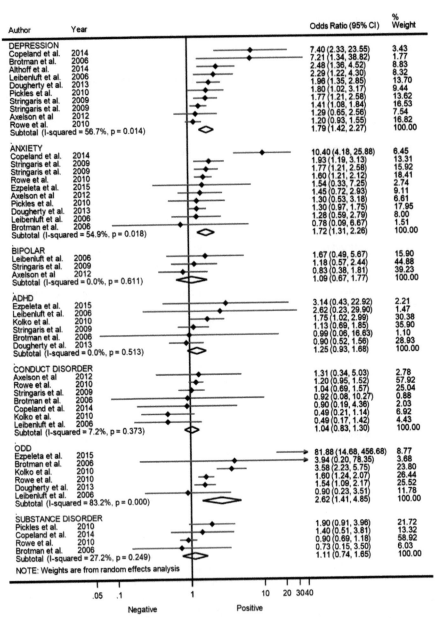

Fig. 2. Forest plot of irritability as a predictor of future psychiatric disorders. Points represent the estimated odds ratio of each study; the lines bisecting the point correspond to the 95% CIs. Pooled effect sizes are represented by diamonds. Weights for each study are given in the right column. (*Adapted from* Vidal-Ribas P, Brotman MA, Valdivieso I, et al. The Status of Irritability in Psychiatry: A Conceptual and Quantitative Review. J Am Acad Child Adolesc Psychiatry 2016;55(7):561; with permission.)

most supportive evidence, discussed later, is that irritability and depression are linked due to shared risk factors (**Fig. 3A**). Secondly, irritability in youth could cause depression through the environment that it elicits, such as school failure or relationship problems with family and peers (**Fig. 3B**). Other aspects of ODD, however, such as arguing and defying adults and annoying and blaming peers, thought to be more associated with academic failure and with interpersonal problems within the family and peers,[20] are less likely associated with depression than the ODD dimension of irritability.[13,14] Finally, insofar as young children may have difficulties in articulating internalizing symptoms but demonstrate externalizing behaviors when depressed, their caregivers observe only the latter in their child. Therefore, for some children, it also is possible that irritability may appear to be an early manifestation of depression (**Fig. 3C**). Although this model does not exclude a shared factor link, more studies and data from multiple informants on irritability and depressive symptoms are needed to support it.

SHARED DEVELOPMENTAL RISKS BETWEEN IRRITABILITY AND DEPRESSION

The evidence that suggests that the association between irritability and depression is explained mostly by shared risk factors includes studies examining genetic risk, family history of depression, shared temperamental and personality characteristics, and negative parenting styles.

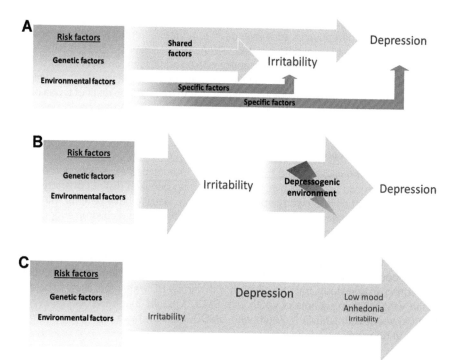

Fig. 3. Potential models to explain the association between irritability and depression. (*A*) Model of shared factors between irritability and depression. (*B*) Model of irritability as a cause of depression. (*C*) Model of irritability as an early manifestation of depression.

Genetic Risk

The stronger evidence for shared risks factors between irritability and depression comes from genetically informative studies. The association between irritability and depression is explained mostly by shared genetic factors both cross-sectionally[15] and longitudinally.[15] In a sample of adolescent twins followed, on average, for 2 years, Stringaris and colleagues[15] found that irritability was associated genetically with future depression more strongly than headstrong/hurtful behaviors were. In contrast, head-strong/hurtful behaviors shared more genetic variance with delinquent behavior at follow-up than with depression. The remaining variance between irritability and depression was explained by nonshared environmental factors. Similar results were found in more recent studies.[21,22] The impact of irritability on future emotional symptoms seems larger than the impact of emotional symptoms on subsequent irritability.[21]

It has been suggested that the genetic covariance between irritability and emotional symptoms varies over time and is especially evident from early to middle puberty (ages 13–14). Recent studies have adopted a developmental approach to examine more in depth the genetic covariance of irritability with distinct psychiatric correlates across time, including ADHD and depression. Irritability in children with ADHD, which often is referred to as emotional or mood lability or dysregulation in the literature, is common both in epidemiologic and clinical samples.[23] More than 35% of children with ADHD in a population-based study showed marked irritability.[24] Chronic irritability in ADHD is also associated with higher rates of depressive symptoms, cross-sectionally[25] and longitudinally,[26] as well as family history of depression.[25] Moreover, as happens with depression, the overlap between irritability and ADHD is mostly accounted by shared genetic risks rather than environmental factors.[27] Using a longitudinal design, a recent study found higher levels of irritability in boys than girls in childhood, whereas girls presented higher irritability than boys in adolescence.[28] In addition, irritability with a childhood onset was higher in boys and associated with ADHD genetic risk, and higher rates of ADHD and depression. In contrast, irritability with an adolescent onset was more common in girls and associated with genetic risk of both depression and ADHD, and higher rates of depression.[29] These findings suggest that there might be distinct types of irritability (eg, ADHD-like irritability and depression-like irritability) that could be distinguished if a development perspective is considered.

Family History of Depression

Family studies can also be used to examine how phenotypes aggregate across generations, which can be suggestive of genetic overlap between distinct phenotypes. It is well known that children with depression are more likely to have parents with a history of depression themselves.[30] Studies examining parents and children suggest that a family history of depression is also associated with irritability in the offspring.[31,32] In addition, the relation between maternal history of depression and adolescent depression is partly mediated by the presence of irritability in childhood.[33] These findings are also supported by studies examining categorical diagnoses of irritability (DMDD/SMD) in offspring. In one study, parents of youth with narrow BD were more likely to be diagnosed with BD than parents of youth with SMD,[34] results that have also been replicated in youth with DMDD.[35] Youth with DMDD, instead, have more parents with a history of depression.[36]

Early Temperament and Personality

Irritability can be viewed as a temperamental or personality characteristic. Indeed, proneness to anger, as irritability is defined, is a dimensional trait that can be found in the population[37] and has shown to be stable over time,[38] ranging from normality to pathology. Anger is thought to develop at approximately 6 months of age out of a broader emotion of distress.[39] According to the model of emotions by Buss and Plomin,[40] anger is considered under the early-temperament of "emotionality," with the other being "activity" and "sociability." Taking this model of temperamental dimensions, Stringaris and colleagues[41] used data from the community-based sample of the Avon Longitudinal Study of Parents and Children to examine the relation between early temperaments of emotionality—characterized by emotion dysregulation—and activity at 38 months and later psychiatric outcomes at 91 months. The investigators found that both temperaments predicted ODD; however, whereas temperamental activity predicted comorbidity between ODD and ADHD, emotionality predicted comorbidity with internalizing psychopathology. This replicated previous studies in which irritable-like temperament was associated with internalizing symptoms.[42]

In adults, irritability is subsumed under the higher-order personality trait of neuroti-cism. In a longitudinal twin study, Kendler and colleagues[43] found a strong relationship between neuroticism and the liability to suffer an episode of major depression. Moreover, close to 55% of the genetic liability to major depression appeared to be shared with neuroticism. The extent to which irritability (as part of neuroticism) may be linked to depression due to such genetic factors remains to be established.

Parenting Styles

Early temperaments can also have an impact on parenting styles, and vice versa, leading to the development of irritability[44–46] and depression.[42,47] Using a longitudinal monozygotic twins difference design, Oliver[48] examined the nonshared environmental links between negative parenting behavior, conduct problems, and irritability. The investigator found that the cross-sectional association between negative parenting and irritability explained by nonshared environment in early childhood and late childhood was larger than the association between negative parenting and conduct problems. Most importantly, while irritability in early childhood predicted negative parenting in middle childhood, the latter predicted higher levels of irritability in late childhood. In other words, the association between irritability and negative parenting styles seems bidirectional. From other studies not involving irritability, it is known that negative parenting has also been associated with an increased risk of depression in children,[44,47] with irritable temperament increasing the depressogenic effect of overprotection and lack of emotional warmth.[47]

TREATMENT IMPLICATIONS OF THE LINK BETWEEN IRRITABILITY AND DEPRESSION

The most supported psychological approaches to treat irritability are parent management training and cognitive behavior therapy (CBT). Evidence for the efficacy of these interventions comes from studies focusing on disorders in which irritability is a common symptom, such as ODD or ADHD.[49–51] In past years, modified psychological therapies have been tested for DMDD. For example, there is encouraging preliminary evidence of efficacy of dialectical behavior therapy[52] and interpersonal psychotherapy (IPT)[53] compared to treatment as usual in adolescents with DMDD. Notably, CBT and IPT are the treatments of choice for adolescent depression.[54]

In terms of pharmacological approaches to irritability, until recently, there was only one pharmacological randomized controlled trial with lithium specifically designed for the treatment of severe chronic irritability,[55] which emerged within the context of the controversy around pediatric BD.[56] However, this small trial[55] showed that lithium was ineffective for the treatment of chronic severe irritability in children and adolescents. Given the link between irritability and depression, the logical next step was to test whether a serotonin reuptake inhibitor (SRI) would be effective in treating severe chronic irritability in youth.

Evidence derived from adult samples indicates that SRIs might be efficacious in the treatment of irritability in depression[57] as well as in patients with intermittent explosive disorder[58] and those with premenstrual dysphoria.[59] In a systematic review on the effect of antidepressants on irritability in young people, Kim and Boylan[60] identified 2 uncontrolled studies of SRIs that reported on irritability as an outcome[61,62]; both studies indicated substantial improvement of irritability with SRI treatment.

Using a double-blind randomized controlled design, the authors examined the effect of adding citalopram, an SRI, to open-label treatment with stimulant medication, compared with adding placebo, in children with DMDD. The decision of treating first with stimulant medication was based on the high rates of ADHD in clinical samples of children with DMDD[63] as well as on the evidence of stimulant on decreased irritability in the context of ADHD.[64] The authors found that add-on citalopram was effective in reducing irritability compared with placebo, although this was not translated to decreases in functional impairment.[65]

It remains to properly test what would be the effect of an antidepressant alone in treating irritability. A case series suggests that atomoxetine, a presynaptic inhibitor of norepinephrine considered a second-line treatment for ADHD when comorbid with emotional disorders, might be effective in reducing irritability in those with DMDD without ADHD.[66] Further research is needed to address the treatment of irritability in children, not only in the context of DMDD but also in its episodic form in the context of depression.

SUMMARY AND FUTURE RESEARCH DIRECTIONS

The association between pediatric irritability and depression can be explained by several shared risk factors, including genetic and environmental risks. Both conditions are associated with higher rates of family history of depression, specific childhood temperaments and personality styles, and negative parenting styles. However, there still are more paths to be explored to fully elucidate why irritability and depression are linked.

For instance, pathophysiological models conceptualize irritability in two ways, namely, an aberrant response to threat and an aberrant response to reward.[1] In terms of the former, it is hypothesized that youth with severe irritability show attentional bias toward threatening angry faces[67] and are more likely to interpret ambiguous or neutral faces as more threatening.[68] Youth with severe irritability also present a generalized deficit in emotion recognition, in both faces[69,70] and voices.[70,71] However, these deficits are not specific to irritability; attentional biases toward negative stimuli and deficits in emotion recognition also are evident in people with depression.[72–74] Therefore, emotionprocessing deficits might represent an increased risk for both irritability and depression. There is some evidence to suggest that deficits in emotion processing in nonreferred samples might predict the development of depressive symptomatology,[75,76] and that this also could be the case in youth with chronic severe irritability.[70]

In terms of aberrant response to reward, irritability and anger are considered responses to frustration, which occurs when an individual continues to do an action in the expectation of a reward but does not actually receive that reward.[77] Similarly, there is a large body of literature supporting the notion that alterations in reward processing are a central mechanism in the development and maintenance of depression.[78,79] In a recent neuroimaging meta-analysis, the authors found that reduced striatal activation was evident in depressed participants during reward feedback and reward anticipation, and the latter was particularly strong in young people.[80] Paradoxically, however, depression typically is associated with low reward approaches,[81] whereas irritability is associated with behavioral approach tendencies and the seeking of rewards.[82] That is, both irritability and depression have an impact on the motivation to obtain rewards, although in opposite directions. Therefore, it is still unclear how altered reward as seen in irritability is associated with that seen in depression and even whether both types of reward deficits are related.

Quantitative genetic studies are needed to test the genetic and environmental contributions to the relationship between irritability, depression, and deficits in emotion and reward processing. It is known that irritability and depression[15] as well as depression and reward sensitivity[83] share genetic risk. It is unknown, however, whether irritability and reward sensitivity share genetic and/or environmental factors. Similarly, preliminary data suggest genetic correlations between deficits in emotion recognition and irritability.[84] To the authors' knowledge, however, no data from quantitative genetic studies exist that examine associations between emotion processing and depression. Longitudinal twin designs could help estimate the direction of effects between irritability, emotion/reward processing deficits, and depression and the genetic contributions to these effects.

Several studies have examined current and future correlates of irritability; yet, few studies have examined early predictors of chronic severe irritability.[85] Early stress is a strong predictor of depression,[86] and it has been hypothesized that this is caused partly by the impact that stress has on the neural reward circuit, according to the reward mediation hypothesis.[87] In addition, it has been suggested that aberrant reward processing can generate interpersonal stress in its own right, which in turn would increase the risk of depression, based on the stress generation hypothesis.[87,88] The authors have shown that stress can be both precursor and consequence of aberrant neural response to rewarding stimuli.[89] It would be interesting to test whether early stress also has an impact on the development of irritability, perhaps also by altering the reward circuit, and whether irritability can be a generator of interpersonal stress as well. If this were the case, that could be another pathway by which irritability increases the risk of depression.

Finally, further investigation is needed in parsing distinct components of irritability and their correlates. At the beginning of this review, the reader is introduced to tonic irritability (ie, persistent irritable mood) and phasic irritability (ie, temper outbursts). Both irritability components have been found moderately heritable (54%–61%); yet, only half of the genetic variance of tonic irritability is shared with phasic irritability.[90] This opens the door to investigate whether they also have differential psychopathological correlates, both cross-sectionally and longitudinally. Preliminary unpublished data from a population-based sample of children suggest that irritable mood is more associated with internalizing disorders like depression and anxiety, whereas temper outbursts are more linked to externalizing disorders like ADHD and CD.

In conclusion, irritability and depression are closely linked, mostly due to shared genetic variance. Other factors, related to this genetic variance or not, also play a role in this association. However, there still are many gaps that need to be examined to fully understand the links between irritability and depression and, consequently, to improve the lives of those children with these emotional problems.

DISCLOSURE

Dr. Stringaris receives grant from National Institute of Mental Health Intramural Research Program Project ZIA-MH002957

REFERENCES

1. Brotman MA, Kircanski K, Stringaris A, et al. Irritability in youths: a translational model. Am J Psychiatry 2017;174(6):520–32.
2. Leibenluft E, Cohen P, Gorrindo T, et al. Chronic versus episodic irritability in youth: a community-based, longitudinal study of clinical and diagnostic associations. J Child Adolesc Psychopharmacol 2006;16(4):456–66.
3. Leibenluft E, Charney DS, Towbin KE, et al. Defining clinical phenotypes of juvenile mania. Am J Psychiatry 2003;160(3):430–7.
4. Association AP. Diagnostic and Statistical Manual of Mental Disorders 5th edition. Washington, DC: 2013.
5. Avenevoli S, Blader JC, Leibenluft E. Irritability in youth: an update. J Am Acad Child Adolesc Psychiatry 2015;54(11):881–3.
6. Burton R. The anatomy of melancholy. London: JW Moore; 1857.
7. Freud S. Trauer und Melancholie. Frankfurt (Germany): Fischer Taschenbuch Verlag; 1915.
8. Stringaris A, Maughan B, Copeland WS, et al. Irritable mood as a symptom of depression in youth: prevalence, developmental, and clinical correlates in the Great Smoky Mountains Study. J Am Acad Child Adolesc Psychiatry 2013; 52(8):831–40.
9. Vidal-Ribas P, Brotman MA, Valdivieso I, et al. The status of irritability in psychiatry: a conceptual and quantitative review. J Am Acad Child Adolesc Psychiatry 2016;55(7):556–70.
10. Burke JD, Loeber R, Lahey BB, et al. Developmental transitions among affective and behavioral disorders in adolescent boys. J Child Psychol Psychiatry 2005; 46(11):1200–10.
11. Angold A, Costello EJ, Erkanli A. Comorbidity. J Child Psychol Psychiatry 1999; 40(1):57–87.
12. Stringaris A, Goodman R. Three dimensions of oppositionality in youth. J Child Psychol Psychiatry 2009;50(3):216–23.
13. Stringaris A, Goodman R. Longitudinal outcome of youth oppositionality: irritable, headstrong, and hurtful behaviors have distinctive predictions. J Am Acad Child Adolesc Psychiatry 2009;48(4):404–12.
14. Burke JD. An affective dimension within oppositional defiant disorder symptoms among boys: personality and psychopathology outcomes into early adulthood. J Child Psychol Psychiatry 2012;53(11):1176–83.
15. Stringaris A, Zavos H, Leibenluft E, et al. Adolescent irritability: phenotypic associations and genetic links with depressed mood. Am J Psychiatry 2012;169(1): 47–54.
16. Althoff RR, Kuny-Slock AV, Verhulst FC, et al. Classes of oppositional-defiant behavior: concurrent and predictive validity. J Child Psychol Psychiatry 2014; 55(10):1162–71.
17. Copeland WE, Shanahan L, Egger H, et al. Adult diagnostic and functional outcomes of DSM-5 disruptive mood dysregulation disorder. Am J Psychiatry 2014;171(6):668–74.
18. Stringaris A, Cohen P, Pine DS, et al. Adult outcomes of youth irritability: a 20-year prospective community-based study. Am J Psychiatry 2009;166(9):1048–54.

19. Dougherty LR, Smith VC, Bufferd SJ, et al. Preschool irritability: longitudinal associations with psychiatric disorders at age 6 and parental psychopathology. J Am Acad Child Adolesc Psychiatry 2013;52(12):1304–13.

20. Greene RW, Biederman J, Zerwas S, et al. Psychiatric comorbidity, family dysfunction, and social impairment in referred youth with oppositional defiant disorder. Am J Psychiatry 2002;159(7):1214–24.

21. Savage J, Verhulst B, Copeland W, et al. A genetically informed study of the longitudinal relation between irritability and anxious/depressed symptoms. J Am Acad Child Adolesc Psychiatry 2015;54(5):377–84.

22. Mikolajewski AJ, Taylor J, Iacono WG. Oppositional defiant disorder dimensions: genetic influences and risk for later psychopathology. J Child Psychol Psychiatry 2017;58(6):702–10.

23. Shaw P, Stringaris A, Nigg J, et al. Emotion dysregulation in attention deficit hyperactivity disorder. Am J Psychiatry 2014;171(3):276–93.

24. Stringaris A, Goodman R. Mood lability and psychopathology in youth. Psychol Med 2009;39(8):1237–45.

25. Eyre O, Langley K, Stringaris A, et al. Irritability in ADHD: associations with depression liability. J Affect Disord 2017;215:281–7.

26. Eyre O, Riglin L, Leibenluft E, et al. Irritability in ADHD: association with later depression symptoms. Eur Child Adolesc Psychiatry 2019;28(10):1375–84.

27. Merwood A, Chen W, Rijsdijk F, et al. Genetic associations between the symptoms of attention-deficit/hyperactivity disorder and emotional lability in child and adolescent twins. J Am Acad Child Adolesc Psychiatry 2014;53(2):209–220 e4.

28. Riglin L, Eyre O, Cooper M, et al. Investigating the genetic underpinnings of early-life irritability. Transl Psychiatry 2017;7(9):e1241.

29. Riglin L, Eyre O, Thapar AK, et al. Identifying novel types of irritability using a developmental genetic approach. Am J Psychiatry 2019;176(8):635–42.

30. Klein DN, Lewinsohn PM, Rohde P, et al. Psychopathology in the adolescent and young adult offspring of a community sample of mothers and fathers with major depression. Psychol Med 2005;35(3):353–65.

31. Krieger FV, Polanczyk VG, Goodman R, et al. Dimensions of oppositionality in a Brazilian community sample: testing the DSM-5 proposal and etiological links. J Am Acad Child Adolesc Psychiatry 2013;52(4):389–400 e1.

32. Wiggins JL, Mitchell C, Stringaris A, et al. Developmental trajectories of irritability and bidirectional associations with maternal depression. J Am Acad Child Adolesc Psychiatry 2014;53(11):1191–205, 1205.e1-4.

33. Whelan YM, Leibenluft E, Stringaris A, et al. Pathways from maternal depressive symptoms to adolescent depressive symptoms: the unique contribution of irritability symptoms. J Child Psychol Psychiatry 2015;56(10):1092–100.

34. Brotman MA, Kassem L, Reising MM, et al. Parental diagnoses in youth with narrow phenotype bipolar disorder or severe mood dysregulation. Am J Psychiatry 2007;164(8):1238–41.

35. Fristad MA, Wolfson H, Algorta GP, et al. Disruptive mood dysregulation disorder and bipolar disorder not otherwise specified: fraternal or identical twins? J Child Adolesc Psychopharmacol 2016;26(2):138–46.

36. Propper L, Cumby J, Patterson VC, et al. Disruptive mood dysregulation disorder in offspring of parents with depression and bipolar disorder. Br J Psychiatry 2017;210(6):408–12.

37. Copeland WE, Brotman MA, Costello EJ. Normative irritability in youth: developmental findings from the great smoky mountains study. J Am Acad Child Adolesc Psychiatry 2015;54(8):635–42.

38. Caprara GV, Paciello M, Gerbino M, et al. Individual differences conducive to aggression and violence: trajectories and correlates of irritability and hostile rumination through adolescence. Aggress Behav 2007;33(4):359–74.

39. Katharine MBB. Emotional development in early infancy. Child Dev 1932;3(4): 324–41.

40. Buss AH, Plomin R. Temperament: early developing personality traits. Hillsdale (NJ): Lawrence Erlbaum; 1984.

41. Stringaris A, Maughan B, Goodman R. What's in a disruptive disorder? Temperamental antecedents of oppositional defiant disorder: findings from the Avon longitudinal study. J Am Acad Child Adolesc Psychiatry 2010;49(5):474–83.

42. Kiff CJ, Lengua LJ, Bush NR. Temperament variation in sensitivity to parenting: predicting changes in depression and anxiety. J Abnorm Child Psychol 2011; 39(8):1199–212.

43. Kendler KS, Neale MC, Kessler RC, et al. A longitudinal twin study of personality and major depression in women. Arch Gen Psychiatry 1993;50(11):853–62.

44. Kiff CJ, Lengua LJ, Zalewski M. Nature and nurturing: parenting in the context of child temperament. Clin Child Fam Psychol Rev 2011;14(3):251–301.

45. Lengua LJ, Kovacs EA. Bidirectional associations between temperament and parenting and the prediction of adjustment problems in middle childhood. J Appl Dev Psychol 2005;26(1):21–38.

46. Lipscomb ST, Leve LD, Harold GT, et al. Trajectories of parenting and child negative emotionality during infancy and toddlerhood: a longitudinal analysis. Child Dev 2011;82(5):1661–75.

47. Oldehinkel AJ, Veenstra R, Ormel J, et al. Temperament, parenting, and depressive symptoms in a population sample of preadolescents. J Child Psychol Psychiatry 2006;47(7):684–95.

48. Oliver BR. Unpacking externalising problems: negative parenting associations for conduct problems and irritability. BJPsych Open 2015;1(1):42–7.

49. Barkley RA. Taking charge of ADHD: the complete, authoritative guide for parents. New York: Guildford press; 2013.

50. Comer JS, Chow C, Chan PT, et al. Psychosocial treatment efficacy for disruptive behavior problems in very young children: a meta-analytic examination. J Am Acad Child Adolesc Psychiatry 2013;52(1):26–36.

51. Sukhodolsky DG, Scahill L. Cognitive-behavioral therapy for anger and aggression in children. New York: Guildford press; 2012.

52. Perepletchikova F, Nathanson D, Axelrod SR, et al. Randomized clinical trial of dialectical behavior therapy for preadolescent children with disruptive mood dysregulation disorder: feasibility and outcomes. J Am Acad Child Adolesc Psychiatry 2017;56(10):832–40.

53. Miller L, Hlastala SA, Mufson L, et al. Interpersonal psychotherapy for mood and behavior dysregulation: pilot randomized trial. Depress Anxiety 2018;35(6): 574–82.

54. Zhou X, Hetrick SE, Cuijpers P, et al. Comparative efficacy and acceptability of psychotherapies for depression in children and adolescents: a systematic review and network meta-analysis. World psychiatry 2015;14(2):207–22.

55. Dickstein DP, Towbin KE, Van Der Veen JW, et al. Randomized double-blind placebo-controlled trial of lithium in youths with severe mood dysregulation. J Child Adolesc Psychopharmacol 2009;19(1):61–73.

56. Wozniak J, Biederman J, Kiely K, et al. Mania-like symptoms suggestive of childhood-onset bipolar disorder in clinically referred children. J Am Acad Child Adolesc Psychiatry 1995;34(7):867–76.

57. Fava M, Rosenbaum JF. Anger attacks in patients with depression. J Clin Psychiatry 1999;60(15):21–4.

58. Coccaro EF, Lee RJ, Kavoussi RJ. A double-blind, randomized, placebo-controlled trial of fluoxetine in patients with intermittent explosive disorder. J Clin Psychiatry 2009;70(5):653–62.

59. Dimmock PW, Wyatt KM, Jones PW, et al. Efficacy of selective serotonin-reuptake inhibitors in premenstrual syndrome: a systematic review. Lancet 2000; 356(9236):1131–6.

60. Kim S, Boylan K. Effectiveness of antidepressant medications for symptoms of irritability and disruptive behaviors in children and adolescents. J Child Adolesc Psychopharmacol 2016;26(8):694–704.

61. Garland EJ, Weiss M. Case study: obsessive difficult temperament and its response to serotonergic medication. J Am Acad Child Adolesc Psychiatry 1996;35(7):916–20.

62. Armenteros JL, Lewis JE. Citalopram treatment for impulsive aggression in children and adolescents: an open pilot study. J Am Acad Child Adolesc Psychiatry 2002;41(5):522–9.

63. Deveney CM, et al. A prospective study of severe irritability in youths: 2- and 4-year follow-up. Depress Anxiety 2015;32(5):364–72.

64. Fernandez de la Cruz L, et al. Treatment of children with attention-deficit/hyperactivity disorder (ADHD) and irritability: results from the multimodal treatment study of children with ADHD (MTA). J Am Acad Child Adolesc Psychiatry 2015;54(1):62–70 e3.

65. Towbin K, et al. A double-blind randomized placebo-controlled trial of citalopram adjunctive to stimulant medication in youth with chronic severe irritability. J Am Acad Child Adolesc Psychiatry 2019;59(3):350–61.

66. Benarous X, Ferrafiat V, Zammit J, et al. Effective use of atomoxetine to treat six inpatient youths with disruptive mood dysregulation disorder without attention deficit disorder. CNS Spectr 2020;25(4):455–7.

67. Hommer RE, et al. Attention bias to threat faces in severe mood dysregulation. Depress Anxiety 2014;31(7):559–65.

68. Brotman MA, et al. Amygdala activation during emotion processing of neutral faces in children with severe mood dysregulation versus ADHD or bipolar disorder. Am J Psychiatry 2010;167(1):61–9.

69. Rich BA, et al. Face emotion labeling deficits in children with bipolar disorder and severe mood dysregulation. Dev Psychopathol 2008;20(2):529–46.

70. Vidal-Ribas P, Brotman MA, Salum GA, et al. Deficits in emotion recognition are associated with depressive symptoms in youth with disruptive mood dysregulation disorder. Depress Anxiety 2018;35(12):1207–17.

71. Deveney CM, et al. Affective prosody labeling in youths with bipolar disorder or severe mood dysregulation. J Child Psychol Psychiatry 2012;53(3):262–70.

72. Peckham AD, McHugh RK, Otto MW. A meta-analysis of the magnitude of biased attention in depression. Depress Anxiety 2010;27(12):1135–42.

73. Dalili MN, Penton-Voak IS, Harmer CJ, et al. Meta-analysis of emotion recognition deficits in major depressive disorder. Psychol Med 2015;45(6):1135–44.

74. Naranjo C, et al. Major depression is associated with impaired processing of emotion in music as well as in facial and vocal stimuli. J Affect Disord 2011; 128(3):243–51.

75. Beevers CG, Carver CS. Attentional bias and mood persistence as prospective predictors of dysphoria. Cogn Ther Res 2003;27(6):619–37.

76. Vrijen C, Hartman CA, Oldehinkel AJ. Slow identification of facial happiness in early adolescence predicts onset of depression during 8 years of follow-up. Eur Child Adolesc Psychiatry 2016;25(11):1255–66.

77. Berkowitz L. Frustration-aggression hypothesis: examination and reformulation. Psychol Bull 1989;106(1):59.

78. Admon R, Pizzagalli DA. Dysfunctional reward processing in depression. Curr Opin Psychol 2015;4:114–8.

79. Forbes EE, Dahl RE. Research Review: altered reward function in adolescent depression: what, when and how? J Child Psychol Psychiatry 2012;53(1):3–15.

80. Keren H, O'Callaghan G, Vidal-Ribas P, et al. Reward processing in depression: a conceptual and meta-analytic review across fMRI and EEG studies. Am J Psychiatry 2018;175(11):1111–20.

81. Treadway MT, Zald DH. Reconsidering anhedonia in depression: lessons from translational neuroscience. Neurosci Biobehav Rev 2011;35(3):537–55.

82. Carver CS, Harmon-Jones E. Anger is an approach-related affect: evidence and implications. Psychol Bull 2009;135(2):183.

83. Bogdan R, Pizzagalli DA. The heritability of hedonic capacity and perceived stress: a twin study evaluation of candidate depressive phenotypes. Psychol Med 2009;39(2):211–8.

84. Rappaport LM, Carney DM, Brotman MA, et al. A Population-Based Twin Study of Childhood Irritability and Internalizing Syndromes. J Clin Child Adolesc Psychol 2020;49(4):524–34.

85. Munhoz TN, Santos IS, Barros AJD, et al. Perinatal and postnatal risk factors for disruptive mood dysregulation disorder at age 11: 2004 Pelotas Birth Cohort Study. J Affect Disord 2017;215:263–8.

86. Pizzagalli DA. Depression, stress, and anhedonia: toward a synthesis and integrated model. Annu Rev Clin Psychol 2014;10(1):393–423.

87. Auerbach RP, Admon R, Pizzagalli DA. Adolescent depression: stress and reward dysfunction. Harv Rev Psychiatry 2014;22(3):139–48.

88. Hammen C. Generation of stress in the course of unipolar depression. J Abnorm Psychol 1991;100(4):555–61.

89. Vidal-Ribas P, Benson B, Vitale AD, et al. Bidirectional associations between stress and reward processing in children and adolescents: a longitudinal neuroimaging study. Biol Psychiatry Cogn Neurosci Neuroimaging 2019;4(10):893–901.

90. Moore AA, Lapato DM, Brotman MA, et al. Heritability, stability, and prevalence of tonic and phasic irritability as indicators of disruptive mood dysregulation disorder. J Child Psychol Psychiatry 2019;60(9):1032–41.

Emotional and Behavioral Dysregulation in Severe Mental Illness

Aditi Sharma, MD[a], Jon McClellan, MD[b],*

KEYWORDS

- Schizophrenia • Bipolar disorder • Borderline personality disorder • Aggression
- Dysregulation • Adolescent

KEY POINTS

- Emotional dysregulation and aggression commonly present in severe mental illnesses, including schizophrenia, bipolar disorder, and borderline personality disorder.
- Emotional dysregulation and aggression can be driven by internal or external factors.
- Determining whether primary drivers of dysregulation are internal or external is important for treatment planning.
- Addressing the underlying psychiatric condition is the primary means of treating emotional dysregulation and irritability in severe mental illness. Accurate diagnosis helps guide appropriate treatment.
- For most people, both disorder-specific treatments and broadly applicable interventions to address skills building and problem solving can be helpful.

INTRODUCTION

Individuals suffering from major mental health disorders, including schizophrenia spectrum disorders, bipolar disorder, and borderline personality disorder, often experience difficulties with irritability, aggression and emotional instability over the course of their illness. Such symptoms may represent core signs of the psychopathology of the underlying disorder, and/or nonspecific problems that are independent of, or in reaction to, the person's primary psychiatric disorder. These clinical presentations challenge differential diagnostic assessments, given the overlap between complex symptom constructs such as irritability, paranoia, and mania.

---
[a] Department of Psychiatry and Behavioral Sciences, University of Washington, 4800 Sand Point WAY Northeast, MS OA.5.154, Seattle, WA 98105, USA; [b] Department of Psychiatry and Behavioral Sciences, University of Washington, 8805 Steilacoom Boulevard Southwest, Lakewood, WA 98498, USA
* Corresponding author.
E-mail address: drjack@uw.edu

Child Adolesc Psychiatric Clin N Am 30 (2021) 415–429
https://doi.org/10.1016/j.chc.2020.10.010
1056-4993/21/© 2020 Elsevier Inc. All rights reserved.

childpsych.theclinics.com

Abbreviations	
ADHD	attention deficit hyperactivity disorder
COBY	Course and Outcome of Bipolar Youth
DBT	dialectical behavior therapy
PANSS	Positive and Negative Syndrome Scale
PRN	pro re nata (as needed)

By themselves, emotional outbursts and displays of anger are neither diagnostically specific nor even prima facie evidence of a mental health disorder. Not all irritability is mania or paranoia, and moodiness does not equate to a major mood episode. The recognition of patterns of symptoms, course of illness, and associated risk factors characteristic of complex psychiatric syndromes is key to accurate diagnosis and treatment, and requires more than counting nonspecific symptom criteria on a diagnostic checklist.

Determining factors that either cause or exacerbate emotionality and aggression is key. In the context of a major mental illness, there are 2 broadly defined overlapping profiles for mood and behavioral dysregulation, namely, symptoms that are primarily driven by internal disruptions of thought and mood and those that are a reaction to environmental triggers and interpersonal interactions, often in the context of skills deficits and maladaptive coping. These patterns are not mutually exclusive. A violent outburst in a person with paranoid delusions is often triggered by some external event, and disproportionate emotional distress is more likely in a person predisposed by virtue of insecure attachment and a history of complex trauma. Recognizing which symptoms are mostly internally driven and independent of environmental triggers and which occur in the context of social negotiations and interpersonal conflicts has important implications for treatment.

SCHIZOPHRENIA SPECTRUM DISORDERS

Schizophrenia is characterized by disruptions of thought and behavior, including hallucinations, delusions, disorganized speech and thought patterns, and negative symptoms (paucity of speech and thought, avolition, asociality).[1] Although irritability and mood dysregulation are not part of the Diagnostic and Statistical Manual of Mental Disorders or International Classification of Diseases diagnostic criteria, individuals with schizophrenia spectrum disorders often present with irritable angry outbursts, reactive moods, and erratic bizarre behaviors.

The diagnosis of schizoaffective disorder reflects the challenges inherent to characterizing mood and behavioral dysregulation in the context of psychosis. To meet the Diagnostic and Statistical Manual of Mental Disorders, 5th edition, criteria for schizoaffective disorder, in addition to meeting the full criteria for schizophrenia, the patient must demonstrate evidence of a major mood episode for the majority of the duration of the illness.[1] However, in community settings, schizoaffective disorder is often diagnosed based on erratic moods and aggressive outbursts, making the diagnosis unreliable in both juveniles and adults.[2,3] It is important to distinguish between major mood episodes (distinct periods of disruptions in mood, energy, sleep, and appetite that persist for days to weeks) versus emotional and behavioral outbursts (explosive and volatile shifts in mood and behavior, often triggered by perceived threats, delayed gratification, or social conflicts). Both are serious psychiatric problems, but qualitatively different presentations and require different approaches for treatment.

Aggression in persons with schizophrenia and related disorders is common. Among individuals enrolled in first episode psychosis programs, 1 out of 4 had a history of violent acts, and 1 out of 10 was arrested for violent behaviors within the first 12 months of first contact.[4] Aggressive behaviors are associated with a number of negative outcomes, including serious harm to others, more frequent and prolonged hospitalizations, higher medication requirements, and stigmatization.[5] In persons with schizophrenia, symptoms of irritability and erratic moods often represent underlying paranoia, disorganized thinking, and reactions to delusional beliefs. Maladaptive emotion regulation strategies are correlated with psychotic and depressive symptoms.[6] Aberrant thought processes and perceptual distortions impair reality testing and the ability to problem solve and negotiate social situations. Outbursts can be difficult to predict, because the triggers are often idiosyncratic and specific to the person's internal belief system. As such, acutely psychotic patients, particularly those with paranoid delusions, experience persistent states of suspiciousness and fearfulness, and are at risk for angry volatile violent outbursts.

Risk factors predicting aggression in persons with schizophrenia include substance abuse, impulsivity, psychopathy, and treatment noncompliance.[7] In addition, developmental risk factors, such as prior criminal behavior, past violence, and childhood histories of trauma also predict histories of aggression in affected individuals.[8] These risk factors are also generally predictive of poorer prognosis. It is important to try and address factors such as comorbid substance abuse and relapse prevention to improve symptomatic and functional outcomes, and to reduce the risk of aggression and unsafe behaviors.

Finally, it is important to recognize that approximately 8% of adolescents (ages 13–18 years) and 17% of children (ages 9–12 years) describe psychotic-like symptoms.[9] Youth reporting these symptoms are at risk for general psychopathology, including anxiety, depression, behavioral problems, substance abuse, and self-harm. Children may describe internal experiences as hearing or seeing things or unusual beliefs, particularly in the context of stress, histories of trauma, developmental impairments, or secondary gain. Most youth reporting psychotic-like symptoms never go on to develop a true psychotic disorder.[10] Equating psychotic-like symptoms to psychosis greatly expands the diagnostic construct, and shifts the characterization of anger outbursts and emotionality from maladaptive coping skills to core symptoms of psychopathology that are not under the child's control. This practice potentially promotes the overuse of antipsychotic medications (and medications in general), while shifting treatment planning away from skills-based coaching, cognitive-behavioral therapies, and parent training.

BIPOLAR DISORDER

Bipolar disorder, particularly pediatric bipolar disorder, perhaps best represents the challenge distinguishing outbursts of mood and behavioral dysregulation from distinct mood episodes of a major psychiatric illness. Classic manic-depressive illness is characterized by remarkable shifts in a person's mood states, with demarcated episodes of elevated mood and affect, markedly decreased sleep and increased energy, racing often grandiose and/or paranoid thoughts (including psychosis), and impulsive reckless behaviors.[11] Most patients, even those presenting with classic euphoria, experience symptoms of irritability, dysphoria, and hostility as they cycle through the different stages of mania.[12] Episodes can last months (or years) without treatment. The classic form of the illness has very substantial morbidity owing to suicide and high heritability. Distinguishing bipolar disorder from other conditions, including schizophrenia, major depression, and borderline personality disorder, has been a longstanding challenge given an overlap in symptoms, and the fact that many persons with manic depressive illness do not present with hallmark euphoric mania.[11]

Beginning in the 1990s, in the child psychiatry literature, the definition of bipolar disorder expanded to include youth with pronounced irritability, explosive outbursts and mood dysregulation.[13] Youth diagnosed with pediatric bipolar disorder are highly comorbid with attention deficit hyperactivity disorder (ADHD) and conduct disorders. The course of illness is often atypical in comparison with the classic adult cyclical illness, and variably described as very rapid, brief, recurrent episodes lasting hours to a few days[14,15] or chronic persistent mania, including those with mania as their baseline functioning.[16] In the Course and Outcome of Bipolar Youth (COBY) study, 3 subtypes of patients with bipolar disorder were characterized: elation only, irritable only, and elation with irritability. The majority of patients had both elation and irritability. Youth in the irritable-only subgroup had similar clinical characteristics and family histories of bipolar disorder compared with those with predominant elation.[17] However, most youth with chronic irritability and severe mood dysregulation do not experience more classic manic or hypomanic episodes over time.[18] As discussed at length in Drs. Vidal-Ribas and Stringaris' article on "How and why are irritability and depression linked?" in this issue, irritability in early adolescence was a specific predictor of self-reported depression and anxiety disorders 20 years later, but not bipolar disorder or antisocial personality disorders.[19,20]

During this era, there was a substantial increase in the diagnosis of bipolar disorder in pediatric populations, particularly in the United States, as well an associated increase in the prescriptions of antipsychotic medications.[21] Whether chronic irritability and mood dysregulation in youth represents bipolar disorder is an ongoing debate.[22,23] Many children with ADHD have significant bouts of irritability, talk fast, sleep less, and present with excessive energy and reckless behaviors.[24] Similarly, many individuals with trauma-associated conditions, including borderline personality disorder (as described elsewhere in this article), experience chronic irritability, mood instability, emotional outbursts, disrupted sleep, and impulsive reckless behaviors. At the time when the diagnosis of pediatric bipolar disorder expanded, mood dysregulation and explosive outbursts did not fit neatly into either behavioral disorder (eg, ADHD, conduct disorder) or mood disorder categories.[13] This debate has generated research trying to better characterize clinical profiles of children with chronic irritability and mood instability (eg, Stringaris and colleagues,[25] 2018). In part owing to the controversy surrounding pediatric bipolar disorder, the *Diagnostic and Statistical Manual of Mental Disorders*, 5th edition, emphasized the cyclical nature of mood episodes in bipolar disorder, and introduced disruptive mood dysregulation disorder to characterize youth with chronic irritability and frequent temper outbursts.[1] Whether this reforms the diagnosis of pediatric bipolar disorder remains to be seen.

Regardless of nomenclature, cyclical episodes defined by dramatic changes in a person's mood, thinking, and behavior are qualitatively and clinically distinct from chronic intermittent irritability and explosive outbursts. Classic acute episodes of mania are potential psychiatric emergencies and require immediate medical intervention, with the most effective treatments being mood-stabilizing and antipsychotic medications. Intermittent explosive outbursts also respond to psychotropic agents, but medications are not sufficient or specific. Persons with these problems need to learn new skills. Functional behavioral assessments and cognitive behavioral therapies are needed to address triggers, reinforcers, and maladaptive coping skills.

BORDERLINE PERSONALITY DISORDER

Borderline personality disorder is characterized by persistent patterns of an unstable sense of self, erratic interpersonal relationships, intense and volatile emotionality, impulsivity, anger outbursts, and self-harming behaviors.[1] Explosive outbursts are often related to underlying affective instability, impulsivity, and sensitivity to interpersonal rejection.[26] Childhood adversity, including abuse and neglect, inconsistent, hostile, or invalidating parenting styles and parental loss, predisposes to the development of borderline personality disorder and traits.[27] Although clinicians are typically hesitant to diagnose personality disorders in youth, symptoms of borderline personality disorder often first present during adolescence, and a younger onset is associated with more long-term impairment.[28]

Originally Adolph Stern[29] (1938) used the term "border line group of neuroses" to describe patients that did not fit into either psychotic or psychoneurotic constructs. Although borderline personality disorder is defined as a categorical condition, there is a great degree of clinical heterogeneity, with symptoms that overlap with other disorders and vary developmentally across the lifespan.[30] Children and teens with histories of complex trauma and chaotic social situations often demonstrate traits of borderline personality disorder, including difficulties with sense of self, impulsivity, emotional and behavioral dysregulation, and interpersonal relationships.[31] Irritability in adolescents with oppositional defiant disorder and mood disorders predicts symptoms of borderline personality disorder at 30 years of age.[32] Moreover, symptoms of borderline personality disorder are common in youth diagnosed with ADHD, oppositional defiant disorder, conduct disorder, depression, anxiety, and post-traumatic stress disorder.[33]

There has been considerable debate as to whether borderline personality disorder and bipolar disorder are independent or related conditions.[34] Both disorders are defined by mood swings, impulsivity, and suicidal behaviors. However, by definition, bipolar disorder is defined by distinct mood episodes whereas borderline personality disorder is defined by chronic pervasive affective instability in the context of unstable interpersonal relationships.[1] The 2 disorders are also distinguished by different patterns of family history and treatment response, with medications having a limited impact on mood states in borderline personality disorder.[34]

Emotional and behavioral dysregulation suggestive of borderline personality disorder and traits are perhaps best conceptualized as maladaptive coping strategies for negotiating interpersonal relationships and conflicts in persons predisposed to these difficulties by virtue of impulsivity, insecure attachment and histories of maltreatment and inconsistent, invalidating, and/or coercive environmental exposures.[35] These symptoms and risk factors are common presenting problems in child psychiatry clinics and inpatient settings, regardless of diagnosis, and qualitatively distinct from outbursts and erratic behaviors secondary to an underlying psychotic illness.

DIAGNOSTIC ASSESSMENTS AND RATING SCALES

Using structured diagnostic interviews, such as the Kiddie Schedule for Affective Disorders and Schizophrenia[36,37] or the Structured Clinical Interview for DSM5 Disorders,[38] helps to reliably diagnose severe mental illness. Some symptom rating scales for psychosis include questions addressing mood and behavioral regulation. The Positive and Negative Syndrome Scale (PANSS) is a 33-item, clinician-administered measure assessing positive and negative symptoms and general measures of psychopathology.[39] The PANSS-Excited Component subscale has been used in several studies examining efficacy of specific interventions for aggression in schizophrenia spectrum disorders or bipolar disorder.[40] The Scale for the Assessment of Positive Symptoms is a clinician-administered scale that assesses positive symptoms, and contains 1 item for rating aggression.[41] Some studies of aggression in schizophrenia use the Buss and Perry Aggression Questionnaire,[42] which assesses 4 different domains of aggression (physical aggression, verbal aggression, anger, and hostility).

The Young Mania Rating Scale is an 11-item, clinician-administered measure based on the patient's subjective report of symptoms and clinical observation. The scale was "not intended to be used as a diagnostic instrument," but designed to assess severity of mania once the diagnosis has been established.[43] The Young Mania Rating Scale includes specific (elevated mood, increased sexual interest, pressured speech, racing thoughts) and nonspecific symptoms (increased activity, irritability) of mania and may fail to discriminate between bipolar disorder and other conditions, primarily ADHD.[44]

Diagnostic measures, including interviews and symptom checklists, augment (but do not supplant) a thorough clinical assessment and mental status examination, and may not account for nuance in presentation. Beyond the measurement of specific symptoms, it is also important to assess overall functioning, which is one of the most important markers of success in treatment.

TREATMENT

Both diagnostically specific treatments and broadly therapeutic intervention strategies are important in managing irritability, aggression, and emotional stability in major mental illness. Specific interventions target the core features of the psychiatric disorder, for example, delusions and hallucinations in psychotic disorders, and the marked shifts in mood, energy, and thinking characteristic of mania in bipolar disorder. Psychotherapeutic and educational interventions are generally applicable across a range of diagnoses, addressing common deficits, such as limited coping skills or emotional reactivity.

TARGETED INTERVENTIONS
Schizophrenia Spectrum Disorders

In schizophrenia spectrum disorders, internal disruptions of thinking and perceptions are associated with aggression and violence.[45,46] Targeting positive symptoms and disorganization helps to decrease the risk of aggression. Several first-, second-, and third-generation antipsychotic medications have been shown to be effective in improving overall symptom severity in early onset schizophrenia.[47] These medications carry considerable side effect profiles, including extrapyramidal symptoms, weight gain, and metabolic complications. Short- and long-term monitoring is required for efficacy and safety. Given the side effects, some agents are not considered first-line agents, such as olanzapine given the risk for weight gain,[48] and clozapine given metabolic problems and neutropenia.[47] In the United States, owing to its serious side effect profile, clozapine can only be prescribed by providers enrolled in a Risk Evaluation and Mitigation Strategy program. Not all antipsychotic agents that are approved by the US Food and Drug Administration for use in adults have been found to be effective in pediatric patients. For example, both ziprasidone[49] and asenapine[50] failed to beat placebo in randomized controlled trials.

Antipsychotic medications more generally decrease symptoms of aggression.[51] Clozapine is the most effective antipsychotic medication for reducing aggression and violence in patients with schizophrenia.[52] In randomized trials examining the effectiveness of antipsychotics, anticonvulsants, benzodiazepines, and lithium treatment for aggression and agitation in hospitalized patients with schizophrenia and bipolar disorder, the evidence best supports the usefulness of antipsychotic agents.[40] Although the use of benzodiazepines, either alone or in combination with antipsychotics, is common practice for treating acute aggression and agitation in persons with psychotic disorders, the data supporting these practices are limited.[53] In a large population study, compliance with antipsychotic medication treatment, including long-acting injectable agents, decreased the risk of violent crime in adults with schizophrenia, whereas mood stabilizers did not have an effect unless the patient had a diagnosis of bipolar disorder.[54]

Although mood stabilizers (lithium and anticonvulsants) are commonly used in clinical practice as adjunctive agents in schizophrenia spectrum disorders, evidence for this practice is mixed.[55] Studies of valproate have yielded conflicting results. In one double blind randomized controlled trial of 249 hospitalized subjects, PANSS improvement was greater in subjects receiving valproate as an adjunct to olanzapine or risperidone, and PANSS hostility scores were also significantly better.[56] These results were not replicated in a later, larger double blind randomized controlled trial.[57] In systematic reviews, the evidence supporting the efficacy of lithium,[58] carbamazepine, or oxcarbazepine[59] for the treatment of schizophrenia is at best limited.

In acutely psychotic patients, the use of a medication to manage an acute episode of agitation or violence is warranted. When behavioral deescalation measures have been ineffective or when a situation is too acute to implement these, the use of an "as needed" or pro re nata (PRN) dose of medication may be necessary. During the initial phases of treatment, the need for PRN doses can help to guide adjustments to the scheduled medications. Ultimately, the goal is to stabilize the patient on a well-tolerated and effective scheduled medication regimen, avoiding the regular use of PRN medications or unnecessary polypharmacy.[60]

Bipolar Disorder

Effective treatments for bipolar disorder in adults, including mood stabilizers and antipsychotic agents, address symptoms of irritability, aggression, and emotion dysregulation associated with mania and mixed episodes.[61] In addition, compliance with mood stabilizers and antipsychotics decreases the risk of violence in adults with bipolar disorder in community settings.[54]

Several antipsychotics and lithium are approved by the US Food and Drug Administration for the treatment of mania in youth ages 10 years and older.[62] The COBY study found that lithium, compared with other mood-stabilizing medications, was associated with less parent-reported aggression.[63] Risperidone has also shown efficacy in reduction of aggression in pediatric bipolar disorder.[64]

Although anticonvulsants are commonly prescribed in clinical practice to address mania and aggression in youth with bipolar disorder, there is limited evidence to support this practice. Large randomized controlled trials did not find either oxcarbazepine or valproate to be superior to placebo for the treatment of pediatric bipolar disorder.[65,66] Risperidone is more effective than valproate for mania and irritability in pediatric bipolar disorder.[67] Valproate has limited evidence supporting its use for pediatric aggression, although not specifically for aggression related to pediatric bipolar disorder.[68]

As with schizophrenia, PRN medications may be needed during acute manic episodes, particularly when the person is experiencing psychotic symptoms or severely disrupted sleep, energy, and behavior. Once the mood episode is stabilized, the long-term goal is to find a maintenance medication regimen that is well-tolerated and prevents further relapses.

Borderline Personality Disorder

Medications can be helpful for treating clusters of symptoms associated with borderline personality disorder, but do not address the core underlying psychopathology.[69] Limited evidence supports the use of valproate, topiramate, and lamotrigine for the treatment of anger, impulsivity and mood instability in adults with borderline personality disorder.[69] Selective serotonin reuptake inhibitors may provide some benefit for comorbid depression and anxiety. Atypical antipsychotics have shown modest efficacy in decreasing anger, but it is unclear whether these benefits persist beyond the short to medium term.[70]

Because symptoms of borderline personality disorder are typically not resolved with medication therapies, PRN medications are generally best avoided. In individuals with borderline personality disorder traits, maladaptive coping skills (eg, self-harm, emotional outbursts) are often inadvertently reinforced by well-meaning family members, care takers, and treatment providers. The offering of PRN medications sends the message to the patient that they "need a medication to calm down," reinforcing an external locus of control and the expression of distress as a mode for engaging others. The goal is to support and reinforce the person's sense of agency in controlling their own behavior, rather than fostering dependency on care providers or turning to psychoactive substances to solve one's problems.

Treat Comorbidities and Other Risk Factors

Treating substance use, comorbid disorders, and promoting medication adherence are important components of any treatment program for a person with severe mental illness. In patients with schizophrenia, clinical factors that predict violence and aggression include depression, young age, male sex, number of childhood conduct disorder symptoms, prior aggressive behavior, current illicit drug use, and treatment noncompliance.[71] Risk factors for aggression in patients with bipolar disorder include substance abuse, recent suicide attempt, initial manic episode, and learning disability.[72,73] In borderline personality disorder, comorbid psychopathology, including substance abuse, anxiety, and antisocial traits, is implicated in violence.[74] Substance use is an independent risk factor for violence, regardless of diagnosis.[75] Relapse prevention is a core tenet of treatment for all psychiatric disorders and has additional implications for reducing the risk for aggression in serious mental illness. Although not all risk factors for aggression and violence are modifiable, those that are should be addressed.

PSYCHOEDUCATIONAL AND BEHAVIORAL INTERVENTIONS

For youth with schizophrenia spectrum disorders and bipolar disorder, psychosocial treatments designed to enhance knowledge, improve interpersonal and family functioning, encourage school attendance and occupational success, promote resiliency, and prevent relapse are important adjuncts to medication therapies.[13,60] Even if these treatments do not directly target emotional and behavioral dysregulation, enhancing the patient's compliance with treatment and functional skill sets will predictably improve overall outcomes and decrease the likelihood of behavioral outbursts and violence.

In adolescents with borderline personality disorder, studies of psychotherapy thus far have focused on the outcomes of self-injurious behavior, depressive symptoms, and suicidality,[76] rather than aggression and affective dysregulation. Considering the associated symptoms, aggression can be targeted indirectly by addressing affective instability, impulsivity, and rejection sensitivity.

Symptoms of borderline personality disorder are targeted by various psychotherapies, including dialectical behavior therapy (DBT), mentalization-based therapy, transference focused psychotherapy, cognitive-analytical therapy, and emotion regulation training.[76–78] DBT has the most evidence supporting its use in adults and has shown some benefit in decreasing self-harm and repeated suicide attempts in adolescents.[79,80] Mentalization-based therapy has also demonstrated efficacy in reducing self-harm in adolescents with depression.[81]

Regardless of the diagnosis, it is important to identify risk factors that contribute to reactive patterns of emotional instability and anger outbursts. Functional behavioral assessments are important to identify triggers and reinforcers of behaviors, particularly in the context of social interactions and interpersonal relationships. This information can be used to design behavioral interventions tailored toward a particular patient's needs, with emphasis on teaching the patient how to achieve the desired "function" in a more adaptive way. For example, if a function of a prolonged aggressive episode involves escape, an intervention emphasizing communicating that need may help to decrease protracted episodes of aggression.

Psychosocial treatments designed for specific conditions can be broadly applied in the management of other disorders with shared symptom domains. For example, although DBT skills were specifically developed for persons with borderline personality disorder, they are potentially helpful for addressing self-regulation impairments across diagnoses. The DBT model assumes that individuals can be responsible for their behavior, learn new interpersonal and self-management skills, and emphasizes multimodal treatment including group work.[82] This model has transdiagnostic applicability and has been adapted for use in correctional settings for juveniles and adults, demonstrating efficacy in reducing problem behaviors.[83,84]

SUMMARY

Unsafe volatile behaviors, including mood instability and threats of violence toward self and others, are serious concern across major psychiatric conditions, and an important public health challenge. Major mental health disorders, including schizophrenia spectrum disorders, bipolar disorder, and borderline personality disorder, often present with irritability, aggression, and affective instability. An accurate diagnosis requires the recognition of characteristic patterns of symptoms and course of illness. Determining whether aggressive outbursts and emotion dysregulation are primarily driven by internal stimuli and disorganized thinking, versus reactive maladaptive responses to social negotiations and interpersonal conflicts, is an important distinction for treatment planning. Intervention strategies are either specific to a defined illness or more broadly focused on teaching new skills and modifying learned patterns of maladaptive coping. For most individuals with major mental health disorders, both types of approaches are beneficial.

DISCLOSURE

The authors have nothing to disclose.

REFERENCES

1. American Psychiatric Association. Diagnostic and statistical manual of mental disorders: DSM-5. 5th edition. Arlington (VA): American Psychiatric Association; 2013.

2. Salamon S, Santelmann H, Franklin J, et al. Test-retest reliability of the diagnosis of schizoaffective disorder in childhood and adolescence - a systematic review and meta-analysis. J Affect Disord 2018;230:28–33.

3. Miller JN, Black DW. Schizoaffective disorder: a review. Ann Clin Psychiatry 2019; 31(1):47–53. Available at: https://www.ncbi.nlm.nih.gov/pubmed/30699217.

4. Whiting D, Lennox BR, Fazel S. Violent outcomes in first-episode psychosis: a clinical cohort study. Early Interv Psychiatry 2019. https://doi.org/10.1111/eip. 12901.

5. Li W, Yang Y, Hong L, et al. Prevalence of aggression in patients with schizophrenia: a systematic review and meta-analysis of observational studies. Asian J Psychiatr 2019;47:101846.

6. Liu J, Chan TCT, Chong SA, et al. Impact of emotion dysregulation and cognitive insight on psychotic and depressive symptoms during the early course of schizophrenia spectrum disorders. Early Interv Psychiatry 2019. https://doi.org/10. 1111/eip.12895.

7. Rund BR. A review of factors associated with severe violence in schizophrenia. Nord J Psychiatry 2018;72(8):561–71. https://doi.org/10.1080/08039488.2018. 1497199.

8. Hodgins S. Aggressive behavior among persons with schizophrenia and those who are developing schizophrenia: attempting to understand the limited evidence on causality. Schizophr Bull 2017;43(5):1021–6. https://doi.org/10.1093/ schbul/sbx079.

9. Kelleher I, Connor D, Clarke MC, et al. Prevalence of psychotic symptoms in childhood and adolescence: a systematic review and meta-analysis of population-based studies. Psychol Med 2012;42(9):1857–63.

10. Isaksson J, Vadlin S, Olofsdotter S, et al. Psychotic-like experiences during early adolescence predict symptoms of depression, anxiety, and conduct problems three years later: a community-based study. Schizophr Res 2019. https://doi. org/10.1016/j.schres.2019.10.033.

11. Ghaemi N, Sachs GS, Goodwin FK. What is to be done? Controversies in the diagnosis and treatment of manic-depressive illness. World J Biol Psychiatry 2000;1(2):65–74.

12. Carlson GA, Goodwin FK. The stages of mania. A longitudinal analysis of the manic episode. Arch Gen Psychiatry 1973;28(2):221–8.

13. McClellan J, Kowatch R, Findling RL, et al. Practice parameter for the assessment and treatment of children and adolescents with bipolar disorder. J Am Acad Child Adolesc Psychiatry 2007;46(1):107–25.

14. Geller B, Zimerman B, Williams M, et al. Diagnostic characteristics of 93 cases of a prepubertal and early adolescent bipolar disorder phenotype by gender, puberty and comorbid attention deficit hyperactivity disorder. J Child Adolesc Psychopharmacol 2000;10(3):157–64.

15. Geller B, Tillman R, Craney JL, et al. Four-year prospective outcome and natural history of mania in children with a prepubertal and early adolescent bipolar disorder phenotype. Arch Gen Psychiatry 2004;61(5):459–67.

16. Wozniak J, Biederman J, Kiely K, et al. Mania-like symptoms suggestive of childhood-onset bipolar disorder in clinically referred children. J Am Acad Child Adolesc Psychiatry 1995;34(7):867–76.

17. Hunt J, Birmaher B, Leonard H, et al. Irritability without elation in a large bipolar youth sample: frequency and clinical description. J Am Acad Child Adolesc Psychiatry 2009;48(7):730–9.

18. Stringaris A, Baroni A, Haimm C, et al. Pediatric bipolar disorder versus severe mood dysregulation: risk for manic episodes on follow-up. J Am Acad Child Adolesc Psychiatry 2010;49(4):397–405. Available at: https://www.ncbi.nlm.nih.gov/pubmed/20410732.

19. Vidal-Ribas P, Stringaris A. How and why are irritability and depression linked? Child Adolesc Psychiatr Clin N Am.

20. Stringaris A, Cohen P, Pine DS, et al. Adult outcomes of youth irritability: a 20-year prospective community-based study. Am J Psychiatry 2009;166(9):1048–54. Available at: https://www.ncbi.nlm.nih.gov/pubmed/19570932.

21. Olfson M, Blanco C, Liu SM, et al. National trends in the office-based treatment of children, adolescents, and adults with antipsychotics. Arch Gen Psychiatry 2012; 69(12):1247–56. Available at: https://www.ncbi.nlm.nih.gov/pubmed/22868273.

22. Parry P, Allison S, Bastiampillai T. 'Paediatric bipolar disorder' rates are lower than claimed - a reexamination of the epidemiological surveys used by a meta-analysis. Child Adolesc Ment Health 2018;23(1):14–22.

23. Goldstein BI, Post RM, Birmaher B. Debate: fomenting controversy regarding pediatric bipolar disorder. Child Adolesc Ment Health 2019;24(1):95–6. Available at: https://www.ncbi.nlm.nih.gov/pubmed/31866767.

24. Spring L, Carlson G. The phenomenology of outbursts in children. Child Adolesc Psychiatr Clin N Am.

25. Stringaris A, Vidal-Ribas P, Brotman MA, et al. Practitioner review: definition, recognition, and treatment challenges of irritability in young people. J Child Psychol Psychiatry 2018;59(7):721–39. Available at: https://www.ncbi.nlm.nih.gov/pubmed/29083031.

26. Scott LN, Wright AGC, Beeney JE, et al. Borderline personality disorder symptoms and aggression: a within-person process model. J Abnorm Psychol 2017; 126(4):429–40. Available at: https://www.ncbi.nlm.nih.gov/pubmed/28383936.

27. Gunderson JG, Fruzzetti A, Unruh B, et al. Competing theories of borderline personality disorder. J Pers Disord 2018;32(2):148–67. Available at: https://www.ncbi.nlm.nih.gov/pubmed/29561723.

28. Winsper C, Marwaha S, Lereya ST, et al. Clinical and psychosocial outcomes of borderline personality disorder in childhood and adolescence: a systematic review. Psychol Med 2015;45(11):2237–51. Available at: https://www.ncbi.nlm.nih.gov/pubmed/25800970.

29. Stern A. Psychoanalytic Investigation of and therapy in the border line group of neuroses. Psychoanal Q 1938;7(4):467–89.

30. Videler AC, Hutsebaut J, Schulkens JEM, et al. A life span perspective on borderline personality disorder. Curr Psychiatry Rep 2019;21(7):51. Available at: https://www.ncbi.nlm.nih.gov/pubmed/31161404.

31. Luyten P, Campbell C, Fonagy P. Borderline personality disorder, complex trauma, and problems with self and identity: a social-communicative approach. J Pers 2020;88(1):88–105. Available at: https://www.ncbi.nlm.nih.gov/pubmed/31066053.

32. Hawes MT, Carlson GA, Finsaas MC, et al. Dimensions of irritability in adolescents: longitudinal associations with psychopathology in adulthood. Psychol Med 2019;1–9 [Online ahead of print]. Available at: https://www.ncbi.nlm.nih.gov/pubmed/31637980.

33. Bozzatello P, Bellino S, Bosia M, et al. Early detection and outcome in borderline personality disorder. Front Psychiatry 2019;10:710. Available at: https://www.ncbi.nlm.nih.gov/pubmed/31649564.

34. Sanches M. The limits between bipolar disorder and borderline personality disorder: a review of the evidence. Diseases 2019;7(3):49. Available at: https://www.ncbi.nlm.nih.gov/pubmed/31284435.

35. Gunderson JG, Herpertz SC, Skodol AE, et al. Borderline personality disorder. Nat Rev Dis Primers 2018;4:18029. Available at: https://www.ncbi.nlm.nih.gov/pubmed/29795363.

36. Kaufman J, Birmaher B, Brent D, et al. Schedule for affective disorders and schizophrenia for school-age children-present and lifetime version (K-SADS-PL): initial reliability and validity data. J Am Acad Child Adolesc Psychiatry 1997;36(7):980–8. Available at: https://www.ncbi.nlm.nih.gov/pubmed/9204677.

37. Kaufman J, Birmaher B, Axelson D, et al. Schedule for affective disorders and schizophrenia for school-age children—present and lifetime DSM-5. 2016. Available at: https://www.kennedykrieger.org/sites/default/files/library/documents/faculty/ksads-dsm-5-screener.pdf.

38. First MB, Williams JBW, Karg RS, et al. User's guide for the structured clinical interview for DSM-5 disorders, clinician version (SCID-5-CV). Arlington (VA): American Psychiatric Association; 2016.

39. Kay SR, Fiszbein A, Opler LA. The positive and negative syndrome scale (PANSS) for schizophrenia. Schizophr Bull 1987;13(2):261–76. Available at: https://www.ncbi.nlm.nih.gov/pubmed/3616518.

40. Correll CU, Yu X, Xiang Y, et al. Biological treatment of acute agitation or aggression with schizophrenia or bipolar disorder in the inpatient setting. Ann Clin Psychiatry 2017;29(2):92–107. Available at: https://www.ncbi.nlm.nih.gov/pubmed/28463343.

41. Andreasen N. The scales for the assessment of positive and negative symptoms. Iowa City (IA): The University of Iowa; 1982.

42. Buss AH, Perry M. The aggression questionnaire. J Pers Soc Psychol 1992;63(3):452–9. Available at: https://www.ncbi.nlm.nih.gov/pubmed/1403624.

43. Young RC, Biggs JT, Ziegler VE, et al. A rating scale for mania: reliability, validity and sensitivity. Br J Psychiatry 1978;133:429–35. Available at: https://www.ncbi.nlm.nih.gov/pubmed/728692.

44. Fristad MA, Weller RA, Weller EB. The Mania Rating Scale (MRS): further reliability and validity studies with children. Ann Clin Psychiatry 1995;7(3):127–32. Available at: https://www.ncbi.nlm.nih.gov/pubmed/8646272.

45. Coid JW, Ullrich S, Kallis C, et al. The relationship between delusions and violence: findings from the East London first episode psychosis study. JAMA Psychiatry 2013;70(5):465–71. Available at: https://www.ncbi.nlm.nih.gov/pubmed/23467760.

46. Keers R, Ullrich S, Destavola BL, et al. Association of violence with emergence of persecutory delusions in untreated schizophrenia. Am J Psychiatry 2014;171(3):332–9. Available at: https://www.ncbi.nlm.nih.gov/pubmed/24220644.

47. Sharma A, McClellan J. Early-onset schizophrenia. In: Prinstein MJ, Youngstrom EA, Mash EJ, et al, editors. Treatment of disorders in childhood and adolescence. 4th edition. New York: The Guilford Press; 2019. p. p704–22.

48. Sikich L, Frazier JA, McClellan J, et al. Double-blind comparison of first- and second-generation antipsychotics in early-onset schizophrenia and schizoaffective disorder: findings from the treatment of early-onset schizophrenia spectrum disorders (TEOSS) study. Am J Psychiatry 2008;165(11):1420–31. Available at: https://www.ncbi.nlm.nih.gov/pubmed/18794207.

49. Findling RL, Cavus I, Pappadopulos E, et al. Ziprasidone in adolescents with schizophrenia: results from a placebo-controlled efficacy and long-term open-

extension study. J Child Adolesc Psychopharmacol 2013;23(8):531–44. Available at: https://www.ncbi.nlm.nih.gov/pubmed/24111983.

50. Findling RL, Landbloom RP, Mackle M, et al. Safety and efficacy from an 8 week double-blind trial and a 26 week open-label extension of asenapine in adolescents with schizophrenia. J Child Adolesc Psychopharmacol 2015;25(5): 384–96. Available at: https://www.ncbi.nlm.nih.gov/pubmed/26091193.

51. van Schalkwyk GI, Beyer C, Johnson J, et al. Antipsychotics for aggression in adults: a meta-analysis. Prog Neuropsychopharmacol Biol Psychiatry 2018;81: 452–8. Available at: https://www.ncbi.nlm.nih.gov/pubmed/28754408.

52. Patchan K, Vyas G, Hackman AL, et al. Clozapine in reducing aggression and violence in Forensic populations. Psychiatr Q 2018;89(1):157–68. Available at: https://www.ncbi.nlm.nih.gov/pubmed/28643049.

53. Zaman H, Sampson SJ, Beck AL, et al. Benzodiazepines for psychosis-induced aggression or agitation. Cochrane Database Syst Rev 2017;(12). CD003079. Available at: https://www.ncbi.nlm.nih.gov/pubmed/29219171.

54. Fazel S, Zetterqvist J, Larsson H, et al. Antipsychotics, mood stabilisers, and risk of violent crime. Lancet 2014;384(9949):1206–14. Available at: https://www.ncbi. nlm.nih.gov/pubmed/24816046.

55. Citrome L. Adjunctive lithium and anticonvulsants for the treatment of schizophrenia: what is the evidence? Expert Rev Neurother 2009;9(1):55–71. Available at: https://www.ncbi.nlm.nih.gov/pubmed/19102669.

56. Casey DE, Daniel DG, Wassef AA, et al. Effect of divalproex combined with olanzapine or risperidone in patients with an acute exacerbation of schizophrenia. Neuropsychopharmacology 2003;28(1):182–92. Available at: https://www.ncbi. nlm.nih.gov/pubmed/12496955.

57. Casey DE, Daniel DG, Tamminga C, et al. Divalproex ER combined with olanzapine or risperidone for treatment of acute exacerbations of schizophrenia. Neuropsychopharmacology 2009;34(5):1330–8. Available at: https://www.ncbi.nlm.nih. gov/pubmed/19052541.

58. Leucht S, Helfer B, Dold M, et al. Lithium for schizophrenia. Cochrane Database Syst Rev 2015;(10). CD003834. Available at: https://www.ncbi.nlm.nih.gov/ pubmed/26509923.

59. Leucht S, Helfer B, Dold M, et al. Carbamazepine for schizophrenia. Cochrane Database Syst Rev 2014;(5). CD001258. Available at: https://www.ncbi.nlm.nih. gov/pubmed/24789267.

60. McClellan J, Stock S, American Academy of Child and Adolescent Psychiatry (AACAP) Committee on Quality Issues (CQI). Practice parameter for the assessment and treatment of children and adolescents with schizophrenia. J Am Acad Child Adolesc Psychiatry 2013;52(9):976–90. Available at: https://www.ncbi.nlm. nih.gov/pubmed/23972700.

61. Yatham LN, Kennedy SH, Parikh SV, et al. Canadian Network for Mood and Anxiety Treatments (CANMAT) and International Society for Bipolar Disorders (ISBD) 2018 guidelines for the management of patients with bipolar disorder. Bipolar Disord 2018;20(2):97–170. Available at: https://www.ncbi.nlm.nih.gov/pubmed/ 29536616.

62. Stepanova E, Findling RL. Psychopharmacology of bipolar disorders in children and adolescents. Pediatr Clin North Am 2017;64(6):1209–22. Available at: https:// www.ncbi.nlm.nih.gov/pubmed/29173781.

63. Hafeman DM, Rooks B, Merranko J, et al. Lithium versus other mood stabilizing medications in a longitudinal study of bipolar youth. J Am Acad Child Adolesc

Psychiatry 2020;59(10):1146–55. Available at: https://www.ncbi.nlm.nih.gov/pubmed/31369795.

64. Frazier JA, Meyer MC, Biederman J, et al. Risperidone treatment for juvenile bipolar disorder: a retrospective chart review. J Am Acad Child Adolesc Psychiatry 1999;38(8):960–5. Available at: https://www.ncbi.nlm.nih.gov/pubmed/10434487.

65. Wagner KD, Kowatch RA, Emslie GJ, et al. A double-blind, randomized, placebo-controlled trial of oxcarbazepine in the treatment of bipolar disorder in children and adolescents. Am J Psychiatry 2006;163(7):1179–86. Available at: https://www.ncbi.nlm.nih.gov/pubmed/16816222.

66. Wagner KD, Redden L, Kowatch RA, et al. A double-blind, randomized, placebo-controlled trial of divalproex extended-release in the treatment of bipolar disorder in children and adolescents. J Am Acad Child Adolesc Psychiatry 2009;48(5):519–32. Available at: https://www.ncbi.nlm.nih.gov/pubmed/19325497.

67. Davico C, Canavese C, Vittorini R, et al. Anticonvulsants for psychiatric disorders in children and adolescents: a systematic review of their efficacy. Front Psychiatry 2018;9:270. Available at: https://www.ncbi.nlm.nih.gov/pubmed/29988399.

68. Munshi KR, Oken T, Guild DJ, et al. The Use of Antiepileptic drugs (AEDs) for the treatment of pediatric aggression and mood disorders. Pharmaceuticals (Basel) 2010;3(9):2986–3004. Available at: https://www.ncbi.nlm.nih.gov/pubmed/27713387.

69. Bozzatello P, Rocca P, De Rosa ML, et al. Current and emerging medications for borderline personality disorder: is pharmacotherapy alone enough? Expert Opin Pharmacother 2020;21(1):47–61. Available at: https://www.ncbi.nlm.nih.gov/pubmed/31693423.

70. Mercer D, Douglass AB, Links PS. Meta-analyses of mood stabilizers, antidepressants and antipsychotics in the treatment of borderline personality disorder: effectiveness for depression and anger symptoms. J Pers Disord 2009;23(2):156–74. Available at: https://www.ncbi.nlm.nih.gov/pubmed/19379093.

71. Hodgins S, Riaz M. Violence and phases of illness: differential risk and predictors. Eur Psychiatry 2011;26(8):518–24. Available at: https://www.ncbi.nlm.nih.gov/pubmed/21277752.

72. Fazel S, Lichtenstein P, Grann M, et al. Bipolar disorder and violent crime: new evidence from population-based longitudinal studies and systematic review. Arch Gen Psychiatry 2010;67(9):931–8. Available at: https://www.ncbi.nlm.nih.gov/pubmed/20819987.

73. Khalsa HK, Baldessarini RJ, Tohen M, et al. Aggression among 216 patients with a first-psychotic episode of bipolar I disorder. Int J Bipolar Disord 2018;6(1):18. Available at: https://www.ncbi.nlm.nih.gov/pubmed/30097737.

74. Gonzalez RA, Igoumenou A, Kallis C, et al. Borderline personality disorder and violence in the UK population: categorical and dimensional trait assessment. BMC Psychiatry 2016;16:180. Available at: https://www.ncbi.nlm.nih.gov/pubmed/27255770.

75. Fazel S, Smith EN, Chang Z, et al. Risk factors for interpersonal violence: an umbrella review of meta-analyses. Br J Psychiatry 2018;213(4):609–14. Available at: https://www.ncbi.nlm.nih.gov/pubmed/30058516.

76. Weiner AS, Ensink K, Normandin L. Psychotherapy for borderline personality disorder in adolescents. Psychiatr Clin North Am 2018;41(4):729–46. Available at: https://www.ncbi.nlm.nih.gov/pubmed/30447735.

77. Kaess M, Brunner R, Chanen A. Borderline personality disorder in adolescence. Pediatrics 2014;134(4):782–93. Available at: https://www.ncbi.nlm.nih.gov/pubmed/25246626.

78. Guilé JM, Boissel L, Alaux-Cantin S, et al. Borderline personality disorder in adolescents: prevalence, diagnosis, and treatment strategies. Adolesc Health Med Ther 2018;9:199–210. Available at: https://www.ncbi.nlm.nih.gov/pubmed/30538595.

79. Mehlum L, Tormoen AJ, Ramberg M, et al. Dialectical behavior therapy for adolescents with repeated suicidal and self-harming behavior: a randomized trial. J Am Acad Child Adolesc Psychiatry 2014;53(10):1082–91. Available at: https://www.ncbi.nlm.nih.gov/pubmed/25245352.

80. McCauley E, Berk MS, Asarnow JR, et al. Efficacy of dialectical behavior therapy for adolescents at high risk for suicide: a randomized clinical trial. JAMA Psychiatry 2018;75(8):777–85. Available at: https://www.ncbi.nlm.nih.gov/pubmed/29926087.

81. Rossouw TI, Fonagy P. Mentalization-based treatment for self-harm in adolescents: a randomized controlled trial. J Am Acad Child Adolesc Psychiatry 2012;51(12):1304–13.e3. Available at: https://www.ncbi.nlm.nih.gov/pubmed/23200287.

82. Linehan M. Cognitive-behavioral treatment of borderline personality disorder. New York: Guilford Press; 1993.

83. Trupin EW, Stewart DG, Beach B, et al. Effectiveness of a dialectical behaviour therapy program for incarcerated female juvenile offenders. Child Adolesc Ment Health 2002;7(3):121–7. Available at: https://onlinelibrary.wiley.com/doi/abs/10.1111/1475-3588.00022.

84. Shelton D, Sampl S, Kesten KL, et al. Treatment of impulsive aggression in correctional settings. Behav Sci Law 2009;27(5):787–800. Available at: https://www.ncbi.nlm.nih.gov/pubmed/19784944.

Dysregulation, Catastrophic Reactions, and the Anxiety Disorders

John T. Walkup, MD[a,*], Susan J. Friedland, MD[b],
Tara S. Peris, PhD[c], Jeffrey R. Strawn, MD[d]

KEYWORDS

- Anxiety disorders • Catastrophic reaction • Functional analysis • Avoidance
- Accommodation • Suicidal behavior

KEY POINTS

- Anxiety disorders begin in childhood, and if untreated, can result in accumulated impairment that puts children at risk for poor adaptation and maladaptive behaviors including self-injury, suicidal behavior, and substance misuse.
- Children with anxiety disorders who cannot avoid their anxiety triggers often catastrophize and become dysregulated with meltdowns, tantrums, and rage episodes.
- In children, dysregulated behavior is reinforced and thus, recurs when it leads to successful avoidance of anxiety triggers. This cycle worsens when parents inadvertently reinforce dysregulated behavior when they intuitively engage children when dysregulated and accommodate avoidance behavior.
- Parents of anxious children can decrease anxiety and the risk for poor outcomes by (1) challenging their children to engage in important developmental activities rather than avoid them, (2) rewarding positive coping, and (3) not reinforcing avoidant behavior.

The clinical relationship between fearfulness and tyranny in children is explored... A characteristic pattern of family interaction emerges that underlies the dynamics of tyranny and fears. Ignoring this interaction and treating fear alone was not helpful; introducing changes in family interaction that lead to the control of tyranny affected both fears and tyranny."[1]
— *Fears and tyranny. Observations on the tyrannical child—Barcai and Rosenthal, 1974.*

[a] Ann and Robert H. Lurie Children's Hospital of Chicago, 220 East Illinois Street, Box 10, Chicago, IL 60611, USA; [b] Ann and Robert H. Lurie Children's Hospital of Chicago & Northwestern University Feinberg School of Medicine, Town and Country Pediatrics, 1460 North Halstead Street, Suite 402, Chicago, IL 60642, USA; [c] UCLA Department of Psychiatry and Biobehavioral Sciences, 760 Westwood Plaza, Room 67-439, Los Angeles, CA 90095, USA; [d] College of Medicine, University of Cincinnati, 260 Stetson Street, Suite 3200, Cincinnati, OH 45267, USA
* Corresponding author.
E-mail address: jwalkup@luriechildrens.org

Child Adolesc Psychiatric Clin N Am 30 (2021) 431–444
https://doi.org/10.1016/j.chc.2020.10.011
1056-4993/21/© 2020 Elsevier Inc. All rights reserved.

INTRODUCTION

In this article, the authors describe the pathway from anxiety symptoms to emotional dysregulation and ultimately to the loss of emotional and behavior control that is common in children and adolescents with anxiety disorders.[2] The pathway presumes that children with anxiety disorders have an underlying pathology, and through engagement with their environment, they become increasingly dysregulated and display catastrophic behavior.

Children with anxiety disorders seem to behave inexplicably. Their anxious reactions seem to come "out of the blue" and are markedly disproportional to the triggering experience. The inexplicable nature of the child's catastrophic reactions gives way once the functional pathway from anxiety disorder to loss of emotional and behavior control is fully explicated. The full range of catastrophic behaviors (temper tantrums, rage episodes, meltdowns, and regressive behavior) functions to allow the child to avoid the demands of his or her environment and are reinforced by parental attention. Defining this pathway facilitates treatment to address not only the anxiety symptoms themselves (ie, cognitive behavioral therapy [CBT] and medication) but also the behavioral and interactional processes that lead to catastrophic behavior.

THE ANXIETY DISORDERS

Anxiety disorders are the most common childhood-onset psychiatric disorders.[3] Although the anxiety disorders comprise six conditions, three (separation, social, and generalized anxiety disorders) start in childhood before puberty,[3] have pervasive symptomatology, and are associated with significant ongoing impairment. Although these three conditions have different triggers for anxiety (separation from a caretaker, social and performance situations, and general intolerance of uncertainty, respectively) they share important commonality:

- Children with anxiety disorders are hypervigilant, constantly scanning the external world and the internal world of their minds and bodies.
- Children with anxiety disorders are reactive to novel stimuli and have sustained threat bias; they interpret innocuous or neutral events as threatening.
- Normal developmental tasks or routine activities are triggers for children and adolescents with anxiety disorders.
- Given this, it is no surprise that the dominant mode of coping for children with anxiety disorders is to avoid situations that they see as threatening.
- When children cannot avoid anxiety triggers, their anxiety rises. They often react catastrophically and rage, tantrum, or shut down.
- Parents often accommodate[4] to reduce their child's anxiety and to avoid a catastrophic reaction.
- Anxiety, the impending threat of a catastrophic reaction, and parental accommodation can become patterns of living that can result in chronic impairment.
- Children with anxiety disorders can also be triggered by experiences that make all children anxious, such as vaccinations, test taking, or high-risk activities. Over time, the child with an anxiety disorder generalizes avoidance coping and catastrophic reactions across life experiences: normal developmental tasks/expectations and universally anxiety-provoking experiences.

ANXIETY: THE UNDERLYING NEUROBIOLOGIC VULNERABILITY

The core features of the anxiety disorders include hypervigilance, increased reactivity to novel stimuli, threat bias, and catastrophic behavior. These features are solidly grounded in neuroscience. Anxiety symptoms and behaviors are associated with increased functional activity of the cingulo-opercular and ventral attention networks, and decreased activity within the frontoparietal and default mode networks (discussed later).[5] These cortical networks are intimately connected with the amygdala, which plays a central role in generating fear responses. In its modulatory role, the prefrontal cortex facilitates emotional control of threat stimuli via downregulation of amygdala hyperactivity.[6,7] Importantly, it has been hypothesized that the abnormalities in prefrontal regions impact the lower limbic structures to give rise to emotional dysregulation in children and adolescents.[8,9] Furthermore, improvements in top-down emotion regulation circuitry have been observed with successful treatment, and further demonstrate increased coupling of amygdala-prefrontal circuitry (**Fig. 1**)[8]:

- The cingulo-opercular network plays a role in maintaining alertness (vigilance), not just attention, and may have a role in detecting errors and summoning cognitive control.[5] Overactivity of the cingulo-opercular network has been associated with increased alertness to potential threats (threat bias) and more reactive, even impulsive cognitive control strategies (catastrophic reactions).
- The ventral attention network, which consists of regions within the ventral prefrontal cortex and temporoparietal junction,[10] reorients attention toward salient stimuli, especially when they appear in unexpected locations or after abrupt changes in sensory stimuli. Overactivity of the ventral attention network has been associated with attention bias toward threat or suddenly appearing stimuli (hyperreactivity to novel stimuli).
- The frontoparietal network (the executive control network) is associated with decreased functioning in individuals with anxiety disorders, even when presented with neutral nonthreatening stimuli.[5,11–13] The lack of emotional control can lead to distraction and off-task behavior, and an inability to summon executive control when under threat (loss of emotional and behavioral control).
- The default mode network is active when one is thinking about the self and others, remembering and planning for the future, and in self-regulation. It is less active when engaged in goal-oriented tasks.[14,15] The default mode network

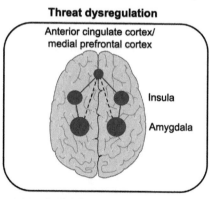

Fig. 1. Proposed neural circuit dysfunction in anxiety disorders represents abnormal patterns of activity and connectivity among structures from default mode, ventral attention, cingulo-opercular, and frontoparietal networks. *Dashed lines* represent inhibitory relationships between nodes and *solid lines* represent positive functional connectivity. (*From* Williams LM. Precision psychiatry: A neural circuit taxonomy for depression and anxiety. Lancet Psychiatry 2016;3(5):472–80. Reprinted with permission from Elsevier.)

encompasses the posterior and anterior cortical midline structures, with major hubs located in the posterior cingulate cortex and precuneus, the medial prefrontal cortex, and the angular gyrus. Reduced functioning of the default mode network as observed in the anxiety disorders has direct implications for overall emotion and behavior regulation.

Dysregulation of these networks provides support for a basic vulnerability youth with anxiety disorders experience in managing their environment. Additionally, these networks dynamically interact within two key anxiety networks that subserve behavioral avoidance and threat processing (see **Fig. 1**).[16] It is hypothesized that dysfunction in these specific networks results in misinterpretation of novel and even neutral stimuli as threatening, and in the context of poor self-regulation, failure to exert effective cognitive control under pressure. Ultimately this network dysfunction sets the stage for a maladaptive pattern of behavior that, in the context of interpersonal relationships, evolves and elaborates over time into catastrophic behavior.

A FUNCTIONAL ANALYSIS OF CATASTROPHIC BEHAVIOR ACROSS DEVELOPMENT

More than 25 years of research documents a range of family factors that shape and maintain child anxiety.[17,18] However, the identification of certain risk patterns does not always translate into effective interventions. The patterns of symptom reinforcement described next (**Table 1**) are central to the evolution and elaboration of anxiety and dysregulated behavior and have direct implications for developing behavioral treatment plans:

- Internal negative reinforcement
- Interpersonal negative reinforcement
- Internal positive reinforcement
- Interpersonal positive reinforcement

Internal and Interpersonal Negative Reinforcement

A behavior that is effective in reducing a noxious stimulus is negatively reinforced and is likely to recur when the noxious stimulus is presented again. A child with an anxiety disorder who is facing an anxiety-provoking experience (ie, a normal developmental task or expectation) but successfully avoids that task will experience a reduction in anxiety (negative reinforcement). When presented with a similar developmental task or expectation in the future, the child is more likely to use an avoidance strategy to reduce their anxiety. Negative reinforcement can be internal, as is seen with children with social anxiety disorder who avoid normal developmental social tasks, such as asking the teacher questions at school or going to a party. Negative reinforcement

Table 1
Patterns of reinforcement of the anxiety disorders

	Positive Reinforcement	Negative Reinforcement
Internally Reinforcing	Provides gratification from positive anxiety cognitions such as perfectionism and caution	Relieves anxious distress
Interpersonally Reinforcing	Pays attention and gives support to anxiety expression	Accommodates avoidance

can also occur at an interpersonal level, when a parent proactively develops a plan with the teacher to not call on the child at school, or declines the invitation to the party. Similarly, children with separation anxiety often cling to their parents to avoid a separation (internal negative reinforcement), and their parents support that avoidance by arranging play dates only at their own home, not arranging sleep overs, or ruling out summer camp experiences (interpersonal negative reinforcement).

Internal and Interpersonal Positive Reinforcement

A behavior is positively reinforced when it is followed by engagement or pleasurable experience. A child with an anxiety disorder who experiences distress when presented with an anxiety trigger often elicits parental attention and support, which unwittingly reinforces or rewards anxiety expression. Even basic comforting during an anxious reaction, although intuitively reasonable, can positively reinforce the anxious reaction and perpetuate anxiety. Even positive attributes of a child, such as caution and perfectionism, if driven by an anxiety disorder, can be internally positively reinforced and thus self-perpetuating. For example, the child who, out of fear, studies much more than he needs to will feel vindicated by a good grade (internal positive reinforcement) and receive positive teacher and parent attention (interpersonal positive reinforcement). However, he may sacrifice sleep, time with peers, and pleasurable activities and over time, may find it difficult to take on academic challenges and succumb to the pressures with a reduction in his sense of well-being.

To summarize, the anxiety-triggering experiences in children with anxiety disorders are normal developmental tasks or expectations. The child's dominant mode of coping is avoidance (internal negative reinforcement), and when they cannot avoid, because the triggers are normal developmental tasks or expectations, they react catastrophically, freeze up, shut down, or rage (internal negative reinforcement). Also, if parents engage or support the child during a catastrophic reaction, they can interpersonally positively reinforce the behavior, and if parents accommodate and support avoidance behavior in response to the catastrophic reactions, the catastrophic reaction is interpersonally negatively reinforced.

Escalation of Emotional and Behavioral Dyscontrol

Early in the course of an anxiety disorder the child feels distressed and seeks to avoid when facing an anxiety trigger. Under the pressure of the child's distress, most parents intuitively move to comfort their child (positively reinforce) and facilitate avoidance (negatively reinforce). Because the triggers are normal developmental experiences/expectations, parents can become confused and anxious themselves (Why can't my child do routine activities like other kids?). At some point parents want to normalize their child's activities. They begin to stop comforting and start demanding that the child engage in normal developmental tasks. Although parents may have some success with "pushing" their child, children who feel "forced" to face their triggers often "up the ante" and intensify the level of their anxious reactions. When parents face the new and more intense anxious reaction they are likely to attend to the child, provide support (positively reinforce), and ultimately capitulate (negatively reinforce) to help the child restore their emotional control.

This is the key dynamic that ultimately results in dysregulated emotions and behavior. Parents use pressure to get the child to take on normal developmental tasks and children, out of fear, develop an increasing repertoire of catastrophic behaviors to maintain their avoidance behavior.

At end stage, parents describe feeling powerless to help their child do what they need to do, often have deviated from their ideal parenting approaches, and feel guilty about the extremes they may have gone to get their child to do even basic routines. But more importantly, many parents "surrender" to the extreme nature of the child's anxious reactions and cede control of the child's daily routine to the child. How long parents hold out before surrendering varies. Some parents whose natural inclination is to be overprotective may not engage in the escalation and "surrender" more quickly, whereas other parents may try to find strategies to get their child to participate in routine developmental tasks and expectations. Similarly, the child who feels "put on" and experiences the pressure from their parents as extreme and unwarranted, feels increasingly alienated from the family, and still remains anxious, but with an expanding repertoire of behaviors to facilitate avoidance.

Once the high-stakes pattern of behavior is established, it is not unusual for the child (and parent) to actually initiate interactions with their most extreme behaviors, a so-called "preemptive strike." For example, Billy is 10 years old and has untreated separation anxiety. Over the years his parents have been able to get him to go to school, but the situation is tenuous. He struggles to get to bed on school nights, and cries and pleads to not go to school in the morning. He is unable to settle without a parent in his room until he is asleep and on awakening in the middle of the night his struggle to fall asleep begins again. Both Billy and his parents are exhausted; getting ready the morning of a school day is an ongoing emotional challenge, he is slow to get up and dress and is too upset to eat. There is a struggle to get in the car (he will not get on the bus), he needs to be escorted to the classroom, becomes tearful, and demands to go home. He does separate, but his parent remains at school until he has securely transitioned to the classroom. In school it is difficult for him to focus on schoolwork, because he is preoccupied with separation fears and does not get all of his work done. When he comes home, he explodes (the preemptive strike) before his parent can even ask him how his day went and how much homework he has. To his parents, his explosion seems "out of the blue" and spontaneous. From a functional point of view, his explosion is not "spontaneous" but is an end-stage learned behavior that serves to facilitate avoidance of the painful after-school accounting with his parents, and of their expectation that he complete his unfinished homework.

As expected, Billy's catastrophic behavior does not really make his anxiety better, and in reality, is only minimally related to the external triggers. The anxiety emanates from within him; it is really not about the trigger, and avoidance does not address the real cause. Children caught in this struggle with their parents remain on edge anticipating the next triggering experience, look unhappy, and proactively engage in catastrophic behavior (a preemptive strike). But most importantly, the child's capacity to cope and adapt to new situations does not develop, and may even deteriorate from baseline. Without successfully taking on new challenges as part of everyday living, a child does not develop robust coping and adaptation skills. Over time the child not only avoids his anxiety triggers, but is also made secondarily anxious by his lack of overall capacity to cope and adapt. For this child, even routine nontriggering experiences become challenging, and the child readily becomes flooded and overwhelmed by the basic tasks of living.

Because the anxiety disorders begin in childhood,[3] the pattern of escalating avoidance behavior often continues over the course of development. Catastrophic behaviors can evolve and become more nuanced and sophisticated during adolescence,[19] and include self-injury to relieve the anxious distress, and even suicidal behavior.[20] Alcohol use is experienced as beneficial for those with social anxiety and generalized anxiety disorder, because alcohol offers temporary relief from self-consciousness or the intolerance of uncertainty.[21] Marijuana produces more variable responses[21]; some young people report temporary symptomatic relief, whereas others experience an exacerbation of anxiety, including panic attacks, and depersonalization and derealization experiences.

The Role of Parents in Children Developing Dysregulation

With respect to children with an anxiety disorder, it is not just a one-way street. It is increasingly clear that children with anxiety have a substantial impact on their parents' behavior, and vice versa. Children with anxiety bring a vulnerability to every interaction with their parents that predisposes them to an evolving pattern of dysregulation. It is also important to note that parents come to the interaction with certain behaviors that either accelerate or mitigate the process.

Observational studies have identified several parenting behaviors that are associated with anxiety in children and adolescent, including parental overcontrol, overprotection, criticism, and rejection.[17,18] Parenting studies are challenging to interpret. First, the concepts overlap and change over time. Currently, there is greater focus on parental accommodation than overcontrol or overprotection.[4] Second, is the "chicken and egg" problem. Correlational studies are difficult to interpret because the direction of the association is unclear, and causation cannot be established.

Another approach to understanding the role of parents is to consider how parents with differing styles might interact with a child with an anxiety disorder. Common to theoretic models of family functioning are two dimensions that sit orthogonally to one another (**Fig. 2**).[22] The dimensions include a flexibility (or control) continuum that ranges from rigid to chaotic, and a cohesion (or support) continuum that ranges

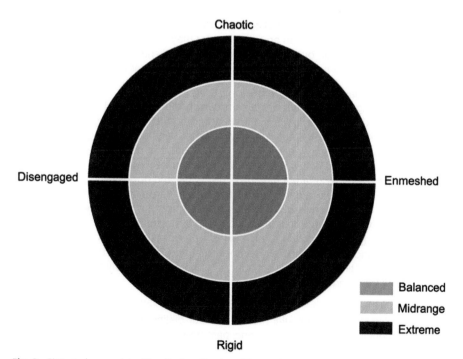

Fig. 2. Circumplex model of family functioning. (*From* Olson D. FACES IV and the Circumplex Model: Validation Study. J Marital Fam Ther 2011;37(1):64-80; with permission.)

from disengaged to enmeshed. Rating scales tailored to tap into these dimensions have yielded common family styles, including balanced, midrange, and extreme styles.[22]

Chaotic and enmeshed parenting style

A parenting style commonly observed in clinical populations of children with anxiety disorders is characterized by enmeshment and a chaotic home structure. This combination leads to a highly reactive interpersonal process that can accelerate the development of dysregulated behavior. Enmeshed parents may more readily attend to their child's expressions of anxiety and also more readily facilitate avoidance. The parents may also be anxious and dysregulated themselves, and model anxiety expression and catastrophic behavior.[23] In the end, enmeshment, escalating dependency, social isolation, and chaotic home structure and relationships can lead rapidly to dysregulated behavior.

Rigid and enmeshed parenting style

A parenting style characterized by intense engagement (enmeshment) and high expectations and standards (rigidity) may lead to more pressure to maintain functioning and control, perhaps even elevated performance. Families with this style of parenting may look highly functional, but might not tolerate change or function well when their child with anxiety has needs that are not readily addressed. In a highly rigid environment, the cost to the child of avoidance and catastrophic reactions may be too great, and while the child is largely adherent to routine expectations, he suffers an ongoing internalized cost from having to perform and behave, and not feeling understood and cared for.

Rigid and disengaged parenting style

A disengaged and rigid parenting style is characterized by the lack of warmth and harsh discipline. Children with an anxiety disorder may express less overt avoidance and catastrophic behavior, but may experience greater alienation and anger about the punitive and insensitive nature of the interaction with the parent.

Chaotic and disengaged parenting style

It is difficult to imagine any parent with an anxious child being uninvolved, because the child's anxiety exerts a great "pull" on the parent to interact. That said, there are often two caretakers involved in the care of a child with anxiety, and although it is uncommon to observe the primary caretaker as uninvolved, it is common to observe the coparent as disengaged. Within the parent dyad, one parent may adopt an enmeshed parenting style, and the coparent may take the "equal and opposite" position—a disengaged role. For example, a parent who is the primary care-taking role moves, understandably, to a more enmeshed style to provide comfort the child. The other parent might disagree with this approach and become more disengaged. Or, a parent who becomes more rigid in response to high levels of expressed anxiety and avoidance behavior might find that the coparent has become more flexible, to the point of being highly accommodating, reactive, or chaotic. This polarization of parenting styles presents a major treatment challenge.

The circumplex model

The circumplex model is a circular model, not four distinct quadrants (see **Table 1**). Family functioning is extremely fluid, and parents who function within the extreme valence do not readily move to the midrange or balanced valences. More commonly, when under stress, parents with an extreme parenting style stay within their valence and move to another extreme parenting style. For example, a parent with an enmeshed and rigid parenting style might, when stressed, shift to a disengaged and rigid style. This shifting parenting style is common in parents of adolescents or young adults with anxiety. Such parents have frequently attended to and accommodated their child's anxiety and avoidance for many years. However, when the child is not moving toward independence, they may feel helpless, become frustrated and angry, disengage, and behave arbitrarily toward their child.

TREATMENT
Assessment and Formulation

A good treatment plan follows a good diagnostic assessment and formulation. There are three levels of anxiety-related symptoms and impairment relevant to treatment planning:

- *Anxiety symptoms and short-term functional impairment.* Early in the course of treatment, a child without other vulnerabilities who experiences intense anxiety symptoms and rapidly engages their parents is expected to do well with straight-forward individual evidence-based CBT treatment. Combining CBT with selective serotonin reuptake inhibitor medication substantially increases the rate of remission in the first 12 weeks of treatment (see later).[24]
- *Anxiety-related accumulated impairment with coping and adaptation.* If over time the child does not get effective treatment, they will express anxiety distress and use avoidance coping. The child will engage the parents, and if they respond intuitively, they will comfort when anxiety is expressed and potentially facilitate avoidance. If the pattern continues, the child will not develop robust and nuanced coping strategies and will begin to experience additional (secondary) anxiety related to the lack of skills to address ongoing challenges. Medication and CBT would be expected to mitigate, but not directly address, this reduced capacity to cope and adapt. Treatment strategies (more akin to rehabilitation) that build or rebuild coping and adaptation strategies are needed. The skills needed depend on when coping and adaptation skills stopped developing, and define the consequent gap between the current level of skill and current developmental demands. A good example is a young adult who has had anxiety since childhood whose coping and adaptation skills arrested in early adolescence but now needs to cope with moving out of the home and function independently as a young adult. This substantial gap in functioning cannot be addressed by CBT (ie, exposure and response prevention) or medication.
- *Anxiety-related accumulated maladaptive behavior, such as substance abuse and suicidal and nonsuicidal behavior.* If over time the child does not get effective treatment, they will accumulate additional impairment and perhaps develop maladaptive behaviors. An extended period of untreated anxiety can lead to demoralization, often confused as depression. In adolescence the child may discover that self-injury can relieve the feeling of being emotionally overwhelmed, or they may resort to the use of substances to "self-medicate." These problems can have a life of their own, and require specific treatment approaches beyond the treatment of the underlying anxiety disorder or accumulated coping and adaptation deficits.

Reinforcement patterns and treatment strategies (**Table 2**) can also be used to formulate how a particular child's anxiety is being maintained and importantly, formulate treatment strategies to address these reinforcement patterns. In **Fig. 3** two contrasting models of anxiety reinforcement are provided. In **Fig. 3**A the anxiety is of

Table 2
Treatment strategies

	Positive Reinforcement	Negative Reinforcement
Internally Reinforcing	Facilitate natural consequences for over-valued anxiety thoughts or behaviors	Provide exposure to anxiety provoking situations and prevent avoidance
Interpersonally Reinforcing	Provide attention and support for engaging in treatment and taking on normal developmental challenges	Facilitate re-engagement in normal developmental tasks that trigger anxiety

moderate severity, with minimal parental involvement in reinforcing anxiety severity. In **Fig. 3B** there is a much greater overall burden of anxiety severity, largely caused by parental reinforcement of anxiety expression and facilitation of avoidance behavior. The child's anxiety would not be that severe were it not for the impact of the parent's engagement and support for avoidance. In **Fig. 3A** treatment with CBT or combination treatment may suffice. In **Fig. 3B** the treatment plan needs a much stronger family

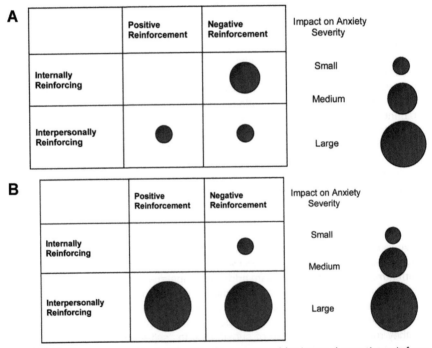

Fig. 3. (*A*) Moderate to severe anxiety, largely maintained by internal negative reinforcement. The young patient is a good candidate for CBT and potentially pharmacotherapy. Family-focused therapy is helpful but more adjunctive, because of the small contribution parents make to reinforcing the expression of anxiety and facilitating avoidance. (*B*) Substantially severe anxiety, largely maintained by parental attention to the expression of anxiety and support of avoidance behavior. Treatment with individual CBT for the child, and substantive parenting work, to refocus parents away from anxiety expression and on supporting the child's efforts to re-engage in activities and not avoid.

focus, and is less likely be responsive to individual CBT and medication alone. The use of a visual aid, such as **Fig. 3**, provides an individualized profile of anxiety severity for the patient and parents, and the starting point for family-based treatment planning.

Treatment of Anxiety Symptoms

The available treatments for the childhood anxiety disorders include CBT and antidepressant medications (selective serotonin reuptake inhibitors and serotonin-norepinephrine reuptake inhibitors).[25–27] These treatments offer the best outcome when they are combined.[28,29] In the Child/Adolescent Anxiety Multimodal Study (CAMS; mean age, 10.7 years),[29] the remission rates (ie, no or minimal symptoms) after 12 weeks of treatment in the combination, medication, and CBT groups were 66%, 46%, and 35%, respectively.[24] At 24 weeks, the remission rates for the combination, medication, and CBT groups were 65%, 49%, and 45%, respectively.[24] The difference in remission rates at 12 and 24 weeks between combination treatment and CBT was statistically and clinically meaningful (~30% and ~20%, respectively) because remission early in CAMS treatment conveyed remission benefit in longer term follow-up.[30] Although everyone supports CBT as the initial treatment, there is a potentially high cost to any treatment approach that does not explicitly include a medication trial in those who do not initially remit to CBT alone.

Psychological treatments

Perhaps the best-known CBT for childhood-onset anxiety disorder is the Coping Cat.[31] The Coping Cat includes 16 to 20 total sessions, two sessions dedicated to family work, and the flexibility to include parents in more sessions to facilitate exposures.[32] The core elements of the Coping Cat include psychoeducation and exposure tasks to address anxious cognitions and improve distress tolerance. On the other end of the spectrum from individual CBT is the Supportive Parenting for Anxious Childhood Emotions (SPACE) Program. This program is focused on reducing accommodation by the parents of children with an anxiety disorder and is similarly effective to individual CBT.[33] Other approaches have taken a transdiagnostic approach, recognizing that anxiety and depression commonly co-occur and the need to develop strategies for regulating distress cuts across disorders.

For youth with accumulated impairment from the lack of early, effective treatment, life skills training, tuned to the gap in skills and developmentally appropriate expectations, can augment CBT. Life skills programs are more generic and include developing effective problem-solving skills, good decision-making, effective communication strategies, developing successful relationships, and building competence and confidence in independent living. A program that aims to address these deficits is the Launching Emerging Adults Program (LEAP) developed by Anne Marie Albano. LEAP is a developmentally informed CBT model that includes parents to addresses anxiety symptoms and facilitates independence in educational settings, work, and relationships unique to this developmental period.[34] An important focus of family involvement is improving family-wide distress tolerance.

Adolescents and young adults who have developed severe anxiety-related dysregulation and more maladaptive patterns of behavior, such as substance misuse, self-injury, and suicidal behavior, may require additional treatment, such as dialectical behavioral therapy[35] or a substance abuse treatment program.

SUMMARY

The anxiety disorders begin in childhood.[3] With the lack of effective treatment, the child loses the capacity to cope and adapt to routine life experiences, may become dysregulated as evidenced by catastrophic reactions, and may go on to develop more severe and maladaptive behaviors.[19,20] Treatment focused on anxiety symptoms, deficits in adaptation and coping, and targeted treatments for maladaptive behaviors offers children, who were not identified early and were not treated effectively, the best hope for a positive outcome.

DISCLOSURE

We would like to acknowledge the support of the Pritzker Foundation for their support of Pritzker Department of Psychiatry and Behavioral Health at Lurie Children's Hospital of Chicago. Dr J.T. Walkup has served on the advisory board of the Tourette Syndrome Association, Trichotillomania Learning Center, and Anxiety and Depression Association of America; and has received royalties from Oxford University Press, Guilford Press, and Wolters Kluwer.

Dr. Peris receives research support from NIMH, PCORI, and the TLC Foundation for Body Focused Repetitive Behaviors. She also receives royalties from Oxford University Press.

Dr. Strawn has received research support from AbbVie, Neuronetics, Lundbeck, Otsuka, PCORI, the National Institutes of Health (NIEHS, NIMH, NICHD) and the Yung Family Foundation. He has provided consultation to Intracellular Therapeutics and the Food and Drug Administration in 2020. He receives royalties from Springer Publishing for two texts and received material support from Myriad. He has also received honoraria from CMEology, Genomind and Neuroscience Educational Institute.

REFERENCES

1. Barcai A, Rosenthal MK. Fears and tyranny: observations on the tyrannical child. Arch Gen Psychiatry 1974. https://doi.org/10.1001/archpsyc.1974.01760090098015.

2. Johnco C, Salloum A, De Nadai AS, et al. Incidence, clinical correlates and treatment effect of rage in anxious children. Psychiatry Res 2015. https://doi.org/10.1016/j.psychres.2015.07.071.

3. Beesdo-Baum K, Knappe S. Developmental epidemiology of anxiety disorders. Child Adolesc Psychiatr Clin N Am 2012;457–78. https://doi.org/10.1016/j.chc.2012.05.001.

4. Lebowitz ER, Marin CE, Silverman WK. Measuring family accommodation of childhood anxiety: confirmatory factor analysis, validity, and reliability of the parent and child family accommodation scale – anxiety. J Clin Child Adolesc Psychol 2019. https://doi.org/10.1080/15374416.2019.1614002.

5. Sylvester CM, Corbetta M, Raichle ME, et al. Functional network dysfunction in anxiety and anxiety disorders. Trends Neurosci 2012;527–35. https://doi.org/10.1016/j.tins.2012.04.012.

6. Etkin A, Büchel C, Gross JJ. The neural bases of emotion regulation. Nat Rev Neurosci 2015. https://doi.org/10.1038/nrn4044.

7. Wilcox CE, Pommy JM, Adinoff B. Neural circuitry of impaired emotion regulation in substance use disorders. Am J Psychiatry 2016. https://doi.org/10.1176/appi.ajp.2015.15060710.

8. Strawn JR, Wehry AM, Delbello MP, et al. Establishing the neurobiologic basis of treatment in children and adolescents with generalized anxiety disorder. Depress Anxiety 2012;29(4):328–39.

9. Gee DG, Gabard-Durnam LJ, Flannery J, et al. Early developmental emergence of human amygdala-prefrontal connectivity after maternal deprivation. Proc Natl Acad Sci U S A 2013. https://doi.org/10.1073/pnas.1307893110.

10. Ptak R. The frontoparietal attention network of the human brain: action, saliency, and a priority map of the environment. Neuroscientist 2012. https://doi.org/10.1177/1073858411409051.

11. Strawn JR, Bitter SM, Weber WA, et al. Neurocircuitry of generalized anxiety disorder in adolescents: a pilot functional neuroimaging and functional connectivity study. Depress Anxiety 2012;29(11):939–47.

12. Shanmugan S, Wolf DH, Calkins ME, et al. Common and dissociable mechanisms of executive system dysfunction across psychiatric disorders in youth. Am J Psychiatry 2016. https://doi.org/10.1176/appi.ajp.2015.15060725.

13. Roy AK, Fudge JL, Kelly C, et al. Intrinsic functional connectivity of amygdala-based networks in adolescent generalized anxiety disorder. J Am Acad Child Adolesc Psychiatry 2013;52(3). https://doi.org/10.1016/j.jaac.2012.12.010.

14. Harrison BJ, Pujol J, Contreras-Rodríguez O, et al. Task-Induced deactivation from rest extends beyond the default mode brain network. PLoS One 2011; 6(7). https://doi.org/10.1371/journal.pone.0022964.

15. Li W, Mai X, Liu C. The default mode network and social understanding of others: what do brain connectivity studies tell us. Front Hum Neurosci 2014;8(1). https://doi.org/10.3389/fnhum.2014.00074.

16. Williams LM. Precision psychiatry: a neural circuit taxonomy for depression and anxiety. Lancet Psychiatry 2016;472–80.

17. Peris TS., Thamrin H., Rozenman M. Family intervention for child and adolescent anxiety: a meta-analytic review of therapy targets, techniques, and outcomes. J Affect Disord n.d.

18. Taboas WR, McKay D, Whiteside SPH, et al. Parental involvement in youth anxiety treatment: conceptual bases, controversies, and recommendations for intervention. J Anxiety Disord 2015. https://doi.org/10.1016/j.janxdis.2014.12.005.

19. Asselmann E, Wittchen H-U, Lieb R, et al. Sociodemographic, clinical, and functional long-term outcomes in adolescents and young adults with mental disorders. Acta Psychiatr Scand 2018;137(1):6–17. https://doi.org/10.1111/acps.12792.

20. Husky MM, Olfson M, He J, et al. Twelve-month suicidal symptoms and use of services among adolescents: results from the National Comorbidity Survey. Psychiatr Serv 2012;63(10):989–96.

21. Vorspan F, Mehtelli W, Dupuy G, et al. Anxiety and substance use disorders: co-occurrence and clinical issues. Curr Psychiatry Rep 2015. https://doi.org/10.1007/s11920-014-0544-y.

22. Olson D. FACES IV and the circumplex model: validation study. J Marital Fam Ther 2011. https://doi.org/10.1111/j.1752-0606.2009.00175.x.

23. Dunne G, Askew C. Vicarious learning and unlearning of fear in childhood via mother and stranger models. Emotion 2013. https://doi.org/10.1037/a0032994.

24. Piacentini J, Bennett S, Compton SN, et al. 24- and 36-week outcomes for the child/adolescent anxiety multimodal study (CAMS). J Am Acad Child Adolesc Psychiatry 2014;53(3). https://doi.org/10.1016/j.jaac.2013.11.010.

25. Strawn JR, Welge JA, Wehry AM, et al. Efficacy and tolerability of antidepressants in pediatric anxiety disorders: a systematic review and meta-analysis. Depress Anxiety 2015;32(3). https://doi.org/10.1002/da.22329.

26. Locher C, Koechlin H, Zion SR, et al. Efficacy and safety of selective serotonin reuptake inhibitors, serotonin-norepinephrine reuptake inhibitors, and placebo

in common psychiatric disorders a meta-analysis in children and adolescents. Jama Psychiatry 2017;74(10):1011–20.

27. Dobson ET, Bloch MH, Strawn JR. Efficacy and tolerability of pharmacotherapy in pediatric anxiety disorders: a network meta-analysis. J Clin Psychiatry 2019; 80(1):17r12064.

28. Beidel DC, Turner SM, Sallee FR, et al. SET-C versus fluoxetine in the treatment of childhood social phobia. J Am Acad Child Adolesc Psychiatry 2007;46(12): 1622–32.

29. Walkup JT, Albano AM, Piacentini J, et al. Cognitive behavioral therapy, sertraline, or a combination in childhood anxiety. N Engl J Med 2008;359(26). https://doi. org/10.1056/NEJMoa0804633.

30. Ginsburg GS, Becker-Haimes EM, Keeton C, et al. Results from the child/adolescent anxiety multimodal extended long-term study (CAMELS): primary anxiety outcomes. J Am Acad Child Adolesc Psychiatry 2018. https://doi.org/10.1016/j.jaac.2018.03.017.

31. Kendall PC. Treating anxiety disorders in children: results of a randomized clinical trial. J Consult Clin Psychol 1994. https://doi.org/10.1037/0022-006X.62.1.100.

32. Beidas RS, Benjamin CL, Puleo CM, et al. Flexible applications of the coping cat program for anxious youth. Cogn Behav Pract 2010;17(2):142–53.

33. Lebowitz ER, Marin C, Martino A, et al. Parent-based treatment as efficacious as cognitive-behavioral therapy for childhood anxiety: a randomized noninferiority study of supportive parenting for anxious childhood emotions. J Am Acad Child Adolesc Psychiatry 2020. https://doi.org/10.1016/j.jaac.2019.02.014.

34. Hoffman LJ, Guerry JD, Albano AM. Launching anxious young adults: a specialized cognitive-behavioral intervention for transitional aged youth. Curr Psychiatry Rep 2018. https://doi.org/10.1007/s11920-018-0888-9.

35. McCauley E, Berk MS, Asarnow JR, et al. Efficacy of dialectical behavior therapy for adolescents at high risk for suicide a randomized clinical trial. JAMA Psychiatry 2018;75(8):777–85.

When No Diagnosis "Fits"
Diagnostically Homeless and Emotionally Dysregulated Children and Adolescents

Deborah M. Weisbrot, MD*, Gabrielle A. Carlson, MD

KEYWORDS

- Diagnostically homeless • Multiple complex developmental disorder
- Schizotypal personality disorder • Multi-dimensionally impaired disorder

KEY POINTS

- Clinical experience supports the concept of children with severe disturbances in most areas of functioning, with psychiatric symptom onset before age 6.
- Diagnostically homeless children experience severe emotional dysregulation and rages and complex, disabling symptoms that do not fit classic diagnostic criteria of mood, anxiety, or psychotic disorders.
- Treatment interventions involve addressing dysfunctional areas; translating diagnostic findings to parents, school, and child; finding an appropriate educational setting; and developing a consistent behavioral program for the child's major problems.

INTRODUCTION

Children with complex developmental problems and severe psychopathologies often do not fit neatly into *Diagnostic and Statistical Manual of Mental Disorders* (DSM) categories.[1,2] Such ill-defined syndromes commonly are encountered, yet little is known about them diagnostically, longitudinally, or in terms of optimal treatment. Understanding their nature is confounded by the multitude of psychiatric diagnoses they receive. Severe deficits in emotional dysregulation, with mixtures of rages, depression, anxiety, and manic symptoms, often are what brings them to attention.[3] They also present with deficits in socialization, language, emotional, and cognitive function.

Challenges in categorizing their symptoms into well-defined diagnoses leads to calling them "diagnostically homeless." Their complex presentations continue to perplex and challenge clinically. Their emotionally dysregulated states occur in combination with executive function deficits and problems in communication (eg, thought disorder and disordered pragmatic language) and socialization. Many display features of autism spectrum disorder (ASD) but of insufficient degree to warrant formal diagnosis.

Putnam Hall-South Campus, Renaissance School of Medicine, Stony Brook University, Stony Brook, NY 11794-0790, USA
* Corresponding author.
E-mail address: deborah.weisbrot@stonybrookmedicine.edu

Child Adolesc Psychiatric Clin N Am 30 (2021) 445–457
https://doi.org/10.1016/j.chc.2020.10.015
1056-4993/21/© 2020 Elsevier Inc. All rights reserved.

childpsych.theclinics.com

Although the authors of the *DSM*[4-7] have attempted to operationalize symptoms for prototypic disorders, many with serious psychopathology still are inadequately described by existing diagnoses. In an attempt to develop specific criteria that exclude extraneous symptoms, many features of known conditions are not included. For example, neither emotion dysregulation nor cognitive deficits is included as a symptom of ASD.

Diagnostically homeless children tend to be excluded from studies, because they do not meet full criteria for ASDs, mood disorders, or schizophrenia. Alternatively, they may be included in samples with attention-deficit hyperactivity disorder (ADHD) or bipolar disorder (BD) because of their mood lability. They typically end up receiving "unspecified" designations or the closest fitting diagnosis. Without a firm way to identify these youngsters, their prevalence cannot be determined. Prospective studies of these patients as a unified group are limited because of diagnostic variability or disagreement on diagnosis. Small series have attempted to address these concerns, but outcomes—not surprisingly—are wide ranging.[2,8-10]

This article highlights *DSM* slippage and describes children and adolescents with complex clinical problems and undetermined diagnoses. The authors explore dilemmas inherent in treating children without truly knowing their underlying disorder and the problems for their parents, given that meaningful psychoeducation is difficult, and no Web site or book exists to help them understand their child. The authors suggest ways in which clinicians can elicit information, which leads to potentially effective interventions.

WHO ARE THE DIAGNOSTICALLY HOMELESS? FOUR CASE EXAMPLES
Case 1: Rule Out Bipolar Disorder

Devon is 6 and a half years old. He was referred for psychiatric evaluation due to severe hyperactivity and impulsivity. He did not speak until he was 3 years old. He had a high pain threshold, leading to many accidents and emergency room visits. He exhibited little but not absent interest in peers and rages when changes were imposed on him. He has tried driving cars, asked strangers to be his father, and walked into people's houses uninvited. He once ran in front of car and explained that he was "faster than cars." He jumped off trees and almost drowned when he tried to swim across the Atlantic. When he was rescued, he said that he heard a voice telling him to do it.

In school, teachers reported severe hyperactivity, inattention, defiance, strange beliefs, and an inappropriate affect. He was bossy with peers and had difficulties with transitions. His IQ scores were in the average range. He had deficits in both receptive and pragmatic language. Devon's family history included depression, substance abuse, and ADHD but no history of BD or ASD.

On examination, he demonstrated poor eye contact. He was jumpy, hyperactive, and easily overstimulated. His mood was cheerful. He said he felt "mad" when his toys were misplaced and that "nothing" makes him scared. He thought he could swim across the ocean, because "my grandpa said I could swim like a fish!" He was inordinately interested in dogs and said he thought he was a dog. He admitted to "voices in his head" telling him to do "bad things, like hit people" but knew this was his imagination. His style of play was obsessional and rigid.

Case 2: Not Your Typical Anxieties

Steven, 11 years old, was referred "to rule out BD" and to evaluate hyperactivity, explosiveness, rages, and nightmares. He was adopted in infancy. He did not want to be held, did not speak until he was 22 months old, and had sensory integration problems, hyperactivity, and impulsivity. He experienced severe separation anxiety and nightmares.

Steven was not just a little nervous about life, peers, or grades. He worried about things like a "Rock & Roll Vampire" sucking out his brains, eating his insides, and turning him into a bat. He dreamed about being in an origami factory and seeing a skeleton-like figure cutting himself. This figure, whom he called the "Paper-Cut Man," jumps on people and cuts them and lurked under his bed. On the good side, Steven had SpongeBob, who protected him from these monsters. Steven worried that bad people were chasing him and saw "scary stuff" out of the corner of his eyes. He did not display other evidence of psychosis. He said his mood was "mostly scared." Due to his fears, Steven insisted that his parents stay with him at night. He would make noises often in response to what he was thinking or hearing.

Case 3: Threats of Violence

Lauren, age 12, previously was diagnosed with ADHD but later presented for psychiatric evaluation due to chronic, withdrawn, depressed, and defiant behaviors. She was described as a "loner" who never related well to children. Lauren described "torture" by peers, which sounded paranoid. Her conversation was extremely circumstantial and rambling. She initially said she "sometimes" hears voices in her head but later said that it usually was her own voice. Referred because she made threats on-line toward school staff, referencing the Columbine shooting, when asked, Lauren said she thought it was funny. She also said, however, that she would like to "see the life drain out" of students whom she felt were bothering her.

Case 4: Broken Brain–Induced Rages

Richard, age 8, was referred by his school district due to concerns that he had a diagnosis of schizophrenia. He had symptoms of ASD, including language delays, disinterest in peers, disinhibited behaviors, and motor coordination difficulties. He had difficulties with transitions and perseverated about his "brain being broken." He was severely oppositional and had prolonged rages. His parents described him as irritable, shifting rapidly from happy to angry to sad. When upset, he would bang his head against the wall, pick his skin compulsively and then lick the blood. He reported hearing the voice of "a scary man" telling him to do bad things, such as hurt people, steal things, and "break stuff." He said he did not mind the voice very much. The voice did not pervade his life the way hallucinations do in schizophrenia.

Children such as these are not rare. It is unclear whether they have a developmental disorder, such as autism, the prodrome of a psychotic or mood disorder, an atypical mood disorder, or an idiosyncratic personality. They have more extensive difficulties than those seen in ADHD, generalized anxiety disorder (GAD), or obsessive-compulsive disorder (OCD). Clinically, they either are forced into a category they somewhat resemble (such as mania in Devon's case or ASD in Steven's case) or are given multiple additional diagnoses of comorbid conditions or an "unspecified" label, the severity of which goes unacknowledged.

SHIFTING NOSOLOGIES AND ATTEMPTS TO DESCRIBE DIAGNOSTICALLY HOMELESS YOUTHS

Over the years, a variety of diagnoses have been proposed to describe diagnostically homeless children. These proposed categories described abnormalities in social relationships, affective regulation, and thought/communication reality-testing domains. They included

1. Multiple complex developmental disorder (MCDD)[11–13]
2. Multidimensionally impaired disorder (MDI)[14]
3. Schizotypal and schizoid personality disorders[15]

Neither MCDD or MDI has been validated or included in the *DSM*, and schizotypal personality disorder cannot be diagnosed in children. Each of these diagnoses is summarized.

Multiple Complex Developmental Disorder—A Trifecta of Severity: Emotional Dysregulation, Anxiety, and Psychosis

Cohen and colleagues[16] emphasized the need to study children struggling with impaired social sensitivity and deficits in regulation of affect, anxiety, and thought problems. They did not have classical autism. He described a condition, included in *DSM* (Third Edition) but later dropped from subsequent *DSM*s, called childhood-onset pervasive developmental disorder. Diagnostic criteria included impaired social relationships and several of the following: sudden, excessive anxiety (including unexplained rages); constricted or inappropriate affect; resistance to change; unusual motor movements; poor speech prosody; and self-mutilation but no psychotic symptoms or marked loosening of associations. Subsequently, a diagnosis of MCDD[12,17] was proposed to describe this disorder in greater detail (**Table 1**).

MCDD criteria included problems with mood dysregulation and anxiety symptoms, for example, unusual fears and phobias as well as panic episodes, and a high frequency of idiosyncratic anxiety reactions and episodes of behavioral disorganization, such as sustained periods of uncontrollable giggling, laughter, or silly affect inappropriate in the situation's context.[12] MCDD highlighted anxiety as a major factor in these children's difficulties and included descriptions of intermittent irrational thinking and impairments in social behavior.[18] Van der Gaag, Buitelaar and colleagues argued that MCDD better represented the group of children who not only had ASD symptoms but also experienced dysregulated affective states, psychotic symptoms, and anxiety.[11,12,19]

Many autistic children struggle with intense anxiety, emotional dysregulation, and possible impairments of reality testing.[20–23] Anxiety is not a major component of ASD in *DSM*, despite anecdotal and recent series suggesting that anxiety may be an integral component, thereby provoking significant debate and confusion.[2,4,24]

Nonetheless, when anxious, affectively dysregulated children with ASDs also present with psychotic-like symptoms, they appear to have more than autism.[12,13] Weisbrot and colleagues[25] examined the frequency of anxiety symptoms in preschool and 6-year-old to 12 year-old clinic children with and without pervasive developmental disorders (PDDs)/ASD symptoms. They found that anxious elementary school–aged children with PDDs had more severe delusions, auditory and visual hallucinations, and grossly disorganized behaviors than nonanxious children, even after controlling for severity of PDDs/ASD symptoms. Similarities in anxiety symptom presentation and their association with psychotic symptoms in both children with and without autism support the possibility of a unique diagnostic entity composed of autistic, anxiety, and psychotic symptoms as well as supporting the overlap in causal mechanisms for anxiety and psychotic symptoms. Alternatively, complex and sometimes heterogeneous clinical presentations of diagnostically homeless children still may be in keeping with *DSM* (Fifth Edition) (*DSM-5*) concepts of spectrum disorders,[7] which do not require absolute homogeneity within diagnostic categories or absolute boundaries between disorders.

Table 1
Comparison of criteria for multiple complex developmental disorder and multidimensionally impaired syndrome

Multiple Complex Developmental Disorder	Multidimensionally Impaired Syndrome
• Impaired regulation of affective state and anxiety: ○ Intense anxiety, tension, irritability ○ Unusual fears and phobias ○ Recurrent panic episodes ○ Episodic behavioral disorganization or regression ○ Significant emotional variability ○ Idiosyncratic anxiety reactions	Nearly daily periods of emotional lability disproportionate to precipitants
• Impaired social behavior and sensitivity: ○ Social disinterest, detachment, or avoidance ○ May appear friendly and cooperative but superficial ○ Inability to initiate or maintain peer relationships ○ High degree of ambivalent attachment to adults ○ Profound limitations in the capacity of empathy	Impaired interpersonal skills despite desire to initiate social interactions with peers
• Impaired cognitive processing: ○ Thought problems including irrationality, intrusions on thought process, magical thinking, neologisms or nonsense words, illogical bizarre ideas ○ Confusion between reality and fantasy life ○ Perplexity and easy confusability ○ Fantasies of personal omnipotence, paranoid preoccupations, overengagement with fantasy figures, grandiose fantasies of special powers, and referential ideation	Poor ability to distinguish fantasy from reality, as shown by ideas of reference and brief perceptual disturbances during stressful periods or while falling asleep

From Cochran DM, Dvir Y, Frazier JA. "Autism-plus" Spectrum Disorders. Child Adolesc Psychiatric Clin N Am 2013;22(4):614; with permission.

Multidimensionally Impaired Disorder—A Good Fit for Atypical Psychosis?

MDI disorder is a syndrome of atypical psychosis, alternatively termed, psychosis not otherwise specified (NOS), or, more recently, in *DSM-5*, unspecified schizophrenia spectrum and other psychotic disorder.[7] This diagnosis was developed during the National Institutes of Mental Health (NIMH) Early Onset Schizophrenia Project to describe children older than 5 years not meeting criteria for schizophrenia but with brief, recurrent, and stress-induced psychotic symptoms.[14,26–28] Similar to the previously described diagnosis of MCDD, these children had problems with affect regulation, rages, and impaired interpersonal skills but had more psychotic-like symptoms like ideas of reference and brief perceptual disturbances during stress or while falling asleep. They also had cognitive deficits, as evidenced by

ADHD-like symptoms and multiple deficits in information processing. They lacked formal thought disorders, however, and were distinguishable from children with schizophrenia. Prognosis was better; only 27% met criteria for schizoaffective disorder at 2-year follow-up. Some (5 of 32) were nearly asymptomatic, although the remainder still were significantly impaired and continued to have severe behavioral problems and mood disorders, albeit not schizophrenia. The bottom line is that these children are more common than children with schizophrenia. Unfortunately, a clear designation of their outcome is hampered by an absence of useful diagnostic criteria.

Schizotypal Personality Disorder—A Good Home for Diagnostically Homeless Children?

The diagnostic category of schizotypal personality disorder is another potential home for mood-dysregulated, anxious, and quasipsychotic diagnostically homeless children. This diagnosis emerged from genetic studies of schizophrenia to describe individuals presenting as less overtly psychotic and fairly stable. It not always is easy, however, to distinguish schizotypy from prodromal schizophrenia.

A study of schizotypal personality disorder in childhood noted a high frequency of social anxiety in these socially isolated children who also displayed magical thinking, bizarre preoccupations, odd speech, and poor rapport.[29,30] They previously received multiple diagnoses, including autism, borderline personality disorder (BPD), and schizoaffective disorder. Their presenting problems primarily were in social and academic areas. They were operationally given the MDI diagnosis instead of schizotypal personality disorder in the NIMH study of childhood schizophrenia because personality disorders were not supposed to be diagnosed in children.

Meticulously gathered, prospectively collected data on offspring of schizophrenic mothers and a socioeconomically matched comparison group were gathered by Dr Barbara Fish, a pioneer in the study of risk for childhood schizophrenia.[2,31,32] Schizotypal children were identified who most likely had schizophrenic mothers, deteriorating IQs, symptoms of social isolation, constricted affect, digressive speech, suspiciousness, and excessive social anxiety. Global assessment scores in schizophrenia spectrum disorder offspring were lower in childhood and adulthood than comparison offspring. These negative symptoms set them apart more than psychotic symptoms. Long-term follow-up demonstrates symptom persistence. Even in a shorter-term follow-up study sample, where mood disorders ultimately manifested, schizotypal symptoms persisted, leading to schizoaffective and atypical bipolar diagnoses (**Table 2**).[31,33–35]

Table 2
Features of schizotypal personality disorder in childhood
Excessive social anxiety and paranoid fears.
Lack of close friends or confidants other than relatives.
Inappropriate or constricted affect.
Ideas of reference
Unusual perceptual experiences
Suspicious; eccentric, Odd thinking:
Vague, circumstantial, overelaborate or stereotyped speech.

Data from Weisbrot DM, Carlson GA. "Diagnostically Homeless": Is it ADHD? Mania? Autism? What to do if no diagnosis fits. Current Psychiatry 2005;4(2):24-42.

Although aggression has not been studied extensively in schizotypal children, data that there are suggest peer victimization is the major explanation for resulting reactive aggression. Children who are perceived as different or weird are "tortured" by other children.[36]

Is it a Language Disorder or a Thought Disorder? A Challenging Distinction

Teasing apart a thought disorder from a language disorder is another challenging situation frequently encountered in the evaluation of children with severe emotional disturbance. Youngsters with social oddities, pragmatic language problems, and psychotic-like symptoms may be mistaken for having a schizophrenia-related disorder. There also can be overlap between these 2 conditions.

Primary information-processing deficits may be involved in the persistence of formal thought disorder in children over age 7.[37] These children become emotionally dysregulated when their illogical thinking distorts their perception of daily social events, which they find threatening. They then have difficulty communicating their fears in a clear manner, leading to affective dysregulation so characteristic of children with MCDD and autism. Semantic differences may confound the discussion of these children's problems further; what is called a pragmatic language problem in the autism literature is called mild thought disorder in the schizophrenia literature.[38,39]

Could Borderline Personality Disorder be a Home for the Diagnostically Homeless?

BPD is characterized by severe affective dysregulation, impulsivity, impaired interpersonal relationships, self-destructive behaviors, suicide attempts, and identity problems as well as stress-related paranoid ideation and severe dissociative symptoms. Until recently, mental health professionals were reluctant to diagnose BPD in children and adolescents, despite personality disorders potentially having developmental trajectories beginning in childhood and borderline traits often becoming prominent in adolescence or even earlier. Descriptions of borderline symptoms in childhood include intense anxiety, reality disturbances, severely impulsive behaviors, and impaired interpersonal relationships.[40] Youth with BPD often experience intense anger, rages, emptiness, shame, and self-loathing. They tend to be extremely sensitive and emotionally reactive and, once upset, it takes them longer to recoup to their emotional baseline. They are hypersensitive to social cues and are prone to interpret things negatively.[6,15,41]

There are significant diagnostic and treatment implications related to diagnostically homeless children being denied this diagnosis. For example, a child may end up with an inaccurate diagnosis of an Axis I disorder, such as BD, ADHD, or a depressive disorder. Even if a comorbid mood disorder or ADHD is diagnosed, the failure to identify BPD could lead to a delay in psychotherapeutic treatments (eg, dialectical behavior therapy) and administration of inappropriate medications. The child's severe and chronic emotional dysregulation symptoms may, in turn, become more entrenched in young adulthood.

ASSESSMENT AND TREATMENT CONSIDERATIONS

Diagnostically homeless children need to be assessed for what they have, rather than trying to "fit" them into current diagnostic schema. Assessment starts with a comprehensive rating scale that includes mood, behavior, and autism-like and psychotic-like symptoms. Although the Achenbach System of Empirically Based Assessment[42,43] has been used for years, other systems, including the Behavior Assessment System

Table 3
Useful questions in the assessment of children's social skills
• Does the child appear disinterested or clueless about social situations? (Seen in ASD; MCDD; MDI; NVLD*****)
• Does the child exhibit atypical or bizarre social interactions? (Seen in Schizotypal Personality Disorder; Schizophrenia)
• Does child appear shut down or inhibited in unfamiliar settings, with greater comfort at home or with familiar people? (Seen in Social Anxiety Disorder)

*Attention Deficit Hyperactivity Disorder; **Autism Spectrum Disorder; ***Multiple Complex Developmental Disorder; ****Multidimensionally Impaired Disorder; *****Nonverbal Learning Disability.

Data from Weisbrot DM, Carlson GA. "Diagnostically Homeless": Is it ADHD? Mania? Autism? What to do if no diagnosis fits. Current Psychiatry 2005;4(2):24-42.

for Children[44] and the Child and Adolescent Symptom Inventory[45] also can be helpful screens. They quantify depression, anxiety, ADHD, and aggressive behavior. Of particular interest, however, are items that focus on the thought problems, atypicality, and social problems domains. The authors also use the Children's Atypical Development ment Scale,[46] which asks about autism-like symptoms as well as symptoms like mood changes quickly without apparent reason, excessively preoccupied with violent stories, gets angry for little apparent reason, extreme reactions to minor inconveniences or irritations, laughs or cries for little apparent reason, has unusual fears not typical for his/her age group (weather or computer games), and overreacts to pain (eg, a minor injury causes a catastrophic reaction). **Tables 3** and **4** outline useful questions for social and language assessment and some of the possible differential diagnoses to consider.

The goal of assessment is not to diagnose something that has no name but to rule out recognizable conditions, which means a thorough assessment of mood, behavior, language and thought, relationships, and stress is necessary (**Fig. 1**). The major point to be made with children considered diagnostically homeless is to recognize the diagnosis simply is not clear. The child has more than ADHD, oppositional defiant disorder, or GAD. For example, having severe outbursts and quirky behavior is not BD. Sometimes a diagnosis becomes clear over the course of follow-up. Devon (see case 1), when observed for a period of time, had ASD, ADHD, and Tourette disorder. What was interpreted as psychosis and delusional was his concrete thinking. He tried to swim across Long Island Sound not because he was grandiose but because if he could swim like a fish, there is no reason he could not make it across a body of water. His "delusion" that he was related to dogs was because he had a barking tic and hyperacusis, which were intermittent and, in his mind, meant he must have canine blood.

Treatment of diagnostically homeless children includes interventions for youth who have rages or are easily overstimulated, executively dysfunctional, and language impaired. In the absence of an optimal classification for diagnostically homeless children, or ways to reliably measure their symptoms, and without a foundation of treatment studies specifically focused on this population, they are at risk receiving unnecessary treatments or deprived of potentially valuable therapies. Ultimately, it becomes necessary to pick the current diagnosis that best addresses the symptom complex causing the most impairment and address that first.

Treatment interventions for the diagnostically homeless child involve (1) identifying and addressing each domain of dysfunction; (2) translating and demystify the findings to parent, child, and educators and being open with parents if their child falls between

---

Table 4
Useful questions in the assessment of children's language skills

1. How did early language develop?
   - Age at first language use?
   - Age at first use of short sentences?
   - Nonverbal communication skills?
   - Communication for sharing?

2. Were language problems present?
   - Communication delayed at first, and then progressed normally? (Seen in developmental language disorder)
   - Communication began normally, then stopped? (Seen in ASD**)
   - Was/is language atypical? (Seen in ASD; NVLD*****)
   - Was/is language bizarre or paranoid? (Seen in Schizotypal PD)
   - Pragmatic language problems? (Seen in ASD; ADHD*; MDI****, MCDD***, NVLD)

3. Are other kinds of language problems present?
   - Problems with written language?
   - Any hearing deficits?
   - Problems with auditory processing?

4. Has the child mastered a range of communication skills?
   - Rules of conversation (eg, who is likely to be interested in what and when to expand shift or end a conversation?).
   - Awareness of nonverbal cues/
   - Understanding emotional language (how to express feelings and the different ways people experience themselves)?
   - Able to make consistent eye contact?
   - Awareness of how social settings affect communication, such as voice volume (whisper in the library, shout on the soccer field) and speech style (slang with peers, formal style for classroom recitation)
   - Body proximity (how to avoid invading someone's space)
       Decoding facial expression (eg, what it means when someone rolls his eyes) and other communications, for example, teasing or sarcasm.

5. Have the following assessments been performed?
   - Audiological; neuropsychological assessments for hearing and/or auditory processing problems.
   - Assessment by a speech/language pathologist for expressive and receptive language deficits?
   - An extensive speech sample for pragmatic language deficits.
   - Psychological assessment for written language problems?
   - Note: assessment tools include: Test of Language Competence (TLC)[47]; The Clinical Evaluation of Language Fundamentals (CELF-5)[48]; The Detroit Test of Learning Aptitude (DTLA)[49] and others

---

*Attention Deficit Hyperactivity Disorder; **Autism Spectrum Disorder; ***Multiple Complex Developmental Disorder; ****Multidimensionally Impaired Disorder; *****Nonverbal Learning Disability.
Data from Weisbrot DM, Carlson GA. "Diagnostically Homeless": Is it ADHD? Mania? Autism? What to do if no diagnosis fits. Current Psychiatry 2005;4(2):24-42.

diagnostic cracks; (3) recommending and advocating for the provision of educational settings and resources that allow the child to work most effectively; and (4) collaborating with educators and parents in the development of a consistent behavioral program to respond to the child's major problems. Longitudinal follow-up also is essential. It usually is necessary to commit to some diagnosis to satisfy insurers and agencies (including schools). When *DSM-5* eliminated PDD-NOS, it removed

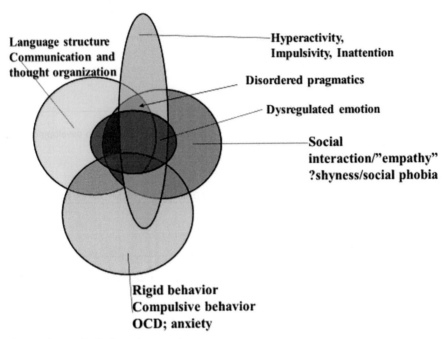

Language structure
Communication and
thought organization

Hyperactivity,
Impulsivity, Inattention

Disordered pragmatics

Dysregulated emotion

Social
interaction/"empathy"
?shyness/social phobia

Rigid behavior
Compulsive behavior
OCD; anxiety

Fig. 1. Diagnostically homeless overlaps.

one of the options such children had. *Psychosis NOS* or, currently, *other specified psychosis* is a more frightening designation that has been used. Some of these children undoubtedly have found their way into studies of *juvenile BD* characterized by severe outbursts (see Lauren Spring and Gabrielle A. Carlson's article, "The Phenomenology of Outbursts," elsewhere in this issue). They do not have adult-like bipolar I disorder, however. It is the authors' experience that although it may be expedient to write something down, simultaneous honesty with parent is accepted gratefully.

SUMMARY

Many children with serious psychopathology fall between the cracks of existing *DSM-5* diagnoses. Clinical experience supports the existence of children with severe disturbances in nearly all areas of functioning, with psychiatric symptom onset before age 6. They are emotionally dysregulated, experience severe rages, and have social and language deficits, peculiar anxieties, and psychotic-like symptoms. They do not meet criteria for ASDs but are most similar to children with autism. They comprise a significant minority of psychiatrically referred children.

The question remains as to whether there is a symptom cluster to differentiate a distinct group of these patients, permitting diagnostic clarity and facilitating treatment research. Until then, their individual mood and anxiety problems, possible psychosis, language deficits, possible thought disorders, and socialization problems need to be assessed and treated. Diagnostic instruments and interviews need to include extensive developmental assessments for ASD, emotion dysregulation, prodromal psychotic symptoms, Axis I disorders, and cognitive and language functioning. Hopefully, future research will lead to accurate homes for diagnostically homeless orphans and the development of effective treatment interventions.

DISCLOSURE

Royalties from Cambridge University Press and Honoraria for American Academy of Child and Adolescent Psychiatry Presentations.

REFERENCES

1. Weisbrot DM. Diagnostically homeless children: is it ADHD, mania, or autism? What to do when no diagnosis fits. Curr Psychiatry 2005;4(2):24–46.
2. Frazier JA, Carlson GA. Diagnostically homeless and needing appropriate placement. J Child Adolesc Psychopharmacol 2005;15(3):337–42.
3. Carlson GA. Presidential address: emotion dysregulation in children and adolescents. J Am Acad Child Adolesc Psychiatry 2020;59(1):15–9.
4. American Psychiatric Association. Diagnostic and statistical manual of mental disorders. DSM-III. 4th edition. Washington, DC: American Psychiatric Association; 1987.
5. American Psychiatric Association. Diagnostic and statistical manual of mental disorders. 4th edition. Washington, DC: American Psychiatric Press; 1994.
6. American Psychiatric Association. Diagnostic and statistical manual of mental disorders - DSM-IV-TR. Arlington (VA): Psychiatric Publishing, Inc; 2000.
7. American Psychiatric Association. Diagnostic and statistical manual of mental disorders, DSM-5. 5th edition. Arlington (VA): American Psychiatric Association; 2013.
8. Kestenbaum C. The borderline child at risk for major psychiatric disorder in adult life: seven case reports with followup. In: Robson KS, editor. The borderline child. New York: McGraw-Hill; 1983. p. 49–82.
9. Lofgren DP, Bemporad J, King J, et al. A prospective follow-up study of so-called borderline children. Am J Psychiatry 1991;148(11):1541–7.
10. Wolff S. Loners: the life path of unusual children. London: Routledge; 1995.
11. Van der Gaag RJ, Buitelaar J, Van den Ban E, et al. A controlled multivariate chart review of multiple complex developmental disorder. J Am Acad Child Adolesc Psychiatry 1995;34(8):1096–106.
12. Buitelaar JK, van der Gaag RJ. Diagnostic rules for children with PDD-NOS and multiple complex developmental disorder. J Child Psychol Psychiatry 1998;39(6): 911–9.
13. Buitelaar JK, Van der Gaag R, Klin A, et al. Exploring the boundaries of pervasive developmental disorder not otherwise specified: analyses of data from the DSM-IV Autistic Disorder Field Trial. J Autism Dev Disord 1999;29(1):33–43.
14. McKenna K, Gordon CT, Lenane M, et al. Looking for childhood-onset schizophrenia: the first 71 cases screened. J Am Acad Child Adolesc Psychiatry 1994;33(5):636–44.
15. Meijer M, Treffers PD. Borderline and schizotypal disorders in children and adolescents. Br J Psychiatry 1991;158:205–12.
16. Cohen DJ, Paul R, Volkmar FR. Issues in the classification of pervasive and other developmental disorders: toward DSM-IV. J Am Acad Child Psychiatry 1986; 25(2):213–20.
17. Towbin KE, Dykens EM, Pearson GS, et al. Conceptualizing "borderline syndrome of childhood" and "childhood schizophrenia" as a developmental disorder. J Am Acad Child Adolesc Psychiatry 1993;32(4):775–82.
18. Sprong M, Becker HE, Schothorst PF, et al. Pathways to psychosis: a comparison of the pervasive developmental disorder subtype multiple complex developmental disorder and the "at risk mental state. Schizophr Res 2008;99(1–3):38–47.

19. Ad-Dab'bagh Y, Greenfield B. Multiple complex developmental disorder: the "multiple and complex" evolution of the "childhood borderline syndrome" construct. J Am Acad Child Adolesc Psychiatry 2001;40(8):954–64.

20. Garralda ME. Hallucinations in children with conduct and emotional disorders: II. The follow-up study. Psychol Med 1984;14(3):597–604.

21. Ulloa RE, Birmaher B, Axelson D, et al. Psychosis in a pediatric mood and anxiety disorders clinic: phenomenology and correlates. J Am Acad Child Adolesc Psychiatry 2000;39(3):337–45.

22. Gillberg C, Billstedt E. Autism and Asperger syndrome: coexistence with other clinical disorders. Acta Psychiatr Scand 2000;102(5):321–30.

23. Muris P, Steerneman P, Merckelbach H, et al. Comorbid anxiety symptoms in children with pervasive developmental disorders. J Anxiety Disord 1998;12(4):387–93.

24. Leyfer OT, Folstein SE, Bacalman S, et al. Comorbid psychiatric disorders in children with autism: interview development and rates of disorders. J Autism Dev Disord 2006;36(7):849–61.

25. Weisbrot DM, Gadow KD, DeVincent CJ, et al. The presentation of anxiety in children with pervasive developmental disorders. J Child Adolesc Psychopharmacol 2005;15(3):477–96.

26. Kumra S, Jacobsen LK, Lenane M, et al. Multidimensionally impaired disorder": is it a variant of very early-onset schizophrenia? J Am Acad Child Adolesc Psychiatry 1998;37(1):91–9.

27. Stayer C, Sporn A, Gogtay N, et al. Multidimensionally impaired: the good news. J Child Adolesc Psychopharmacol 2005;15(3):510–9.

28. Fitzgerald M. Multidimensionally impaired disorder. J Am Acad Child Adolesc Psychiatry 1998;37(11):1125–6.

29. Nagy J, Szatmari P. A chart review of schizotypal personality disorders in children. J Autism Dev Disord 1986;16(3):351–67.

30. Wolff S, Chick J. Schizoid personality in childhood: a controlled follow-up study. Psychol Med 1980;10(1):85–100.

31. Carlson GA, Fish B. Longitudinal course of schizophrenia spectrum symptoms in offspring of psychiatrically hospitalized mothers. J Child Adolesc Psychopharmacol 2005;15(3):362–83.

32. Hans SL, Auerbach JG, Auerbach AG, et al. Development from birth to adolescence of children at-risk for schizophrenia. J Child Adolesc Psychopharmacol 2005;15(3):384–94.

33. Asarnow JR, Tompson MC, Goldstein MJ. Childhood-onset schizophrenia: a followup study. Schizophr Bull 1994;20(4):599–617.

34. Asarnow JR. Childhood-onset schizotypal disorder: a follow-up study and comparison with childhood-onset schizophrenia. J Child Adolesc Psychopharmacol 2005;15(3):395–402.

35. Harrow M, Grossman LS, Silverstein ML, et al. Thought pathology in manic and schizophrenic patients. Its occurrence at hospital admission and seven weeks later. Arch Gen Psychiatry 1982;39(6):665–71.

36. Raine A, Fung AL, Lam BY. Peer victimization partially mediates the schizotypy-aggression relationshiop in children and adolescents. Schizophr Bull 2011;37(5):937–45.

37. Cochran DM, Dvir Y, Frazier JA. Autism-plus" spectrum disorders: intersection with psychosis and the schizophrenia spectrum. Child Adolesc Psychiatr Clin N Am 2013;22(4):609–27.

38. Solomon M, Ozonoff S, Carter C, et al. Formal thought disorder and the autism spectrum: relationship with symptoms, executive control, and anxiety. J Autism Dev Disord 2008;38(8):1474–84.

39. Eussen ML, de Bruin EI, Van Gool AR, et al. Formal thought disorder in autism spectrum disorder predicts future symptom severity, but not psychosis prodrome. Eur Child Adolesc Psychiatry 2015;24(2):163–72.

40. Hecht KF, Cicchetti D, Rogosch FA, et al. Borderline personality features in childhood: the role of subtype, developmental timing, and chronicity of child maltreatment. Dev Psychopathol 2014;26(3):805–15.

41. Petti TA, Vela RM. Borderline disorders of childhood: an overview. J Am Acad Child Adolesc Psychiatry 1990;29(3):327–37.

42. Achenbach TM, Rescorla LA, University of Vermont, et al. Manual for the ASEBA school-age forms & profiles. Burlington (VT): University of Vermont, Research Center for Children, Youth, & Families; 2001.

43. Achenbach TM. Manual for the child behavior checklist/4-18 and 1991 profile. Burlington (VT): University of Vermont, Department of Psychiatry; 1994.

44. Reynolds CR, Kamphaus RW. Behavior assessment system for children, third edition (BASC™-3). Toronto: Pearson; 2015.

45. Gadow KD, Sprafkin J, Gadow KD, et al. Child symptom inventories manual. Stony Brook (NY): Checkmate Plus.; 1994.

46. Stein MA, Szumowski E, Sandoval R, et al. Psychometric properties of the children's atypical development scale. J Abnorm Child Psychol 1994;22(2):167–76.

47. Wiig E, Secord W. Test of language competence - expanded edition (TLC-expanded). London (UK): Pearson; 1989.

48. Wiig EH, Semel E, Secord WA. Clinical evaluation of language fundamentals | fifth edition (CELF-5). London (UK): Pearson; 2013.

49. Hammill DD. Detroit test of learning aptitude (DTLA). London (UK): Pearson; 1998.

Appendix

The following case descriptions, anonymized and compiled, represent examples of the problems described in the articles of the *Child and Adolescent Psychiatric Clinics of North America*, Part I devoted to mood-dysregulated children and adolescents. Note that cases are sometimes those that were already described in the articles.

CONDUCT DISORDER AND OUTBURSTS (CHAPTER 3)

Stephon is a 16-year-old incarcerated male adolescent referred for psychiatric evaluation by corrections officers at the state adolescent prison where he has resided for the past 8 weeks. Cell-block officers are concerned about Stephon's unpredictable explosive temper, aggression, and dangerousness and note that he recently was placed in "lockdown" for "beating down" his cell mate. He is referred to the prison's consulting psychiatrist for a "dangerousness" evaluation.

On interview, Stephon was cocky and self-confident, minimizing the circumstances that led to his incarceration ("I caught some charges...."). When asked his understanding of why he was being evaluated by the psychiatrist, he replied: "I likes to make people do things they don't want to do. If they doesn't do it, I goes off on 'em." After a pause he adds, "ain't my problem if they gets beat, they shoulda done what I wanted."

Stephon's developmental history is significant for birth to an unwed, cocaine-addicted 19-year-old biological mother living on welfare. He never met his biological father. Removed from his mother's care at age 7 years because of her continued substance abuse, he was placed in a succession of temporary foster care settings. All placements were terminated because of Stephon's impulsive, explosive violent temper, stealing, and lying. He was eventually returned to his mother's care at age 10 years. Largely unsupervised, he failed at school, stopped attending school at age 11 years, hung out on "the street" with much older adolescents, was active in a local street gang, and was first arrested at age 12 years for automobile theft.

A violent, explosive temper was noted early by age 5 years. Stephon's temper could be precipitated by mild frustration, by disappointment, or by his wanting to avoid unpleasant tasks. As he grew into adolescence, his violence took on characteristics of predatory aggression, whereby he used violence to obtain things he wanted from others. During his psychiatric interview, Stephon noted: "I is a good fighter....I hurts them and they can't hurt me....I can't be beat." He became a leader in the street gang hierarchy and at age 13 years was arrested again on a weapons charge. Released by the juvenile court on probation, he promptly returned to life on the streets. By mid-adolescence, 2 more arrests occurred, and he received probation for both. Arrested yet again for the armed robbery of a convenience store, he was sentenced by the court to 1 to 2 years in the state prison for adolescents and young adults.

The above case study is reprinted from Carlson GA, Chua J, Pan K, et al. Behavior Modification Is Associated With Reduced Psychotropic Medication Use in Children With Aggression in Inpatient Treatment: A Retrospective Cohort Study. J Am Acad Child Adolesc Psychiatry 2020;59(5):632-641; with permission.

AUTISM AND OUTBURSTS (CHAPTER 6)

Ten-year-old Tyreke was referred for outbursts whereby he bangs his head on the desk and walls, stabs himself with a pencil, and says that he wants to die when he gets agitated. What gets him upset is the noise other children make, the things they say, doing math, and any change in his routine. He has major outbursts that occur at least once a week, sometimes more often. There are minor tantrums in between. His outbursts can last for up to an hour. Although he is irritable in school (unless he is by himself), he is not a problem at home, where he is an only child. He has been having outbursts since at least third grade. Stimulants were tried in the past and were thought to cause motor and vocal tics. A low dose of risperidone caused weight gain and was stopped. Guanfacine, which he takes currently, helps tics but not much else.

Tyreke's developmental history is consistent with an autism spectrum disorder, with a combination of language delay, scripted language and play, and an obsessive interest in Thomas the Tank Engine. He had sensory sensitivities, including sound, food texture, and smell. His full-scale IQ is 82. Language testing showed similar scores. It is difficult to get Tyreke to talk about abstract things like feelings. He has some trouble focusing, but he is not hyperactive nor especially impulsive. He has performance anxiety and worries about making a mistake because then he'll have to do the work over. He doesn't like other children looking at him or talking about him. Parent and teacher rating scales are similar for autism and anxiety symptoms except that teachers feel he is depressed (based on what he says when agitated) and aggressive, but his parents do not.

ATTENTION-DEFICIT AND OPPOSITIONAL DEFIANT DISORDERS (CHAPTER 7)

Jared, age 7, was referred by his second-grade teacher and parents for help with his terrible "meltdowns." He was referred for inpatient treatment the day he refused to do math, refused a break when offered, swiped his school work onto the floor, and was removed from the class to the principal's office. There he threw the snacks he was offered onto the floor, tried to flip a table, and hit/punched and kicked the adults. Instead of being sent home (the usual response), he was sent to the psychiatry emergency room and was hospitalized. He was diagnosed with attention-deficit/hyperactivity disorder, oppositional defiant disorder, and a language delay. Medication treatment, the behavior modification program, and parent training commenced.

Initially compliant with rules, one evening, Jared became rude, provocative, and increasingly aggressive when he didn't like the evening movie. He refused his "time out" for aggression in the hall and had to take it in a room out of the milieu (under staff observation but without interaction). After 45 minutes of screaming and cursing, he calmed down and rejoined the milieu. His second outburst 2 days later was managed with a "time out" only and was over in 10 minutes. There was no third outburst. He was discharged for outpatient follow-up after 15 days. He was placed in a 12:1+1 class.

Jared was readmitted at age 11 for similar behaviors and medication readjustment. At this time, the behavior program had been discontinued. When his first agitation episode occurred, according to nurse's notes, "he was given almost every activity and privilege he demanded. When the evening movie was about to begin, he became more agitated, refused alternative choices, ignored attempts at de-escalation and distraction, took a clay object from the activity room and smashed it, brandishing one of the shards at staff. A 2-minute physical hold was initiated. He accepted an oral PRN of risperidone 0.5 mg but continued to be demanding, irritable and rude. He finally fell asleep at 9:30 PM."

The first episode lasted 3.5 hours. Jared had 4 more medicated outbursts and 2 more "holds" over his 15-day stay. These outbursts varied in duration from 40 minutes to 2 hours. At discharge, he was placed on home instruction pending and out of district placement.

TOURETTE DISORDER AND OUTBURSTS (CHAPTER 8)

Kyle is an 8-year-old boy who, for the past year, has experienced increasing episodes of explosive anger, typically in response to frustration, usually occurring at home, mostly directed at his mother. Episodes had increased to multiple episodes nearly every day from perhaps once to twice a month about a year earlier. Worsening coincided with increased tics. Recently, Kyle had an outburst at school resulting in his out-of-school suspension. He was formerly a good student, without medical problems, and he is well liked by school staff and peers. He is more socially withdrawn presently.

Since early childhood, Kyle experienced motoric hyperactivity, impulsivity, inattention, poor frustration tolerance, anxiety, somatic symptoms (headaches, stomachaches, enuresis), sensory sensitivities, and tics, including eye rolling, facial twitching, snorting, throat clearing, head turning, and repeated touching. Fears of harm coming to his parents or to himself, horrifying recurrent images, and fears of acting on aggressive urges are accompanied by compulsive checking, repeated requests for reassurance, and need to repeat certain gestures until it feels "just right." OT evaluation demonstrated sensitivity to and difficulties processing multisensory input and misperception of certain social exchanges as threats. A brief trial with guanfacine IR 1.5 mg daily caused worsened aggression and mood lability. Sertraline 25 mg daily caused increased irritability, nausea, and headaches; attempts to increase sertraline resulted in worsening of aggression and was stopped. Aripiprazole 2.0 mg daily led to decreased tics and anxiety and reduced frequency and intensity of anger outbursts.

Kyle attained normal developmental milestones but experienced repeated trauma during early childhood, including witnessing his father physically and verbally assaulting his mother. He suffered with repeated abrupt separations, and most recently, a forced evacuation and loss of his family's home due to fire.

Family history is significant for sociopathy, alcohol and substance abuse on the paternal side, and for obsessive compulsive disorder, anxiety, and affective disorder on the maternal side.

POSTTRAUMATIC STRESS DISORDER AND OUTBURSTS (CHAPTER 9)

Mark is an 8-year-old boy, brought into an outpatient clinic by his foster mother because he grabbed a knife and threatened another foster child in the home. Mark has a history of "abuse," but no one has details because he is in his fifth foster home with his third Child Protective Services worker. Mark is sent for evaluation in the emergency department and is admitted to a children's psychiatric inpatient unit. There, he continues to be aggressive, requiring seclusion and intramuscular (IM) neuroleptic medication as needed to manage behaviors. Despite multiple episodes of seclusion and as-needed injections, Mark's behavior does not improve. The inpatient staff feel threatened, worn out, and angry. When asked if there is a pattern to the explosive behavior, they report that it is a result of Mark "not getting his way" or being told "no." When asked to wait his turn by a favorite staff member, Mark becomes angry and begins crying, yelling, and disrupting the activity. After the area is cleared for safety, the child psychiatrist on the unit approaches Mark and urges him to take some deep breaths and talk about what is on his mind. He is asked what it means when someone he likes tells him "no," and he relates that it means that the person is saying that they don't like him, like other kids more, and will never care about what he wants. The favorite staff member reassures Mark that he still likes him. Mark is able to deescalate to the point where he can recognize his cognitive distortions. Later, in a calmer state, Mark agrees to "practice" being told "no" and makes a safety agreement. Eventually, Mark is able to recognize that feelings of rejection remind him of violence and abandonment experienced earlier in life.

ANXIETY PLUS OUTBURSTS (CHAPTER 13)

Nikki is a 9-year-old girl with a 1 year history of anxiety symptoms. She worries about school, her mother and father, "being good at stuff," whether her friends like her, and "what's going to happen next." Her anxiety is present for several hours daily and is particularly impairing around times of separation. In addition, she has difficulty controlling her anxiety, and she frequently requires reassurance. She has anxiety-related somatic symptoms (headaches and stomachaches) as well as initial insomnia, anergia, and psychomotor agitation. She denies any suicidal ideation, intent, or plan, and there are no manic or psychotic symptoms. She denies depressed mood.

Nikki's anxiety is accompanied by frequent irritability and occasional violent outbursts, particularly in anxiety-provoking situations. For example, when Nikki's mother and father were leaving to go to the grocery store, Nikki was to have stayed home with her 16-year-old sister. However, she ran into the driveway screaming and threw herself onto the family minivan and attempted to block her family from leaving. Each evening, after her mother or father reads her a story, she screams as her mother or father leaves the room and clings to them while screaming, often barricading herself in her room to prevent her mother and father from leaving. At times, she will become physically aggressive toward her parents but has not been physically aggressive to those outside the family. She struggles to attend school, and numerous school staff must be involved in coaxing her from the car each morning; however, she is often crying and at times yelling as she pleads not to go to school.

Nikki attained normal developmental milestones. She had a recent physical examination from her primary care physician, and her vital signs were within normal limits, with for a body mass index being in the 94%ile.

Nikki's family history is significant for a depressive disorder in her mother and panic disorder in her father as well as bipolar disorder in her paternal grandmother. In terms of social history, Nikki has 2 older siblings, and her mother has a history of postpartum depression and anxiety and is in remission.

ANXIETY/DEPRESSION AND OUTBURSTS (HOW AND WHY ARE IRRITABILITY AND DEPRESSION LINKED?, PABLO VIDAL-RIBAS AND ARGYRIS STRINGARIS)

Consuelo is a 12-year-old girl with a 2-year history of anxiety and depressive symptoms and doesn't feel she is good enough at anything. She asks her parents a lot of questions and needs frequent reassurance from her mother because she feels she can't do anything right. When stressed, she is restless and has headaches and stomachaches. She is also exhausted at the end of the day but can't fall asleep. She is tired in the morning because she hasn't slept. She denies any suicidal ideation, intent, or plan, and there are no manic or psychotic symptoms. However, she doesn't enjoy much, which is a change for her as her parents say that she used to be a very happy kid, especially when playing sports.

Consuelo's irritable and at times explosive behavior has gotten worse over the past 6 months. Her parents have noticed that when she is feeling pressured, she wants her parents to "make it better," but when they try, for example, by talking to the teacher or the coach, Consuelo gets even more angry! She studies to the point of exhaustion because she has trouble concentrating and then "freezes up" taking tests, so her grades have dropped. If her parents back off, she gets upset, but when they push her to get through her homework, she explodes. If they ask her if she wants help with homework, she explodes. She can't decide about anything. Her soccer coach tried to cheer up her "sour mood," but she got angry with him in front of the other kids and he "benched" her. When her coach put her back in the game, she then was zoned out, didn't perform well, and then walked off the field and started crying.

Consuelo says she doesn't understand why she explodes and wants help. She told the doctor she was "stressed out" but not anxious and got angry when the doctor suggested she was depressed, saying she wasn't a "depressed" person. She was annoyed when asked about suicidal behavior.

According to Consuelo, neither play therapy nor relaxation therapy was helpful. She has never been treated with any psychotropic medications or psychiatrically hospitalized. Her family history is significant for depression in her father and social phobia in her mother. She is an only child. Developmental history was unremarkable.

DEPRESSION AND OUTBURSTS (CHAPTERS 10 AND 11)

Cody is in eighth grade, referred because of multiple absences and being sullen and uncooperative with teachers. Indicative of Cody's problems, Mrs N reported that he initially refused to come for the evaluation, became physically and verbally aggressive toward his mom, and threatened to kill himself if she didn't leave him alone. She felt she had to call the police to calm him down. The police came and made it clear that if he didn't cooperate, he'd be cuffed and taken to the emergency room. He ultimately calmed down, and he came to the evaluation and was cooperative.

At home, Cody has become disrespectful, doesn't make eye contact, refuses to go out, and displays violent behavior. His mom describes him as very depressed looking. He stays in bed all day long, refuses to take care of his hygiene (hasn't showered for over a month), and "his room is a mess." His mood switches quickly, and he can get easily upset about random things. Mrs N has a child-lock in her car in order to prevent Cody from getting out from a moving car when angry and upset. Cody has kicked the back seat so hard when angry that Mrs N often has had to turn around and drive back home. Mrs N knows he isn't using drugs because he has no social life and doesn't go out enough to get drugs.

On the Children's Depression Inventory, Cody scored 35, very depressed. He is sad many times, feels like crying all the time, does many things wrong, is bad many times, can't concentrate, has to push himself to do school work, is tired all the time, doesn't feel like eating, feels alone, never has fun, isn't sure if anyone even loves him. He said he "thinks about suicide but he wouldn't do it." Although he denied suicidal ideation, he often wished he was dead and said he hated his life when upset. He did not have these thoughts or feelings otherwise and denied ever doing anything to hurt himself. On the other hand, he said he feels weird and different from other kids and is reminded of that fact by his peers. Cody denied psychosis (hallucinations or delusions) and drug or alcohol use.

BIPOLAR DISORDER AND OUTBURSTS (CHAPTER 12)

Maddy is a 15-year-old girl diagnosed with depression and attention-deficit/hyperactivity disorder, who was hospitalized a year ago for suicidal ideation. Her mood and attention problems briefly improved with an antidepressant followed by a psychostimulant, but she stopped taking both medications after a month or two when she started to experience brighter moods.

Maddy's most recent presentation was for explosive irritability, increased energy, and decreased need for sleep. Due to her irritability, she had lost friends and found everything her parents said to be a nag. Several times during the week, Maddy said she punched walls as a way to suppress her urge to hit her mom for seemingly benign requests to participate in household chores. Maddy was frequently observed by her parents to shift from a calm baseline to agitated or excited within a very short span of time, with excitability manifesting as uncharacteristic pressured, productive, and loud speech. Previously a straight A student, she was now struggling to complete assignments, being distracted by thoughts racing in her mind about her newfound interest in making furniture and how it could make her a star in her own television show for do-it-yourself projects. She stole her mother's credit card to buy tools that she stayed up all night trying to use. In taking a careful history, it appeared that milder forms of these symptoms had occurred, sometimes being followed by long periods of depressed mood and anhedonia.

BORDERLINE PERSONALITY DISORDER AND OUTBURSTS (CHAPTER 12)

Jenny is a 17-year-old girl readmitted to an acute care psychiatric hospital for a history of suicidal behavior, self-harm, aggression, and reports of hallucinations. Since age 12, Jenny has been hospitalized multiple times and has also spent time in juvenile detention. She has been variably diagnosed with schizoaffective disorder, bipolar disorder, major depression with psychotic features, polysubstance abuse, plus other mood, anxiety, and behavior disorders.

Jenny's developmental history is significant for early neglect and social chaos. Her biological mother was reportedly diagnosed with bipolar disorder and polysubstance abuse. The identity of her biological father is not known. Jenny was removed from her biological mother's care at 3 months of age due to neglect and placed in multiple foster homes until she was adopted at 2 years of age. Jenny reports a history of sexual abuse by an older male relative of her adoptive family when she was 11 years of age. She is also suspected to have engaged in prostitution while living on the streets.

She has had difficulties with poor impulse control, oppositional behaviors, and anger outbursts since elementary school. These behaviors have escalated over time. Her adoptive parents report that her outbursts typically occur when she does not get her own way, or when they try to set limits. Her moods are highly volatile, with screaming, threats of violence, acts of self-harm, and assaulting others over slight provocations (eg, being told she cannot leave the house or see her boyfriend). She has not gone to school for several months and runs away from home for days at a time.

She describes hallucinations that mostly occur in the evening around bedtime, or when she is upset or in conflict. Her descriptions of visual and auditory hallucinations are vivid and detailed. She has named the voices and says that they sometimes are helpful, but at other times say mean things or tell her to cut herself. She says the voices do not sound like a real person, but the experiences happen more when she is having flashbacks of past traumatic experiences. She otherwise does not demonstrate overt signs of thought disorder or responding to internal stimuli.

While hospitalized, Jenny engaged in frequent aggressive, assaultive, and self-harming behaviors, at times requiring mechanical restraints multiple times a day. In a typical episode, she would become demanding and threatening over a conflict or rule and escalate to kicking, hitting, and spitting at staff, and/or self-harming behaviors (eg, wrapping clothing around her neck, pounding her head on the floor). These outbursts would generally last 1 to 2 hours and required a great deal of staff time trying to negotiate, redirect, and/or intervene with her behaviors. She routinely received IM doses of antipsychotic agents as needed for unsafe behavior and was staffed 2:1 given her acuity.

Moving?

Make sure your subscription moves with you!

To notify us of your new address, find your **Clinics Account Number** (located on your mailing label above your name), and contact customer service at:

Email: journalscustomerservice-usa@elsevier.com

800-654-2452 (subscribers in the U.S. & Canada)
314-447-8871 (subscribers outside of the U.S. & Canada)

Fax number: 314-447-8029

Elsevier Health Sciences Division
Subscription Customer Service
3251 Riverport Lane
Maryland Heights, MO 63043

*To ensure uninterrupted delivery of your subscription, please notify us at least 4 weeks in advance of move.